POWER and PORK

I0306719

POWER and PORK

A JAPANESE POLITICAL LIFE

AURELIA GEORGE MULGAN

E PRESS

ASIA PACIFIC PRESS
THE AUSTRALIAN NATIONAL UNIVERSITY

E PRESS

Co-Published by ANU E Press and Asia Pacific Press
The Australian National University
Canberra ACT 0200, Australia
Email: anuepress@anu.edu.au
Web: http://epress.anu.edu.au

© Aurelia George Mulgan 2006
This work is copyright. Apart from those uses which may be permitted under the *Copyright Act 1968* as amended, no part may be reproduced by any process without written permission from the publisher.

The views in this book do not represent those of Asia Pacific Press or the Crawford School of Economics and Government at The Australian National University.

National Library of Australia Cataloguing in Publication entry

Mulgan, Aurelia George.
Power and Pork: a Japanese political life

Bibliography.
Includes index.
ISBN 0 7315 3757 2 (pbk)
ISBN 1 920942 33 5 (online)

1. Japan - Politics and government. 2. Japan - Economic policy. I. Title.

320.952

Editors: Bridget Maidment, Clare Shamier, Paula Sutton
Cover design: Annie Di Nallo Design

First edition © 2006 ANU E Press and Asia Pacific Press

CONTENTS

Tables		vi
Abbreviations		vii
Acknowledgments		viii
Preface		ix
1	Beyond generalisation	1
2	Becoming a politician	6
3	Accommodating electoral reform	50
4	Exercising power as a *nôrin giin*	73
5	Exercising power as a *nôrin zoku*	121
6	The identical twins of Nagata-chô	150
7	Electoral vicissitudes	204
8	Conclusion	251
Bibliography		254
Index		267

TABLES

2.1	Farm household composition/votes cast for Matsuoka by municipality in Kumamoto (1) in 1990 Lower House election	22
2.2	Farm household composition/votes cast for Matsuoka by municipality in Kumamoto (1) in 1993 Lower House election	32
2.3	Ranking of Kumamoto Prefecture as an agricultural producer/by commodity and farm/forestry households (per cent), 2000	39
3.1	Farm household composition/votes cast for Matsuoka by municipality in Kumamoto (3) in 1996 Lower House election	58
3.2	Farm household composition/votes cast for Matsuoka by municipality in Kumamoto (3) in 2000	62
4.1	Matsuoka's Committee Memberships et cetera	111
6.1	Top 10 contributors to Matsuoka Toshikatsu New Century Politics and Economics Discussion Association	164
6.2	Top 10 individual contributors to Matsuoka Toshikatsu New Century Politics and Economics Discussion Association	165
6.3	Companies contributing to Matsuoka and Muneo	167

ABBREVIATIONS

AFF	Agriculture, Forestry and Fisheries
ALIC	Agriculture and Livestock Industries Corporation
BSE	Bovine Spongiform Encephalopathy
DPJ	Democratic Party of Japan
FTA	Free Trade Agreement
JCP	Japanese Communist Party
JETRO	Japan External Trade Organization
JSP	Japanese Socialist Party
LDP	Liberal Democratic Party
LIPC	Livestock Industry Promotion Corporation
MAFF	Ministry of Agriculture, Forestry and Fisheries
MAF	Ministry of Agriculture and Forestry
METI	Ministry of Economy, Trade and Industry
MMD	Multi-member district
MOF	Ministry of Finance
MoFA	Ministry of Foreign Affairs
NHK	Nippon Hôsô Kyôkai [Japan Broadcasting Association]
OB	'Old Boy'
PARC	Policy Affairs Research Council
SMD	Single-member district
WTO	World Trade Organization

ACKNOWLEDGMENTS

The author would like to thank the National Library of Australia in Canberra for providing such pleasant sanctuary for several months in 2005, which enabled much of the research for this book to be undertaken. Under the conditions of a Harold White Fellowship, the library granted privileged access to the Japanese Collection. Library staff of the Collection, especially Mayumi Shinozaki, greatly assisted in obtaining materials from the Diet Library in Japan and in furnishing all kinds of research advice and support. An Australian Research Council Discovery Project funded valuable research assistance, which was provided by Reiko Take. Maree Tait showed great patience and forbearance during the inevitable hiccups in the publication process, while Richard Mulgan, the unsung hero of this book, deserves my undying gratitude for the index and his usual illuminating intelligence, which contributed to the book in so many different ways. The book is dedicated to Sasha and Basil, who will always be greatly missed.

PREFACE

Japanese politics is notoriously opaque and corrupt. It is not surprising, therefore, that few studies reveal what really goes on in the political lives of Japanese politicians. This is one such exposé, the fruits of intensive research into the public and private political life of a member of the Japanese Diet. It is an extraordinary tale about an ordinary politician who exemplifies what it is to be a parliamentarian affiliated with the ruling Liberal Democratic Party (LDP) in present-day Japan. Like the majority of LDP Diet members, the politician at the centre of this study represents a clutch of special interests whilst striving to direct public resources to his own electorate. As a veteran lawmaker and policy specialist, he also belongs to a 'policy tribe' (*zoku*), regularly intervening in the making of government policy and interfering in the administrative affairs of the bureaucracy. As a tribe Diet member he has achieved the ultimate goal, appointed as minister over the very ministry that he sought to influence from within the ranks of the LDP.

This is a story of political failures and of political successes, a tale of ambition furthered by the successful delivery of benefits to constituents and supporters, but also thwarted by suspicion and allegations of corrupt behaviour. In the brave new world of Japanese politics opened up by Prime Minister Koizumi, such an independent policy entrepreneur and political fixer appeared to represent the old style of LDP politician, whose days might be numbered. However, his accession to the ministry in the first Abe cabinet, as this book went to press, underlines the resilience of traditional patterns of Japanese politics. Only the final chapter of this political life, which is yet to be written, will reveal the extent to which a new model of Japanese politics has taken root.

1

BEYOND GENERALISATION

The story told in this book—or rather the inside story of a political life that now spans 16 years—is neither pure biography nor pure scholarly treatise. It falls somewhere in between. It is not pure biography because it is only concerned with political phenomena. It focuses on the political career, connections, performance and activities of one of Japan's Diet politicians, Matsuoka Toshikatsu, Liberal Democratic Party (LDP) member in Japan's House of Representatives for Kumamoto No. 3 district. His private life outside politics is only of incidental interest to this account.

Some may feel that such an approach will omit potentially the most absorbing and interesting details, but, as this book will show, not only is politics Matsuoka's life, but there is intriguing detail aplenty in his political machinations. The book delves into Matsuoka's early life and career, but only to provide important background details and to help explain Matsuoka's decision to enter politics. His reputation in Nagata-chô[1] for liking women, for greeting female Diet members with unwelcome comments and for sweet-talking hostesses in high-class nightclubs[2] are the only comments that will be made about his private predilections. As for Matsuoka's personality, this is not explicitly the focus of analysis, but sometimes glimpses of it are revealed—in the descriptions of his relations with bureaucrats, businessmen, other politicians, local government figures and organisational leaders, and also in accounts of what he said and what he did—in words that are his and theirs, not mine. Indeed, Matsuoka has both a public persona and a private personality, and the two do not necessarily match. The persona he presents to the outside world is that of someone who is highly principled, and who works tirelessly on behalf of his constituents, supporting groups and various important causes. Privately, as a dealmaker and political fixer, by all accounts, he is completely different, aggressive[3] with a reputation for shouting, bullying and violence.

The book is not pure scholarly treatise because it eschews generalisation, or, at least, uses generalisation only in order to elucidate Matsuoka's activities and behaviour, rather than as the main explanatory device. The book does not, therefore, adopt a scientific approach in the sense that this terminology is normally used. At the same time, the study aspires to be labelled political science because it analyses political events and facts, and seeks to understand the nature of Japanese politics not through generalisation, but rather through the rich description of individual example.

The methodological approach adopted in this book is that of 'thick' description, a research technique borrowed from social anthropology,[4] which weds 'the tools of modern social science…to the artful narrative skills of the humanities'.[5] Thick description is an approach that goes beyond generalisation and is designed to yield insights that cannot be captured by universal statements about particular phenomena. It is possible for an individual story to provide deeper and more rounded understanding than any generalisation can offer. Such an approach is underpinned by the realisation that, even amongst scholars of Japanese politics, we often know the general contours of an institution or practice, but we do not have a sufficiently precise or inside knowledge of it.

Concentrating on an individual politician is unusual in studies of Japanese politics unless the work is strictly biographical or focuses on dominant leaders or iconic figures, such as Tanaka Kakuei.[6] Little has been written about the political lives of individual, ordinary Diet members[7] in spite of the fact that, as everywhere, politics in Japan is shaped by human factors. As Curtis wisely observed: 'Decisions made by individuals…are the direct cause of what happens in politics'.[8]

While the book offers an account of the political life and activities of an individual Japanese politician, it is hoped that the analysis will generate insights into Japanese politics as a whole. Such insights are not explicitly manipulated into generalisations in the study; they have to be gleaned from the material that is presented. Nonetheless, understanding how one person thinks, acts and operates may produce greater understanding of how a political system functions and even how it is changing. If the book furnishes information that leads others to make more reliable, illustrative generalisations about the behaviour of Japanese politicians, then it will have succeeded in its modest ambitions. It is certainly hoped that this book will provide a counterweight to the tendency amongst political scientists to try to reduce the detail and variation in political phenomena to numbers or to highly selective illustrations of deductive theory.

The particular nature of this study posed some unique research problems, particularly in obtaining crucial, inside information. This was gathered from a diverse range of sources: the mass media—newspapers, records of TV interviews, and industry and investigative journals—as well as from more scholarly articles and texts, and from Matsuoka's own website. Recourse was also made to the Internet where too much credence was not placed in the potentially libellous words of those engaging in 'threading', where individual members of the Japanese public can vent their spleen about public figures, such as Matsuoka, with the advantage of anonymity. At no time was the subject of the study interviewed (for reasons that may become obvious to readers); but those who had interviewed him were. The study was done at arm's length—as a book such as this has to be done.

Why write such a book? The project arose out of earlier research that revealed the importance of individual politicians in Japanese electoral politics and the influence individual Diet members in the ruling LDP can exert over government policy. The book began its life as an examination of the 'government versus ruling party' phenomenon in Japan: the LDP has often maintained different positions on policy from the bureaucracy and, more spectacularly in recent times, from the prime minister. What this early research revealed was a great deal of evidence that individual politicians—ordinary backbenchers in the LDP—customarily drive government policy in Japan. Indeed, the policy directions of the LDP frequently seem to be propelled by the policy activities of individual Diet members, who specialise in particular aspects of policy in a highly decentralised policymaking apparatus that corresponds to, and is supported by, the bureaucratic ministries and agencies. As a policymaking body, the LDP is a decentralised organisation that supports the policy specialism of Diet members and allows them to exercise their own, individual policy influence from the bottom up.

The traditional paradigm of Japanese politics is encapsulated in this structure of individual Diet member-dominated politics. The LDP is, as Machidori depicts it, 'a decentralised party dominated by Diet members'.[9] With their own individually-centred, secure voting bases, LDP politicians have operated as their own persons, not completely beholden to their own party, but wielding power, including policymaking power, in their own right—independently of both the bureaucracy and the government leadership of the day.

In policymaking, individual LDP politicians—ordinary backbenchers—instead of being the usual parliamentary fodder that they are in other parliamentary cabinet systems, wield direct power over policymaking and over the administrative affairs of the bureaucracy. They actually encroach on the role of government in two important ways. They are routinely involved both in the making of government policy and in processes of government administration. The former amounts to intervention (*kainyû*) in government policymaking, and the latter to interference in areas of administrative competence.

Policymaking by ordinary LDP backbenchers takes place within the committees of the LDP's Policy Affairs Research Council (PARC). This allows the LDP to shape its public policy independently. Moreover LDP politicians have behaved like individual policy entrepreneurs rather than as members of a party bound by a commonly agreed policy view imposed by cabinet and the party leadership. At the same time, individual backbenchers have exerted influence over specific administrative decisions made by bureaucracy through direct contacts with ministry officials, completely bypassing the minister. To analyse politics operating in such a way, an approach had to be adopted that implicitly recognised Japanese politics as fundamentally individually-based (*kojin honi*) rather than party-based (*seitô honi*).[10] The LDP has lacked policy coherence; it has not moved as one, but as a vast conglomeration of individual politicians, each pushing his or her own barrow.

Why focus on Matsuoka? Why does he deserve 'thick' or 'rich' description? He is so illustrative of a certain genre of Japanese politician, he might have been created as a composite of the characteristics of the archetypal Japanese politician. He exhibits all the features for which the LDP is famous and with which it has traditionally been associated. First, he is strongly representative of special interest groups as one of the most notorious, influential, outspoken and colourful representatives of farm, forestry and rural–regional interests in Japan. Second, he has been implicated in a number of corruption scandals involving so-called 'money politics' that have somewhat dented, but not destroyed, his political standing. Third, he is a fervent advocate of pork barrelling, and puts enormous effort into bringing public works back to his own constituency. Fourth, he has operated within the LDP's policymaking process as a direct and indirect representative of the special interests that are central to his electoral support coalition.

The methodology used in this book—that of thick description—can also offer insight into the nature of political change in Japan. Does the traditional model of an LDP politician still apply? Is Japan experiencing the much-heralded 'regime shift'[12] that was flagged more than 10 years ago? How did Matsuoka adjust to the reforms instituted by the Koizumi administration? Has the prototypical LDP politician been eclipsed, or has he merely found new ways of maintaining his influence?

NOTES

1. This is the area of Tokyo where both the National Diet and the headquarters of the LDP (Jimintô honbu) are located.
2. These comments were made by a fellow LDP Diet member, and quoted in 'Han Koizumi Giin no "Yoru no Kao"' ['The "Night Face" of a Diet Member Opposed to Koizumi'], Shûkan Shinchô, 13 December 2001, p. 161.
3. Itô Hirohide, 'Heisei Jiken Fuairu: Nôrin Jigyô Hojokin o Dokusen Suru Matsuoka Toshikatsu Shûin Nôsuiiinchô no Eikyôroku' ['Heisei Scandal File: The Influence of House of Representatives Agricultural, Forestry and Fisheries Committee Chairman Matsuoka Toshikatsu Who Monopolises Agricultural and Forestry Works Subsidies'], Seikai, Vol. 22, No. 6, June 2000, p. 65.
4. Clifford G., 1973. *The Interpretation of Cultures*, Basic Books: New York.
5. Rhodes R. A. W. and Weller P., 2001. *The Changing World of Top Officials: Mandarins or Valets?*, Open University Press, Buckingham and Philadelphia, p. 7.
6. See, for example, Tachibana Takashi, 1976. *Tanaka Kakuei Kenkyû: Zenkiroku* [*Tanaka Kakuei Research: A Total Record*], Kôdansha, Tokyo; Schlesinger, J.M., 1999. *Shadow Shoguns: The Rise and Fall of Japan's Postwar Political Machine*, Stanford University Press, Stanford; Hôsaka Masayasu, 1993. *Yoshida Shigeru to iu Gyakusetsu* [*The Paradox of Yoshida Shigeru*], Chûô Kôron Shinsha, Tokyo; Shiota Ushio, 1996. *Kishi Nobusuke*, Kôdansha, Tokyo; Arai Shunzô, 1982. *Bunjin Saishô Ôhira Masayoshi* [*The Cultured Prime Minister Ôhira Masayoshi*], Shunjûsha, Tokyo.
7. One exception is Curtis, G.L., 1971. *Election Campaigning, Japanese Style*, New York, Columbia University Press.
8. Curtis G.L., 1999. *The Logic of Japanese Politics: Leaders, Institutions and the Limits of Change*, Columbia University Press, New York, p. 4.
9. Machidori Satoshi, 2005. 'The 1990s reforms have transformed Japanese politics', *Japan Echo*, June 2005, pp. 38–43.
10. Tatebayashi Masuhiko., 2004. *Giin Kôdô no Seiji Keizaigaku: Jimintô Shihai no Seido Bunseki* [*The Political Economy of Diet Members' Activities: An Analysis of the System of LDP Rule*], Yuhikaku, Tokyo: p. 7.
11. Tatebayashi Masuhiko., 2004. *Giin Kôdô no Seiji Keizaigaku: Jimintô Shihai no Seido Bunseki* [*The Political Economy of Diet Members' Activities: An Analysis of the System of LDP Rule*], Yuhikaku, Tokyo: pp. 4-5.
12. Pempel T J., 1998. *Regime Shift: Comparative Dynamics of the Japanese Political Economy*, Cornell University Press, Ithaca.

2

BECOMING A POLITICIAN

Born in February 1945, Matsuoka Toshikatsu was the eldest son of an ordinary farming household in Aso Town in Aso County in Kumamoto Prefecture.[1] His family home (*jika*) remains there to this day, in the locality that has been central to his whole political life. The setting is quintessentially rural in the Japanese style. As Matsuoka himself puts it, 'in my boyhood, I grew up as a high-spirited young lad revelling in mother nature at the foot of Mt Aso, which is an active volcano in Kyushu.'[2]

EDUCATION AND EARLY CAREER[3]

After graduating from junior high school, Matsuoka lodged in Kumamoto City while attending the prestigious Seiseikô Prefectural High School in the city. He studied hard as well as learning karate, which he continued into his college days. One of his classmates at junior high school and high school said of him

> [h]e was not a 'clever student' but worked quite hard preparing for the university entrance examination. In a big contrast to now, he was a quite inconspicuous student. In his high school days, he joined a cheer group and karate club and pretended to be a straight-laced person, but I often saw him chasing girls. I had a question about his behaviour.[4]

After graduating from high school, Matsuoka tried for two years to enter the National Defence Academy. Matsuoka's choice of university was informed by his father's career as a professional soldier (a former member of the military police).[5] However, Matsuoka's academic record was not good enough for him to make it into the defence academy. He explained his failure by saying that although he passed the first-stage examination, he failed the second-stage examination, which was a medical test.⁶ He stopped taking the academy entrance examination, but continued to work hard and was accepted into Tottori University.

While Matsuoka was at Tottori University, he was not an especially remarkable student. A classmate said

> ...at that time, the wave of the student movement rushed here later than in the cities. When we were in our fourth year in spring, our school was blockaded. There were large numbers of apolitical students, and Matsuoka was one of them. He said, 'I want to be a soldier', and was popular in the karate club. He always wore jeans and shirts, and he attended classes faithfully, and his academic record was good.⁷

Matsuoka subsequently graduated with a BA in Forestry from the Faculty of Agriculture. He sat and passed the national public servant exam (*kokka kômuin shiken*) in Forestry and then the interview for entry into the Ministry of Agriculture and Forestry (MAF),⁸ as it was known in the 1960s. Because he passed the Level One national public servant exam, he became a 'career' (*kyaria*) government official. However, because Matsuoka was a graduate in Forestry, his post in the MAF was that of technical official or specialist (*gikan*), not a generalist, or policy administrator (*jimukan*). Unlike *gikan*, *jimukan* typically had university qualifications in law, economics (including agricultural economics) or administration. They were the 'élite course' bureaucrats who climbed to high-ranking executive positions within the ministry, such as administrative vice-minister, director-general of the forestry, fishery or food agencies, or director of a MAF bureau. Matsuoka belonged to the non-élite technical stream along with others who had qualifications in forestry, civil engineering, veterinary and livestock sciences, and other technical fields.⁹ His employment as a *gikan* meant that he could only occupy positions within the ministry designated for technical specialists. Moreover, he could not aspire to the top administrative posts within the MAF. Technical bureaucrats could only rise as high as bureau assistant director (*jichô*) and for this reason were often considered 'quasi-career' bureaucrats.¹⁰ Matsuoka's inability to reach the highest levels of the ministry was to provide a spur to his political ambitions.

Having joined the MAF in 1969, Matsuoka spent 19 years there. This was an occupation, in his own words, that was 'connected to regional areas, particularly agricultural and forestry industries, and agricultural mountain villages'.¹¹ A lot of his time in the ministry was spent in the Forestry Agency,¹² reputedly a strong-hold of technical bureaucrats with specialist qualifications in forestry. At one time Matsuoka became head of the Planning and Coordination Office of the Local Forestry Bureau in Akita Prefecture, and chief of the Teshio County Forestry Management Station in Hokkaido Prefecture. Hokkaido is Japan's most important prefecture in terms of the area of forestland (over 5 million hectares) and in terms of its forestry industry (logging). By 1988, Matsuoka had attained his highest post in the ministry as head of public relations in the Planning Division of the Forestry Agency.

Besides the Forestry Agency, Matsuoka also worked in the Minister's Secretariat, which is officially tasked with acting as a planning and general coordinating body for the entire ministry. In the Minister's Secretariat, he was head of the Planning Office. While working there, he was in charge of reformulating the 'Basic Plan for New Industry Cities and Industrial Infrastructure Special Regions' (*Shinsangyô Toshi to Kôgyô Seibi Tokubetsu Chiiki*), shifting the emphasis from prioritising production to prioritising livelihood-related facilities (*seikatsu kanren shisetsu*).[13] The new plan proposed the further conversion of agricultural land, fishing grounds and forests for the promotion of construction and infrastructure, including the development of 10,200 hectares of land for industrial use, 110 public housing units, 160 hectares for other residential usage, and water pipes, roads, railways, ports and other transportation facilities, and health and education facilities.[14]

Like all bureaucrats in the Ministry of Agriculture, Forestry and Fisheries (MAFF),[15] as it became in 1978,[16] Matsuoka was rotated every two to three years, enabling him, like other officials, to climb the *gikan* career ladder based on seniority rather than ability. He was also seconded for a while to the National Land Agency's Regional Development Division where he served as assistant director of the Mountain Village Heavy Snowbelt Development Division (a position reserved for *gikan*).[17] It was not uncommon for MAFF bureaucrats to be seconded for a time to positions in the National Land Agency, because agriculture and forestry are both land-based industries. Matsuoka was also attached to the Secretariat of the National Land Agency for a time. While at the Land Agency, he drafted the Peninsula Promotion Law (*Hantô Shinkôhô*), a private member's bill (Diet members' legislation) incorporating depopulated area countermeasures.[18]

After a climb up the career ladder in the MAFF to a moderately high position, Matsuoka launched himself into politics. Bearing in mind that the national bureaucracy prefers to recruit law graduates from prestigious Tokyo

University (Tôdai), and that many MAFF *gikan* with degrees in civil engineering and agricultural sciences come from the prestigious Kyoto University, one MAFF 'Old Boy' (OB) speculated that because a 'Tottori University graduate is a "lesser being" in the MAFF, Matsuoka may have realised that the limits of his ambitions within the organisation'.[19] Tottori University was derisively described as a 'Mickey Mouse university' by a Matsuoka critic on a public website.[20] Another MAFF OB commented

> …because Matsuoka failed to enter a university for two years, younger University of Tokyo graduates were promoted to important positions ahead of him. For that reason, it seems that he became a House of Representatives member, reversing his status with the big one, thus relieving years of pent-up feelings in one go.[21]

Matsuoka displayed the characteristics of a 'status incentive politician', described as 'individuals who became politicians because of a need for prestige'.[22] They are often former administrators.[23]

Whatever his true motivations, in 1988 Matsuoka decided to pursue a career in national politics, resigning from his last post as public relations officer at the Forestry Agency and returning home to Kumamoto. This was despite the fact he would have been eligible for a pension had he worked for just one more year.[24] It would seem that he risked everything to enter politics. As he stated publicly, he was likely to go broke by resigning from his job to enter politics.[25]

PREPARING TO ENTER POLITICS

Matsuoka's preparations for the political arena had begun while he was still working in the MAFF. He gained a reputation for proactive subsidy allocation to local areas. Such subsidies could later be converted into votes and political funds—vital for electoral success. There were numerous tales of his 'heroic' exploits during his administrative career[26] as a devisor of 'new works'. One former MAFF OB recounted that while Matsuoka was tackling measures to deal with heavy snow and problems with state-owned land—during his time in the Land Agency and Forestry Agency—he started new public works projects and acquired know-how for securing the necessary budget for these projects through his connections with politicians.[27]

Matsuoka also took the first steps along the path to acquiring political funds by acting as a broker, which was to prove both lucrative and potentially dangerous for his subsequent political career. In 1986, while a public relations officer in the Forestry Agency, Matsuoka tried to use his ministry position and contacts to obtain campaign finance to launch himself into politics. He offered to sell information to Sasaki Kichinosuke, one of the 'Kings of Real Estate' during Japan's bubble economy of the 1980s. At their peak, Sasaki's total assets were said to be worth ¥900 billion.[28] At the time, Sasaki was making arrangements for about ¥60 billion in bank loans for a tender to purchase a Forestry Agency site in Tokyo, located next to the Hotel Okura in Toranomon, Minato Ward. On the site was a demolished building. The person in charge at one of the banks, Kokumin Bank, which Sasaki had approached, brought Matsuoka along to meet Sasaki.[29] As Sasaki recalls

> Matsuoka said 'I have the survey map of the Forestry Agency tender site, and I want you to buy the survey map'. As an experiment, I asked how much the price would be to buy the survey map. But reversing the question, Matsuoka asked 'if you want the survey map, say the amount'. When we asked to copy the map, Matsuoka said 'Since the map is real, you cannot copy it'. Because we had this sort of an exchange, I became rather suspicious. I thought that if I had made a successful bid for the land already, the survey map was important. But if my bid were unsuccessful, the map was nothing but a piece of paper. A large number of companies were participating in the tender for the site of the demolished building, and the issue was picked up by the television program, 'NHK Special', thus attracting public attention. However, in the final analysis, because Kokumin Bank suddenly terminated financing to us, a rival company, Company A, made a successful bid. I heard that the survey map had been taken to Company A.[30]

In December 1987, Matsuoka (who was Forestry Agency public relations officer at the time) called Sasaki hoping to act as a broker for a local businessman in Kumamoto. As Sasaki recounts

> Matsuoka said 'to tell you the truth, I have a utterly shameless request but a local construction company, Company T, needs temporary funds. It owns a mountain, and so do you think there would be a good buyer in Tokyo'? I had a feeling at the time from what Matsuoka was saying that they wanted to borrow ¥30 million. The next day, the president of Company T came to Tokyo and to my company with Matsuoka. As I thought, the negotiation to sell the mountain became a negotiation to borrow money. Matsuoka pestered me, saying 'because the company is a devoted supporter, can you help the company somehow'? I replied, 'to help the company, Mr Matsuoka, please become a co-signer for the loan'. I lent ¥30 million for three months at the annual interest rate of 15 per cent. I accepted the mountain as security for the loan. However, the bill was not honoured on the day it was due to be settled. The deadline was changed about 36 times and the bill was even renewed. I found out later that the mountain was located next to Namino Village, Kumamoto Prefecture, where Aum Shinrikyo[31] followers were living in a group. Therefore, the mountain did not seem to have any value. Later, the loan was paid off, but Company T went virtually bankrupt.[32]

In preparation for launching himself into politics, Matsuoka tried not only to amass the necessary funds but also to cultivate useful political connections. He formed close relationships with leading members of the agriculture and forestry 'tribe' (*nôrin zoku*). They were Nakagawa Ichirô from Hokkaido (who served as Minister of Agriculture and Fisheries in 1977–78, was a candidate for the LDP presidency and prime ministership in 1982, and who later hanged himself in January 1983)[33] and Tamaki Kazuo, Director-General of the Management and Coordination Agency in the Nakasone administration. Both Nakagawa and Tamaki were members of the Seirankai (Young Storm Society), an overtly nationalist body with extreme right-wing views. In fact, all the *nôrin zoku* at the time, including Watanabe Michio (who took over from Nakagawa as Minister of Agriculture, Forestry and Fisheries in 1978–79), were members of Seirankai.[34] Matsuoka was intimately connected to Nakagawa through MAFF study groups and in other contexts, and he was also close to Tamaki.[35]

An ex-MAFF official, who was Matsuoka's former boss, revealed some of the background to Matsuoka's bid to enter politics.

> Ever since he was sent to Land Agency, he wanted to be a politician. About 15 years after he entered MAFF, he and two Tokyo University graduate career officials who entered the MAFF in the same year as him, were somehow unpopular with their boss. Matsuoka became disgusted and started talking about resigning. At the time, he was associated with Nakagawa Ichirô and Tamaki Kazuo [former Director-General of the Management and Coordination Agency, and a leader of Seirankai]. He intended to receive their support when he ran for an election. Unfortunately, just before Matsuoka ran for an election, Tamaki died suddenly [in 1987].[36]

Arai Satoshi, former MAFF official and Democratic Party of Japan (DPJ) member of the House of Representatives, recalled that when Matsuoka was a sub-section chief in the ministry, Arai formed a study group with him. However,

> [w]hen former MAF Minister Nakagawa, with whom I got acquainted in a study group, ran for election as LDP president, I heard that Mr Matsuoka had got into an election car and supported Mr Nakagawa. I made an international phone call from Sri Lanka, where I had been sent, and said 'isn't that exceeding the duty of a public servant'? Mr Matsuoka replied, 'since the LDP presidential election is not subject to the Public Office Election Law (*Kôshoku Senkyohô*), there is no problem'.[37]

Around about the time he stood for election, Matsuoka commented, 'my master was Nakagawa Ichirô, and my teacher was Tamaki Kazuo'.[38] After both these prominent politicians died in succession, Suzuki Muneo, Nakagawa's secretary, who stood successfully for his Lower House seat, stepped into the breach. He supported Matsuoka and thus took Nakagawa and Tamaki's places.[39]

A report also surfaced of Matsuoka's conducting pre-election campaigning while he was a government official. A political affairs reporter recounted

> [j]ust before Matsuoka ran for an election, when he was the Forestry Agency public relations officer, he returned to Aso Town, Kumamoto on the weekends and conducted an election campaign. At the time, pre-election campaigning by a government official became problematic in another municipality. Questions were raised about Mr Matsuoka's action.[40]

Matsuoka was already showing a propensity to bend the rules if it meant furthering his political career.

THE REQUISITES FOR SUCCESS

Having resigned from the MAFF in 1988 with a view to contesting a seat in the next House of Representatives election, Matsuoka needed three vital prerequisites. These were a local support base (*jiban*), name recognition (*kanban*) and money (*kaban*).

Local support base

Matsuoka's choice of constituency was preordained. He would stand for election in the 5-seat district of Kumamoto (1). Kumamoto was his home prefecture and his home town, Aso Town, in Aso County, was located in Kumamoto (1). Matsuoka's electoral challenge was thus mounted from his hometown, Aso Town. The Aso region would become his main voting base (*shujiban*).[41] This was the logical place on which to focus and construct his political support base. He could utilise all kinds of local connections, including family and social ties, links to local politicians and businesses, and associations with various social and economic groupings in the area, tapping into the forces of localism that generated 'hard' votes based on personal connections and loyalties. By mobilising these kinds of local community and blood ties, which were traditionally strong in rural areas, Matsuoka could secure support for a political career.

Matsuoka organised a personal support group (*kôenkai*) as his primary vote-gathering machine and body of grassroots supporters. Its main branches were in Aso Town and Kumamoto City. It was called the 'Matsutomokai'—the 'Matsu Friends Association'. It began with a few thousand members, but expanded as Matsuoka's campaign gathered steam. It functioned as Matsuoka's electoral organisation, campaign machine and political funding body all rolled into one.

Having chosen to stand for Kumamoto (1), Matsuoka sought official electoral endorsement from the LDP, which was his natural party given his background in agriculture and forestry. The majority of Japanese farm and forest owners, and rural dwellers supported the LDP, and, in those days, ex-MAFF bureaucrats always became LDP Diet members. However, Matsuoka's bid for endorsement by the LDP was unsuccessful.[42]

Part of the problem of securing the backing of the LDP was the fact that Matsuoka would be standing for one of the five seats in Kumamoto (1) at the same time as four sitting LDP members. In the previous Lower House elections (in 1986), the district of Kumamoto (1) returned four members from the LDP, plus one politician from the Kômeitô. The losers were candidates from the Japan Socialist Party (JSP) and Japan Communist Party (JCP). If the LDP endorsed yet another candidate, it would potentially split the LDP vote to the point where perhaps three or less LDP candidates would be successfully elected. Furthermore, the LDP had performed poorly in the 1989 Upper House election, and so it was likely that the total LDP vote would be down in the subsequent Lower House election.

The party also had a rule about endorsing only those candidates who had a strong local organisation and/or organisational support, name recognition and money. These attributes were far from assured in Matsuoka's case. At the same time, from Matsuoka's perspective, LDP endorsement was neither a necessary nor a sufficient condition for his candidature. Under the Lower House multi-member district (MMD) system, standing as an Independent would not necessarily be a barrier to success, because in the personalised, candidate-centred elections in which LDP candidates from different factions competed against each other, the party label was of secondary importance.

Matsuoka would run on the basis of a mobilised personal vote, not on the basis of his party affiliation. His own individual support group could step into the breach as an organisation providing local backing. In this respect, Matsuoka was no better or worse off than LDP candidates, who similarly relied on their own *kôenkai* to connect with voters. His personal support group would provide an organisational setting in which he could conduct various campaign activities directed at local voters in specific regions and occupational fields connected to his own interests and expertise.[43] Even as a member of the LDP, Matsuoka would not necessarily have had name recognition in the broader electorate, which he would have had to establish independently of the party.

Lacking LDP endorsement, Matsuoka formally stood as an Independent. This was not unusual for first-time candidates (and others) who could not secure the backing of the ruling party. Matsuoka's move to style himself as a 'conservative Independent' in the 'conservative kingdom' (*hoshu ôkoku*) of Kumamoto Prefecture further confirmed this. Independents like Matsuoka were simply LDP candidates who had been unsuccessful in securing the party's endorsement. As everyone knew, they were LDP in all but name.

Name recognition

Matsuoka was already well known in the Aso region because that was where his family home was and where his mother still lived. His pre-campaign activities, conducted whilst still in the ministry and in the period between when he resigned from the MAFF and the election, were directed at getting his name more widely known across the district. He organised meetings with local voters to publicise his candidature and to expand and consolidate his political support base. He painstakingly built a support base and fought an uphill battle against rival candidates, especially the well-established LDP candidates.

Matsuoka had good connections in Kumamoto City where he had attended Seiseikô High School, which had an influential alumni association. Reputedly its OB connections were abnormally influential in elections.[44] It was alleged that 'behind Matsuoka's latent power was this Seisei power, and, it is said, his connections with Seisei-line *yakuza*.'[45] On the other hand, having been a MAFF bureaucrat bestowed a certain degree of status and respectability as well as policy knowledge and a natural link to large numbers of farm and rural voters. Matsuoka claimed to be 'famous both in name and in reality for being an expert in agriculture, forestry and fisheries, which are especially the foundation of the country.'[46]

The electoral district of Kumamoto (1) was semi-urban. In 1990, it had 413 persons per square km of population density compared with a national average of 332 persons, and it had five cities including Kumamoto City.[47] At the same time, it had five counties (including Aso County), encompassing more rural farming districts. The semi-urban character of the constituency meant that Matsuoka's election-campaign strategy could not be geared solely to rural dwellers, including agricultural and forestry voters.

According to the 1990 census, there were 23,121 farm households in the electorate, which made up 6.4 per cent of the total number of households.[48]

At an average of 3.7 voters per farm household,[49] this comprised a potential support base of 86,268 farm votes, which was only 13.4 per cent of the total cast vote in Kumamoto (1).[50] Almost every farm household vote would need to be secured in order to win a seat based solely on the farm household vote. This made it necessary for Matsuoka to cast the net for potential supporters much wider.

Money

As previously noted, Matsuoka had already tapped into funds from business during his years in the MAFF by offering his services as a broker with company executives in exchange for money. As a declared candidate, however, Matsuoka established six organisations to gather political funds. Altogether, they collected a substantial total of ¥131 million for his first election bid.[51] The first and most important of these was his personal support group, the Matsuoka Toshikatsu *kôenkai*, which was under the legal jurisdiction of the Kumamoto Prefecture Election Administration Commission, and which gathered ¥63 million.[52]

The balance of officially reported funds was provided by five political funding groups under the administration of the Ministry of Home Affairs. These groups were the Matsuoka Toshikatsu New Century Politics and Economic Discussion Association (Matsuoka Toshikatsu Shinseiki Seikei Konwakai), the 21st Century Discussion Association (21 Seiki Konwakai), the Green Friends Association (Ryokuyûkai) and the Matsuoka Toshikatsu Policy Research Association (Matsuoka Toshikatsu Seisaku Kenkyûkai). The Policy Research Association recorded the highest amount at ¥26 million.[53]

Matsuoka had direct financial support from a key political backer in Tokyo, who was already an LDP member of the Diet and who wanted to build his own loyal following amongst LDP Diet members—a vital prerequisite for becoming a faction leader and holding high political office, including the prime ministership. Chairmanship of a faction guaranteed one's candidature for the party's presidency. This politician was Suzuki Muneo,[54] Nakagawa's successor as Matsuoka's patron. Suzuki made a good substitute for the LDP faction that would have selected Matsuoka for party endorsement and provided him with political funds, had Matsuoka been an official candidate of the LDP.

Political revenue and expenditure reports for 1990 reveal that Matsuoka's political funding groups received direct donations from Suzuki's own political funding groups. The Matsuoka Toshikatsu Policy Research Association received ¥5 million from the Osaka Food Distribution Research Association (Ôsaka Shokuhin Ryûtsû Kenkyûkai), the Green Friends Association received ¥10 million from the 21st Century Policy Research Association (21 Seiki Seisaku Kenkyûkai) and the 21st Century Discussion Association received ¥10 million from the Hokkaido Development Research Association (Hokkaidô Kaihatsu Kenkyûkai).[55] The total sum obtained from Suzuki was ¥25 million, and, despite Matsuoka's not securing LDP endorsement, Suzuki also came to Matsuoka's side during the campaign.[56]

THE CAMPAIGN

The February 1990 Lower House election followed the July 1989 Upper House election in which the LDP was 'defeated', meaning that it lost its Upper House majority for the first time since 1955. The 'defeat' was caused by three main factors: voters' rejection of the consumption tax (introduced in April 1989); the Recruit scandal tainting a large number of LDP Diet members, including many of its prominent leaders and cabinet members; and among farmers, a wholesale rejection of the Takeshita government's December 1988 agreement to liberalise the beef and citrus markets. The 1989 election became one of the JSP's biggest post-war election victories, with many women candidates scoring victories over standing LDP members. The same political wave carried over to the February 1990 election. Matsuoka was able to turn his non-endorsement by the LDP into an electoral advantage by mounting an anti-LDP offensive, campaigning against the newly introduced consumption tax and tapping into farmers' dissatisfaction with the government's agricultural policy: an issue on which the party remained vulnerable.

Matsuoka ran a typically candidate-centred campaign. He presented himself as 'an independent political entrepreneur with his own local organisation and his own marketing strategies'.[57] He was able to take advantage of an electoral system in which the individual basis of the vote (*kojin honi*) was extraordinarily strong, and the party basis (*seitô honi*) was extraordinarily weak.[58] His campaign slogan was 'Momotaro (peach boy) of the Heisei era'. Momotaro was a hero in Japanese folklore who destroyed the marauding *oni* (ogres). Matsuoka's catchphrase was 'Momotaro in Heisei destroys the demons'.[59] Another prominent Matsuoka campaign slogan was 'I am mounting a crusade against misgovernment' [*akusei taiji ni idomu*].[60]

Unlike most former MAFF officials with political ambitions, Matsuoka did not get the backing of the MAFF for his campaign.[61] He did not fit the normal pattern of an ex-MAFF official seeking national political office. Not only was he from a low-ranking university in a regional backwater, but also he did not occupy a particularly high position in the MAFF when he resigned from the ministry. Perhaps, most importantly, as a *gikan*, Matsuoka was not an OB from the Structural Improvement Bureau (now Rural Development Bureau), with links to the land improvement industry. This industry represented a vast and lucrative agricultural public works enterprise, which was a very important source of votes, organisational support and political funds for MAFF land improvement *gikan* who entered politics. Matsuoka did not have the advantage of this kind of leg-up into the political world. He did not possess the right qualifications to call himself a civil engineering *gikan*,[62] and not being a *jimukan*, he could not base his campaign on being an 'organisational representative' (*soshiki daihyô*) of the MAFF in the Diet. He did not, for example, have an ex-MAFF administrative vice-minister heading up his campaign organisation, nor did he have campaign functionaries who were MAFF OBs.[63] In spite of all these disadvantages, Matsuoka tried to use his known MAFF connections to good effect in the election campaign.

In Kumamoto (1), eight candidates were competing for five seats. It was known as a closely contested constituency.[64] The JSP candidate, Tanaka Shôichi, was campaigning on an anti-consumption tax ticket, targeting housewives.[65] On that basis, he was thought to have scored a lead over the conservative camp. Kitaguchi Hiroshi from the LDP was a former director of Kumamoto City Agricultural Cooperative (Nokyo) and apparently had the agricultural, forestry and fisheries votes sewn up; while Noda Takeshi, another LDP Diet member and former Minister of Construction, reputedly obtained 'hard' votes from Kumamoto City and other urban areas in the electorate.[66] He was a leader of the commerce and industry 'tribe' (*shôkô zoku*), having been chairman of the LDP's Commerce and Industry Division and chairman of the Lower House Commerce and Industry Committee. Noda was well versed in fiscal, tax and economic policy and was also prominent in the LDP's Special Coal Countermeasures Committee, a salient fact in Kyushu given that at the time coalmines were being shut down in the prefecture.

The rest of the candidates were supposed to be fighting it out for the remaining votes. This group included the JCP and Kômeitô candidates, and the other two LDP candidates, Uozumi Hirohide and Matsuno Raizô. Uozumi had infiltrated the commerce and industry vote: he was a large stockholder in a road paving company, former chairman of the prefectural Chamber of Commerce and Industry, the former mayor of Kumamoto City as well as a former prefectural assembly politician. Moreover, he was knowledgeable in all prefectural issues associated with agricultural policy and regional development. He was a long-time rival of Matsuoka's, having attended Kumamoto High School, a rival school to Seiseikô High School. He also differed from Matsuoka in having made his way into politics through mayoral and prefectural office, compared with Matsuoka who was an ex-bureaucrat seeking a career in national politics.[67]

Matsuno was a prominent and long-standing LDP politician from Kumamoto, with a good base of support in both regional areas and in the cities, where he had been chairman of a brewing company. Matsuno had been in the Diet almost continuously since 1947, elected in only the second election after the war. He was so senior in the LDP that he had been minister of almost everything. He usually received the backing of the agricultural cooperatives and had been a former Minister of Agriculture and Fisheries in Prime Minister Sato's administration. He had also been Minister of Transport, Director-General of the Defence Agency, and also chairman of the LDP's PARC and Executive Council.[68] However, pre-election coverage of his campaign by the media suggested that the Matsuno camp was in crisis mode because of the powerful rollback in support for Matsuno in local regions.[69]

VICTORY!

When the results of the election were finally declared, Matsuoka scraped in at the bottom of the victors' list (in fifth position), but for him, the most important thing was that he had won a seat in the Lower House (see Appendix). Matsuoka described the electoral contest and his subsequent victory in the following terms: 'despite being an unknown candidate, I won by a narrow margin in the most famous, closely contested constituency in the whole country, after defeat seemed certain'.[70] The media reported that Matsuoka had put up a good fight.[71]

The final result saw the LDP lose two of its seats in Kumamoto (1) with the usual ranking of candidates in the electorate completely overturned. The two victorious LDP candidates were ranked lower than the two opposition party members. The biggest vote-winner was the JSP candidate,[72] followed by the

Kômeitô candidate, followed by Noda and Uozumi, with Matsuoka coming in behind the two LDP candidates.

In winning a seat in Kumamoto (1), Matsuoka was victorious over Matsuno and Kitaguchi, both well-established LDP Diet members. As he boasted himself, 'I pushed aside senior (*senpai*) Diet members and was successfully elected'.[73] He scored extremely well against Matsuno, beating him by just over 2,300 votes (see Appendix A).[74] In fact, Matsuno's support was broadly distributed across the cities and towns of Kumamoto (1). He won more votes than Matsuoka in all counties and cities of the electorate except Kumamoto City and Aso County, although he was first–place getter in Oguni Town in Aso County. In the total vote count, Matsuno's county vote was lower than Matsuoka's, while his city vote was higher.[75] It was Matsuoka's really solid showing over Matsuno in Aso County (24,905 votes compared with 6,795 votes)[76] that gave Matsuoka victory over Matsuno, because Matsuno beat him in all the other counties and in the total city vote, but by a smaller amount overall. Matsuoka's slightly stronger showing in Kumamoto City also helped. Media commentary concluded that Matsuno's failure was due to the fact that he could not win against the tide of generational change. Matsuno himself also attributed his defeat to the changing of generations.[77]

Matsuoka won by a smaller margin of around 360 votes over Kitaguchi, a second-generation politician and son of a Nokyo politician who had previously held the seat. Kitaguchi was also a Nokyo man through and through and had some of the agricultural cooperatives in the prefecture mobilising an organisational vote on his behalf. Despite this, Matsuoka still edged him out. Kitaguchi was stronger than Matsuoka in the cities including Kumamoto City, but despite winning more votes than Matsuoka in three counties, he scored a lower overall vote in the counties.[78] Once again, it was Matsuoka's extraordinarily strong showing in Aso County that came through for him, producing a much higher county vote. In fact, no other candidate came anywhere near Matsuoka as an electoral performer in Aso County. Nevertheless, taking the larger view, there was not much to separate Matsuoka, Matsuno and Kitaguchi, who won 12.6 per cent, 12.2 per cent and 12.5 per cent of the total vote respectively.[79]

Matsuoka's electoral victory could be attributed to a number of key factors. First, in standing as a conservative Independent, he was not directly identified with the LDP, and in fact could use his independent stance to attack it for policies that were highly unpopular at the time. Choosing an Independent candidate who failed to get LDP endorsement was one of the ways that traditional LDP voters could cast a protest vote against the party. Constituents could still vote conservative without going the whole way and voting for an opposition party member in the full knowledge that a candidate standing as a conservative Independent would invariably join the LDP after the election. This option 'regularly produced the defeat of LDP incumbents even as the LDP retained power'.[80] Matsuoka gained anti-LDP protest votes over issues such as agricultural trade liberalisation (in rural areas) and the consumption tax and money politics scandals (amongst city voters).

Second, as already emphasised, Matsuoka had an extremely solid electoral base in the Aso region. It was really Aso County with its 24,905 votes that delivered Matsuoka his seat. Matsuoka won more votes in Aso County than in any other single county (see Table 2.1). This county alone out of the five counties in Kumamoto (1) provided Matsuoka with just under a third of his total vote (see Table 2.1). He was the most popular candidate in 11 out of 12 towns in Aso County (see Table 2.1). Only in Oguni Town did he cede first place to another (Matsuno, as noted above). Matsuoka was also popular in Ozu Town and Kikuyo Town in Kikuchi County, where he was ranked first and second-highest vote-winner amongst all the candidates (see Table 2.1). In fact, Matsuoka was the most popular candidate overall in the rural counties, winning top place as vote-getter, with a total of 41,690 votes, or 51.5 per cent of his total vote tally (see Table 2.1). Of course, the Aso vote helped put him into this position, beating both Matsuno and Kitaguchi, who were also relatively strong in the counties.[81] Aso County provided more than half of Matsuoka's total vote tally in the counties (41,690). Matsuoka clearly had a strong, geographically concentrated *jiban* in the Aso region. The media reported at the time that Matsuoka's campaign centred on the fact that he was from the Aso region.[82]

Third, Matsuoka was a MAFF OB, which would have gained him the votes of both current and retired MAFF officials residing in his electorate as well as support from primary industry voters in the electorate. In addition, given his career background, some of his votes were undoubtedly generated by his alignment with and knowledge of agriculture and forestry matters and interests. He certainly stressed this in his campaign. He described himself in the Diet handbook as follows

> [b]ecause of my background as a bureaucrat, I have my own original ideas about policies such as the consumption tax and agricultural and fisheries policies and so on. If you show considerable spirit to the citizens, then they will want to make friends with you and will become attached to you.[83]

Matsuoka's success in the rural counties was indicative of strong sectoral (that is, agricultural and forestry) support. Because Matsuoka had been in the MAFF, farmers and forest owners regarded him as having useful personal connections with currently serving ministry officials, which could be mobilised to secure policy benefits such as subsidies for agricultural and forestry projects as well as other material benefits in the form of income support and border protection for agricultural products.

Six ex-MAFF bureaucrats were successful in the 1990 elections, four of whom represented electorates in Kyushu—including Matsuoka. He joined 115 other former bureaucrats-turned-politicians in the Diet in 1990, representing 15.2 per cent of the total Diet membership.[84] Amongst this group, 15 were from the MAFF.[85] It was not unusual for former bureaucrats to enter Diet politics. In fact, the central government bureaucracy was one of the main recruiting grounds for national politicians.

Matsuoka, however, was not your usual bureaucrat-turned-politician. Almost all former bureaucrats were so-called 'élite course' bureaucrats: that is, they had graduated from the Law Faculty of the University of Tokyo, and had served in their ministries as *jimukan*, not *gikan*. Matsuoka was from lowly Tottori University Faculty of Agriculture, with a degree in Forestry. The MAFF OBs in the Diet in 1990 were all Tôdai Law Faculty graduates, except for one who was a graduate of Tokyo University's Faculty of Agriculture, and another who was a graduate of Kyoto University's Faculty of Agriculture.[86] As already noted, Matsuoka had an unconventional background for an ex-MAFF Diet member. Perhaps that is why he was so ambitious and had so much to prove.

Even though Matsuoka was the most popular candidate in the rural counties in his electorate, and even though he was the candidate with the least number of votes in the cities (apart from the JCP candidate)[87]—both of which underlined his rural orientation—he could not consider himself to be solely a representative of farm and forestry interests. Indeed, no significant statistical correlation was observable between the percentage of farm households in a particular municipality and the percentage of the total vote Matsuoka received from that municipality.[88] This was due to two main factors. Matsuoka could not break into the electoral power bases of the other LDP candidates, in particular, rural counties (Kitaguchi was relatively strong in all counties except for Aso County, but was particularly strong in Tamana County, while Matsuno was quite strong across all counties but particularly in Kikuchi County).[89] Matsuoka was,

Table 2.1 Farm household composition/votes cast for Matsuoka by municipality in Kumamoto (1) in 1990 Lower House election

Name of Municipality	No. of Farm Households	Farm Households as % of Total in Municipality/ies	Votes Cast for Matsuoka	% of Total Vote Cast	% of Matsuoka's Total Vote	Placing among 8 Candidates
Cities	6,002	2.3	39,183	9.7	48.5	7th
Kumamoto City	2,237	1.1	34,704	11.4	42.9	5th
Arao City	475	2.6	1,078	3.1	1.3	8th
Tamana City	1,151	8.7	1,427	5.2	1.8	7th
Yamaga City	1,035	9.9	1,161	5.6	1.4	7th
Kikuchi City	1,104	13.6	813	4.4	1.0	7th
Counties	17,119	17.2	41,690	17.5	51.5	1st
Hotaku County	2,383	18.6	2,084	7.2	2.6	7th
Hokubu Town	601	11.0	945	9.0	1.2	7th
Kawachi Town	857	43.4	292	5.1	0.4	7th
Akita Town	329	11.9	392	6.2	0.5	7th
Temmei Town	596	23.0	455	6.9	0.6	7th
Tamana County	3,525	16.7	2,853	5.7	3.5	7th
Taimei Town	449	11.1	314	3.4	0.4	7th
Yokoshima Town	600	42.3	209	5.1	0.3	7th
Tensui Town	881	49.9	104	2.0	0.1	7th
Gyokuto Town	337	20.5	318	8.0	0.4	7th
Kikusui Town	209	10.5	598	11.7	0.7	5th
Mikawa Town	445	26.2	154	3.6	0.2	7th
Nankan Town	411	11.9	557	7.2	0.7	7th
Nagasu Town	193	3.8	599	5.7	0.7	7th
Kamoto County	3,144	29.8	3,193	8.4	3.9	7th
Kahoku Town	399	26.3	538	12.8	0.7	4th
Kikuka Town	566	27.7	352	6.1	0.4	6th

Kamoto Town	452	17.7	504	8.5	0.6	7th
Kao Town	441	29.4	394	9.4	0.5	7th
Ueki Town	1,286	43.7	1,405	7.9	1.7	7th
Kikuchi County	3,296	10.4	8,655	13.1	10.7	4th
Shichijo Town	483	33.9	222	5.39	0.3	6th
Kyokushi Town	388	28.2	313	8.4	0.4	6th
Ozu Town	697	9.9	3,184	22.7	3.9	1st
Kikuyo Town	575	8.5	2,181	16.8	2.7	2nd
Koshi Town	354	6.7	1,086	16.2	1.3	7th
Shisui Town	478	14.9	412	5.5	0.5	7th
Nishigoshi Town	321	4.8	1,257	9.9	1.6	5th
Aso County	4,771	20.7	24,905	45.5	30.8	1st
Ichinomiya Town	468	15.0	2,845	41.1	3.5	1st
Aso Town	986	17.1	8,741	64.0	10.8	1st
Minamioguni Town	344	25.7	1,261	36.7	1.6	1st
Oguni Town	420	15.0	1,489	21.7	1.8	2nd
Ubuyama Village	196	41.0	566	41.7	0.7	1st
Namino Village	220	40.7	574	40.6	0.7	1st
Soyo Town	479	33.0	1,822	50.4	2.3	1st
Takamori Town	523	21.4	2,468	43.4	3.5	1st
Hakusui Village	439	35.2	1,388	42.6	1.7	1st
Kugino Town	241	35.2	959	51.1	1.2	1st
Choyo Village	191	10.3	1,869	56.6	2.3	1st
Nishihara Village	264	19.8	923	27.1	1.1	1st
Total/average	23,121	6.4	80,873	12.6	100.0	5th

Sources: Sōmuchō, Tōkei Kyoku, *Heisei 2-nen Kokusei Chōsa Saishū Hōkoku Dai 3-kan Dai 2-ji Kihon Shūkei Kekka Sono 2 Todōfuken, Shichōson-hen 43 Kumamoto-ken*, pp. 288-289 and 294-303; Asahi Shinbunsha Senkyo Honbu, *Asahi Senkyo Taikan: Dai 39-kai Shūgiin Sōsenkyo*, p. 194.

therefore, competing for the farm vote against Matsuno and Kitaguchi: the two well-known politicians with established agricultural connections in the electorate.

The other factor was the large number of votes Matsuoka received from Kumamoto City. This proved to be the fourth and final reason why Matsuoka was victorious in the election. The cities in Kumamoto (1) generated 39,183 votes for Matsuoka, which amounted to 48.5 per cent of his total vote. The most important aspect of the distribution of his support was the cluster of votes Matsuoka won in Kumamoto City, which was his largest single source of support (34,704 votes, or 42.9 per cent of his total vote as shown in Table 2.1).

The fact that Matsuoka's city votes were so concentrated in one place suggests that they were generated less by programmatic appeals (Matsuoka's criticism of the consumption tax) than by more traditional kinds of social networks, in this case, the network centring on Matsuoka's old high school alumni association, which reportedly acted as Matsuoka's biggest vote-collecting machine.[90] One of his old school mates whose support he solicited said of him

> [h]e pretended to be a hard-liner, but I got the impression that he was not a fervent soul. He was definitely not a person who left a good impression. However, since he was running for election, all his classmates supported him. A rally was held at a hotel. Although people said he would lose the election, he was able to get elected, and we were glad that it was worth supporting him. But after he was elected, he did not thank us. I have supported various other people, but it is unusual for there to be no telephone call to express his gratitude.[91]

In summary, Matsuoka was able to carve out an electoral niche for himself in Kumamoto (1) based primarily on his home town and Aso county connections, and his network of old school ties in Kumamoto City, in addition to his identification with and career background in the agricultural and forestry bureaucracy. His victory was constructed on the basis of an electoral coalition, which combined personal and local connections with the backing of special interests in agriculture and forestry, which married an area, or geographic zone (*chiiki wari*), vote to a sector or policy field (*seisaku bunya wari*) vote.[92] In the mix were also programmatic (anti-LDP) appeals, particularly on issues such as the consumption tax and 'misgovernment'. Because the election was held in 1990, Matsuoka was able to exploit the LDP's electoral vulnerabilities at the time.

The *Yomiuri Shinbun* attributed Matsuoka's victory to a big rise in the mood for generational change, given that Matsuno was 73 (which the newspaper identified as a salient feature of the whole election, given that new candidates won 25 per cent of the seats in that election).[93] Matsuoka himself was 45 years old, a relatively young age for entering politics, particularly following a bureaucratic career. The paper referred to Matsuoka's forming his political camp for the first time, appealing for political renewal (*seiji sasshin*) and for generational change in the representation of the electorate (*sedai kôdai*), and his gaining support from his alma mater (his old school in Kumamoto City) and from Kasumigaseki OBs.[94]

Although Matsuoka was not endorsed by the LDP, he was seeking support predominantly from customary LDP voters. The composition of his electoral support cut across both the two main vote divisions in Lower House electoral politics—geographic region and sectoral specialisation. In this fashion, Matsuoka, (albeit standing as an Independent), was able to differentiate himself from other LDP candidates by pointing out his special characteristics as an individual candidate (*tokka*) in terms of both region and policy field.

JOINING THE LDP

After he won his Diet seat, Matsuoka joined the LDP. There was no contradiction in his having attacked the party during the campaign and joining it afterwards. Becoming a member of the LDP after the election was the norm for successful Independents who had sought but failed to get the party's endorsement in the election. The custom was for the party to welcome such Independents into the fold, especially if the size of the LDP's majority in the Diet were an issue. The LDP, internally, contained both government and opposition forces in the form of the mainstream (the factions who supported the LDP president) and the anti-mainstream (those factional groupings who opposed his election). These vertical divisions were reinforced by horizontal ones amongst groups of Diet members representing differing sets of sectional interests, who often opposed their own administration's decisions on policy as part of an entrenched pattern of 'government versus party' conflict in policymaking.

Within the LDP, Matsuoka joined the Seiwakai faction led by Mitsuzuka Hiroshi,[95] and received a ¥20 million in political funds from the faction.[96] However, it was Suzuki (who was not a member of this faction) and not Diet members who belonged to the Seiwakai, who took Matsuoka around and introduced him to each office in the Diet members' office building.[97] Suzuki's 21st Century Policy Research Association also donated ¥50,000 towards Matsuoka's celebration party.[98]

Moreover, the LDP had lost two of its sitting members in Kumamoto (1), and so Matsuoka would bolster its ranks. Last but not least, victory in 1990 was a sufficient demonstration to the LDP that Matsuoka was a vote-winner, especially as Matsuno and Kitaguchi gave up politics altogether. In this way, Matsuoka gained membership of the LDP via the back door. The LDP endorsed his candidature in all subsequent Lower House elections—in 1993, 1996, 2000, 2003 and 2005.

Matsuoka also had to join the party of government in order to place himself in a position to influence policy, particularly measures relating to agriculture and forestry. Being a member of the ruling party was also vital in order to gain direct access to administrative officials who exercised discretionary power over the allocation of public works projects to particular regions and public works contracts to particular companies, and/or who could grant particular administrative dispensations to companies, individuals and organisations. This was virtually impossible for an Independent member of the Diet or a member of the Opposition to do.

WITH A VIEW TO THE 1993 ELECTION...

The election of 1993—which was the first poll that Matsuoka contested as an officially endorsed LDP candidate—was very different from the 1990 poll. On the positive side, Matsuoka started much further ahead than in 1990 when he was a newcomer striving to break into the ranks of Diet members representing the electorate. In the interim since the last election, Matsuoka bolstered his name recognition and consolidated his support base in Kumamoto (1). His first term in the Diet and his membership of the LDP assisted in both respects, attracting greater numbers into his *kôenkai* in search of the favours and benefits that flowed from having a personal, institutionalised connection and channel of communication with a Diet member.[99]

Another factor that augured well for Matsuoka was the LDP's split prior to the election, which left only two LDP candidates standing in Kumamoto (1) instead of the customary four. The LDP might have split at the centre, but for electoral purposes it was always split at the local level, meaning that Matsuoka was facing no more electoral competition than normal. Matsuoka and Noda were the only two LDP-endorsed candidates left. Noda's support base was primarily in the cities.[100] Uozumi had left to join the breakaway Renewal Party,[101] leaving Matsuoka as the strongest LDP candidate in rural areas with connections to primary industry voters. One of the advantages of standing as a sector-specific candidate was that Matsuoka could tap into support from the interest groups operating in the agricultural and forestry sectors. Representing their interests was a way of harnessing their support over the entire electoral district and receiving hard support, meaning reliable votes, in return.[102]

Over time, Matsuoka built up stronger connections to the agricultural cooperative organisations in his district, including the Kumamoto Prefecture Farmers' Political League (Kumamoto-ken Nôgyôsha Seiji Renmei, or *nôseiren*) as well as at the centre, including the national political arm of the agricultural cooperatives, the National Council of Farmers' Agricultural Policy Campaign Organizations (Zenkoku Nôgyôsha Nôsei Undô Soshiki Kyôgikai), or National Council (Zenkoku Nôseikyô) for short. This grouping was newly established in 1989. Matsuoka addressed gatherings of local representatives of the agricultural cooperatives and attended large functions of the Nokyo faithful organised by the National Council.

Matsuoka also developed links with specific producer groupings such as dairy farmers through the prefectural dairy farmers' political league (*rakunô seiji renmei*). He addressed national conventions of dairy farmers and made a point of attending the joint councils of local and central dairy farmers' political leagues. He also built a close relationship with tobacco farmers through the tobacco cultivation associations (*takabo kôsaku kumiai*). He became an advisor to the National Central Union of Tobacco Cultivation Associations (Zenkoku Tabako Kôsaku Kumiai Chûôkai), which was the mouthpiece for the country's 23,000 tobacco farmers. It campaigned directly against the Ministry of Health and Welfare's plans to cut national smoking rates. The total number of leaf tobacco farmers was small, but they were strongly unified. Moreover tobacco farming was important in Kumamoto Prefecture.[103] Tobacco farmers in Kumamoto (1) thus became a reliable source of votes for Matsuoka.

As an OB of the Forestry Agency, Matsuoka had a natural connection with the forest associations (*shinrin kumiai*) that represented forest owners, who harvested their timber for sale on the domestic market in Japan.[104] In Kumamoto Prefecture, there were approximately 42,000 members of the forest associations. Their activities included logging (trees that were more than 40 years old), marketing logs and providing guidance for members about management of their forestland.[105]

For all of these primary industry groupings, the key issue was the price received for their product on the domestic market. As most commodities were subject to administrative price intervention, LDP Diet members could exert influence on price policymaking, and in so doing, harness the organisational backing of these groups in elections. Between 1990 and 1993, Matsuoka put his foot on the lower rungs of all the appropriate ladders in farm and forestry policymaking in the LDP and in the Diet.[106]

Going into the election, Matsuoka had much stronger organisational support from agricultural and forestry groupings than he did in 1990. This also helped to differentiate him from Noda and Uozumi. Noda, as before, centred his voting base on urban, commercial interests, and Uozumi, while ex-LDP, obtained his primary backing from commerce and industry organisations.

On the negative side, one of the other candidates contesting the election was Hosokawa Morihiro, the very popular ex-governor of Kumamoto Prefecture (1983–91). This was an election in which Hosokawa, who was playing the main role in the Japan New Party[107] boom, was a powerful force eating into the support strata of both the LDP and JSP. Indeed, everyone standing in Kumamoto (1) was affected by the 'Hosokawa typhoon'. Although Matsuoka and Noda, as the only two LDP candidates, tried to use the split in the LDP as some kind of springboard to consolidate their support amongst conservative voters, some groups, hitherto out-and-out supporters of the LDP such as the Kumamoto Prefecture *nôseiren*—Nokyo's political front organisation—and the prefectural medical association, decided to support Hosokawa and the Japan New Party.[108] Matsuoka's face reportedly went very pale when Hosokawa stood as a candidate.[109] The two were reputedly not friendly with each other at all. The constituency that Matsuoka inherited was from Fujita Yoshimitsu who was a distant relation of Hosokawa's. Matsuoka 'feared that in regional areas where local connections and blood relations are important, the people he was counting on for votes would support Hosokawa like tumbling snow'.[110] As one commentator put it, 'Hosokawa was the "emperor". He was a famous governor. He transcended local and blood connections and took votes as if a typhoon had blown through. Matsuoka of course, went for bribery, and furiously distributed rice balls with ¥10,000 notes in them. There were police in his office despite it being during an election period'.[111]

Hosokawa's entry into the race, and his absorption of votes that would have normally flowed to the other parties, meant that the competition amongst the remaining candidates was made all the more severe. Noda, Matsuoka and Uozumi in the conservative camp were slightly ahead, while Tanaka and Kurata Eiki (representing the JSP and Kômeitô respectively) were about even. Noda had very strong support from the conservative camp, whilst Matsuoka had thoroughly infiltrated the rural areas. Unluckily for Tanaka, some of the progressive/reformist vote was also flowing to Hosokawa. Noda Masaharu, standing as an Independent, was appealing for support for his own independent rice policy.[112]

BEER COUPONS OR VOTE-BUYING?

For Matsuoka, money was another issue altogether. In 1991 and 1992 (non-election years) official donations to his political funding groups and personal support organisation dipped—from ¥131 million in 1990 to around ¥110 million and ¥104 million respectively.[113] As 1993 was an election year, funding levels rose again—to around ¥126 million.[114] The largest portion flowed to the Matsuoka Toshikatsu New Century Politics and Economics Discussion Association (¥75 million), with ¥56 million donated to Matsuoka's *kôenkai*.[115]

Kumamoto was infamous for the old-fashioned vote-buying system of finding a ¥10,000 note in a cabbage delivered to voters' houses. Even in the 1990s, this system was still in evidence.[116] Buying votes with money was described as 'commonsense' in some local towns and villages of the prefecture.[117] Kumamoto was also famous for the distribution of soy sauce and pickled leaf mustard at election time.[118]

One month before the 1993 election, Matsuoka ordered ¥25 million worth of beer coupons contained in 10 boxes that were delivered to his Diet office. To obtain the beer coupons, he used Mitsukoshi Department store's reception system that allows unlimited credit for certain special customers.[119] Matsuoka had been introduced to Mitsukoshi by Hirota Daisuke, another user of the system. Hirota, a self-proclaimed construction consultant, was a character with a history.[120] Rumours connected him to a former politician's secretary, Ozaki Mitsurô, who was arrested over a scandal and who was also audited by the National Taxation Office.[121] Hirota was chairman of a company called Miyauchi Denkô, which changed its name to Japan Orando.[122] Matsuoka became a reception client in June 1993, and almost immediately issued his first order for beer coupons. Because it was such an expensive order, the store liaison officer assigned to look after Matsuoka initially thought to himself, 'you must be joking'.[123] However, when he checked with Matsuoka's office, Matsuoka's secretary brushed aside his reservations saying, 'there's no time. We need it next month' (July 1993, when the Lower House election was due to be held).[124] The beer coupons, like tailoring coupons, were extremely convertible, and so Mitsuokoshi did not want to go ahead with the expensive sale.[125]

To make matters worse, when the time came for payment, Matsuoka did not honour the debt. Mitsukoshi repeatedly pressed him for payment, including a threat to go to his office in Kumamoto. The department store finally received a promissory note in lieu of payment. The drawer of the promissory note was Hirota.[126] However, accepting such a note was not store policy, and when the store officer visited Matsuoka's home in Tokyo, he was given cash to cover the debt owed (approximately ¥25 million). The band around the cash was from the Sanwa Bank. At the time, Hirota had been using the Sanwa Bank for his business transactions, and so the Mitsukoshi people assumed that the money had originated from Hirota.[127]

Matsuoka continued to make purchases using the reception system and also continued to default on payment. Hirota again covered Matsuoka's bill with a promissory note. When the store objected to this system of payment, Hirota was enraged. He phoned and visited the Mitsukoshi Head Office along with an influential right-wing figure. He also stopped paying his account. When interviewed, Hirota declined to acknowledge even knowing Matsuoka, saying that all the transactions had been made through a representative of Matsuoka's *kôenkai*, a Mr T, who had subsequently died. Matsuoka himself also denied knowing Hirota and denied making any of the transactions, blaming Mr T. who clearly could not answer for himself.[128]

THE 1993 ELECTION RESULTS

When the results of the 1993 election were declared, Matsuoka was ranked third amongst the five successful candidates (see Appendix). This was two rungs higher than in 1990, but was well behind the very popular Hosokawa, who was way out in front of everyone else with 213,125 votes (see Appendix). Overall, the Japan New Party won 25.16 per cent of the votes, while the LDP secured 20.83 per cent. It was the first time that the LDP had vacated its position as the leading party in Kumamoto (1) since the amalgamation of the conservatives in 1955.[129] For the LDP, the entire election was held in the shadow of the Japan New Party boom.

Hosokawa claimed nearly 40 per cent of the votes in Kumamoto City and was also the most popular candidate overall in the rural counties, which meant that Matsuoka ceded top ranking to Hosokawa in regional areas. In fact, Hosokawa had broad support across all parts of the electorate, suggesting strong personal popularity.

Next in ranking to Hosokawa were Noda Takeshi in second place with 93,824 votes and Matsuoka with 82,620 votes. The split in the LDP and Uozumi's defeat left Matsuoka as the only LDP politician with strong connections to farm and forestry voters across the entire electorate, given that Noda's connections were mainly with business, although he gained respectable support from county voters.[130] The election results thus served to reinforce Matsuoka's sectoral differentiation strategy.

All up, Matsuoka's vote tally was only a small improvement on his 1990 performance, with 12.95 per cent of the total cast vote (see Appendix). This suggested a stable support base, with supporters who were not easily deflected by Hosokawa's appeal. On the other hand, Matsuoka appeared to have hardly gained in electoral performance in spite of his having secured a position in the LDP and in the Diet over the previous three years.

Apart from Hosokawa's taking a large slice of the total vote, the rankings amongst the various party candidates settled back more into the normal order of things with the JSP and Kômeitô candidates coming in behind the LDP candidates (in fourth and fifth position respectively) in contrast to the 1990 election (see Appendix). This was despite the fact that the LDP, as a whole, lost its majority in the Lower House in this election, gaining only 223 seats in the 511-seat Diet. The JSP as a party also went down to a massive defeat with only 70 seats. These figures provided the context in which a non-LDP coalition took power for the first time since 1955.

The two dominant features of Matsuoka's electoral performance—the fact that his total vote tally hardly changed and the fact that distribution of his support was broadly similar to the 1990 election result—suggested that the composition of Matsuoka's vote was hard rather than soft or floating. Matsuoka was not a politician that attracted floating votes. Nor, it would seem, did he have much personal popularity, the kind that produced such high levels of support that Hosokawa enjoyed for instance.

The only movement in Matsuoka's votes was the decline in voting support in the cities (where his overall support fell by 3000 or so votes) and the somewhat larger gain in rural votes (see Table 2.2). In Kumamoto City alone, Matsuoka's

Table 2.2 Farm household composition/votes cast for Matsuoka by municipality in Kumamoto (1) in 1993 Lower House election

Name of municipality	No. of farm households[a]	Farm households as % of total in municipality/ies	Votes cast for Matsuoka	% of total vote cast	% of Matsuoka's total vote	Placing among 8 candidates
Cities	6,002	2.3	36,021	8.4	43.6	6th
Kumamoto City	2,237	1.1	27,249	8.2	33.0	6th
Arao City	475	2.6	2,107	6.4	25.6	6th
Tamana City	1,151	8.7	2,608	10.0	3.2	4th
Yamaga City	1,035	9.9	2,852	14.0	3.5	3rd
Kikuchi City	1,104	13.6	1,205	6.7	1.5	6th
Counties	17,119	17.2	46,599	22.5	56.4	2nd
Hotaku County[b]	N/A	N/A	N/A	N/A	N/A	N/A
Tamana County	3,525	16.7	7,518	15.3	9.1	3rd
Taimei Town	449	11.1	1,621	17.2	2.0	3rd
Yokoshima Town	600	42.3	811	20.1	1.0	3rd
Tensui Town	881	49.9	549	10.9	0.7	4th
Gyokuto Town	337	20.5	613	16.4	0.7	3rd
Kikusui Town	209	10.5	368	7.5	0.4	4th
Mikawa Town	445	26.2	1,362	32.2	1.6	1st
Nankan Town	411	11.9	1,140	15.1	1.4	3rd
Nagasu Town	193	3.8	1,054	10.2	1.3	4th
Kamoto County	3,144	29.8	5,162	14.1	6.2	4th
Kahoku Town	399	26.3	687	16.9	0.8	3rd
Kikuka Town	566	27.7	717	13.5	0.9	4th
Kamoto Town	452	17.7	1,057	19.1	1.3	2nd
Kao Town	441	29.4	728	18.1	0.9	3rd
Ueki Town	1,286	43.7	1,973	11.2	2.4	4th

Kikuchi County	3,296	10.4	8,649	12.6	10.5	4th
Shichijo Town	483	33.9	443	11.2	0.5	3rd
Kyokushi Town	388	28.2	491	13.9	0.6	3rd
Ozu Town	697	9.9	2,882	19.6	3.5	2nd
Kikuyo Town	575	8.5	2,133	15.8	2.6	2nd
Koshi Town	354	6.7	885	7.6	1.1	6th
Shisui Town	478	14.9	642	8.3	0.8	5th
Nishigoshi Town	321	4.8	1,173	8.8	1.4	6th
Aso County	4,771	20.7	25,270	47.7	30.6	1st
Ichinomiya Town	468	15.0	2,711	41.1	3.3	1st
Aso Town	986	17.1	8,300	62.2	10.0	1st
Minamioguni Town	344	25.7	1,297	38.9	1.6	1st
Oguni Town	420	15.0	2,616	39.4	3.2	1st
Ubuyama Village	196	41.0	722	54.7	0.9	1st
Namino Village	220	40.7	821	60.1	1.0	1st
Soyō Town	479	33.0	1,688	49.1	2.0	1st
Takamori Town	523	21.4	2,128	40.3	2.6	1st
Hakusui Village	439	35.2	1,490	47.2	1.8	1st
Kugino Town	241	35.2	975	52.8	1.2	1st
Chōyō Village	191	10.3	1,521	46.4	1.8	1st
Nishihara Village	264	19.7	1,001	29.8	1.2	1st
Total/average	23,121	6.4	82,620	12.95	100.0	3rd

Notes: [a] Farm household data are for 1990. [b] Hotaku County was merged into Kumamoto City on 1st February 1991.

Sources: Sōmuchō, Tōkei Kyoku, *Heisei 2-nen Kokusei Chōsa Saishū Hōkoku Dai 3-kan Dai 2-ji Kihon Shūkei Kekka Sono 2 Todōfuken, Shichōson-hen 43 Kumamoto-ken*, pp. 288-289 and 294-303; Asahi Shinbunsha Senkyo Honbu, *Asahi Senkyo Taikan: Dai 40-kai Shūgiin Sōsenkyo*, p. 153.

vote tally fell by more than 7,000 votes, suggesting that some of his original supporters had switched their allegiance to Hosokawa. The even split between city and rural votes for Matsuoka in the 1990 election expanded to a 10,000 or so vote margin in 1993, with county support clearly starting to predominate. His county vote tally rose by 5,000 or so votes overall. Only in Kamoto County did his votes fall appreciably.

Aso County, on the other hand, went from strength to strength, supplying 25,270 votes, or just over 30 per cent of Matsuoka's total vote (see Table 2.2). This result pointed to continuing solid local support from his home county. Matsuoka was uniformly ranked top vote-winner in all the towns and villages of the county (see Table 2.2) suggesting a hardening of his *jiban*.

THE 'KARATE KINGS' FIGHT IT OUT

Top of the list of losers in the 1993 general election in Kumamoto (1) was Uozumi, who had joined the Renewal Party. Matsuoka reportedly destroyed Uozumi's *jiban* in the election, with the result that Uozumi performed poorly, coming just below last on the elected list by a margin of 2000 votes (see Appendix).[131]

In the 1994 mayoral election for Kumamoto City, Uozumi and Matsuoka engaged in an unscheduled but publicly televised karate fight in the square in front of the election office of one of the candidates. The two men had very different backgrounds and career histories and were also on bad terms and arch rivals. Matsuoka was known as a conservative, whereas Uozumi had followed Ozawa Ichirô, the flag-bearer of reform, and split from the LDP. Matsuoka had become excessively agitated at the gathering in the square, and, when Uozumi tried to restrain him, Matsuoka jumped on Uozumi. They exchanged karate slaps, Matsuoka's glasses went flying and he ended up with a cut lip.[132] What surprised the *Asahi* journalist who investigated the event was that there was so little sympathy for the victim, Matsuoka, who was slapped by Uozumi, and very little criticism of Uozumi, the perpetrator of the slap.[133]

The local media were covering the event, but only the *Asahi* reported it. The response from other media groups was that 'this is within the limits of normal in Kumamoto'.[134] Kumamoto had a reputation as a land of political strife and fierce electoral contests in which all-night vigils were held around fires in steel drums to prevent people from crossing over to the 'enemy's' side. As for Matsuoka, he went on to acquire a reputation for being hot-blooded, argumentative and a hard drinker.[135] What is more, he became known in Nagata-chô as the one man who hit his secretaries with his fists.[136] Other reports also surfaced of Matsuoka's physical altercations with Diet members.[137]

REPRESENTATION OF INTERESTS

Over two elections, Matsuoka had constructed an electoral coalition centring on his *jiban*, agricultural and forestry interests and sections of the city vote. This coalition required careful attention to a range of interests – local, sectional and client-based.

Local interests

Of all the victorious candidates standing in Kumamoto (1) in the 1990 and 1993 elections, Matsuoka had the most regionally concentrated vote, with wide variations in the percentage of the total vote he obtained across the cities, towns and villages that made up Kumamoto (1). Matsuoka's political stronghold was clearly in the Aso region. That was where his primary *jiban* was located. He was consistently the most popular candidate there over two elections. He also gained a substantial percentage of his total vote from Kumamoto City. In fact, these two sources of support comprised just under two-thirds of his vote tally (74 per cent in 1990 and 64 per cent in the 1993 election, as shown in Tables 2.1 and 2.2).

Such regionally concentrated support encouraged Matsuoka to make a strong commitment to a specific locality and to engage in policy activities that delivered 'regionally concentrated policy services' (*chiiki shûchûgata seisaku sâbisu*).[138] This meant directing pork-barrel benefits, particularly public works projects, to his *jiban*.[139]

For politicians such as Matsuoka, the beauty of public works was that they could be guided to a particular place—they were location-specific, and so their beneficiaries could be identified and votes could be harvested in return. Public works were different from general policy benefits that were delivered uniformly to broad sub-categories of voters, such as all rice farmers, or all dairy farmers wherever they were located.

By guiding benefits (*rieki yûdô*)—such as public works—to a specific locality, Matsuoka could secure high support rates within his *jiban* and a strong personal vote.[140] A large part of Matsuoka's policy activities were, therefore, geared to getting public resources directed to particular regions, such as funds for agricultural, forestry and rural infrastructure development, including agricultural and forest roads, and community facilities of various kinds.

At the same time, such projects doubled as support for construction companies in the electorate. They advantaged not only the beneficiaries of the facilities that were built, such as road users, school children, and the patrons of sports and cultural facilities, but also the executives of both large and small construction companies and their employees, many of whom were locally-based, including part-time farmers. In this sense, public works projects had high utility as an electoral strategy geared to the mobilisation of both votes and political funds. In geographic terms they could also be used to service both Matsuoka's urban and rural support bases, because some construction companies were located in Kumamoto City. Public works thus provided, for Matsuoka, a bridge between urban and rural areas.

Public works also served indirectly as industry promotion policies and policies to promote the regional economy. They generated wider spin-offs for industry and business in the area—not just local, but also regional and prefectural—as well as for employment and regional development as a whole. Kumamoto was well known as a region with a culture of dependency on public works,[141] and Matsuoka claimed to be especially concerned with the development of Kumamoto Prefecture as his 'birthplace'.[142]

The incentive generated by the distribution of Matsuoka's voting support was, therefore, for him to engage in politics that benefited local interests in regional areas (*rieki yûdô seiji*)[143] and to become a 'benefit-guiding type of politician' (*rieki yûdôgata no seijika*). Matsuoka continued the 'long-established tradition in which the gaining of subsidies and public works was seen as a good achievement'.[144] By providing voters and local industries in his home region with public works, Matsuoka could secure in exchange his own interests such as votes, funding and support.[145]

Rieki yûdô seiji was also an important means for Matsuoka to build a strong following amongst local politicians in a line from the centre (Matsuoka as a Diet politician) to the periphery (prefectural and municipal politicians, including mayors and assembly members, who were dependent on the flow of funding from the centre for their own public works programs). Such projects helped local politicians obtain support from voters in local elections. They enhanced the image of particular local politicians as effective representatives who could draw government resources back to their local areas. They also demonstrated the utility of the local politicians' links to national Diet politicians and thus their standing and importance. The return for Matsuoka was the role local politicians played as *kôenkai* kingpins, election campaigners and generators of support amongst voters loyal to them. Through this system, it was Matsuoka, the individual Diet member, not the party (LDP), who delivered benefits to his local politician supporters and to localities that were the most important to him politically.

Directing public works projects to his local region required Matsuoka to engage in policy interference, that is, interference in the administrative affairs of the bureaucracy. It meant interceding with and exerting influence over officials in the MAFF (and other ministries such as the Ministry of Construction and the Ministry of Transport), because it was government officials who decided where particular public works projects would be located.

Because Matsuoka had not been a mainstream career official in the MAFF and only had a BA in Forestry from Tottori University, he did not have an automatic foothold for influence within the MAFF when he started out as a politician. His big break was the Uruguay Round Agriculture Countermeasures Expenditure (*UR Nôgyô Taisakuhi*) package. It provided an opportunity for him to start exercising enormous power over the distribution of special agricultural and rural public works projects funded either totally or partially by the package.[146] Officially the policy was designed to compensate farmers for greater exposure to international trade competition as a result of the Hosokawa administration's agreement to the Uruguay Round Agreement on Agriculture (URAA) finalised in March 1994 for initial implementation in the 1995 fiscal year. The total budgetary allocation for the countermeasures package was ¥6.01 trillion over six years (1995-2001). In practice, the package provided a huge financial boost for local public works projects.

Since the nature of the expenditure package represented just a 'grab for money' (*tsukamikin*), how it was to be used was not clear. Matsuoka himself said that 'there is an abundance of funds. There was no choice but to use the money for projects'.[147] With that money, Matsuoka constructed large-scale facilities in his constituency.[148]

Sectional interests

Matsuoka's strong support from rural counties made it inevitable that he would represent agricultural and forestry interests in the party and in the Diet. One of his primary policy-related activities in Tokyo was policy intervention, whereby he would participate in the making of policy decisions for the agriculture and forestry sectors—in the committees of the LDP's PARC, in Diet standing committees (Kokkai *iinkai*), and in Diet members' activities (*giin katsudô*) in the Diet members' leagues (*giin renmei*).

Agriculture and forestry policy included both distributive measures (those involving the allocation of public funds) and non-distributive measures (those relating to the regulation of markets etc.). For the most part, however, agriculture and forestry policy, even when it was distributive, was not locality-focused, but targeted to broad groups of farmers such as rice producers. Agricultural subsidies for price support schemes, for example, applied not only to particular types of producers in Matsuoka's electorate, but also across the entire Kumamoto Prefecture, and in most cases across the entire nation. Such benefits could not be restricted to farmers in certain localities or electoral *jiban*. They contrasted with subsidies for agricultural and rural public works, which, although included under the broad umbrella of agricultural policy in the aggregate (that is, allocating total quantities of funds annually), were essentially geared to an electoral strategy focusing on local, rather than sectional interests.

Moreover, broadly focused agricultural and forestry policies, including subsidies, applied to all members of a group irrespective of whether or not they supported Matsuoka. The incentive was for Matsuoka to try and influence these policies as a means of maximising farm and forestry votes, but he could not withhold benefits from those who did not support him. The connection between voter and politician, in this case, was indirect. In this respect, agricultural policies possessed the characteristics of programmatic policies, even though they applied to a specific group of producers. Nonetheless, though indirect, the electoral benefit from supporting industry-wide policies was potentially substantial.

The combination of high levels of rural support and Matsuoka's career experience in the MAFF practically preordained his political career as a typical agriculture and forestry Diet member (*nôrin giin*), with a primary specialism in agriculture and forestry policy, a niche in which he could bring his specialised knowledge of agricultural and forestry administration to bear. Creating an identity as a *nôrin giin* also served to distinguish Matsuoka from other LDP rival candidates and politicians from Kumamoto (1). Support generated by agricultural and forestry-related policy activities was derived from all those areas of his electorate where people were gainfully employed in agriculture and forestry. It was not specific to his *jiban* in the Aso region.

A sectionalist orientation was also important in enabling Matsuoka to maintain profitable relations with the interest groups operating in this field. These could sometimes be the specific beneficiaries of policy measures and subsidies for their own organisational purposes. In return, producer groups could be mobilised to provide various forms of organisational backing, including campaign assistance, votes and funds.

Representing agricultural and forestry interests enabled Matsuoka to exploit the organisational muscle of farm and forestry-related interest groups. The biggest and most comprehensive grouping, which mobilised the most primary industry votes in the electorate, was Nokyo. Within Nokyo, the Kumamoto Prefecture *nôseiren* was the most important group for Matsuoka; it had branches in each county of his electorate and sub-branches in the municipalities.

Table 2.3   Ranking of Kumamoto Prefecture as an agricultural producer/ by commodity and farm/forestry households, 2000 (per cent)

Product/households	Ranking
Rushes (for tatami mats)	1st (95%)
Corn	1st (35%)
Tomatoes	1st (11%)
Tobacco	2nd
Strawberries	3rd
Ginger root	3rd
Japanese mandarins (*mikan*)	4th
Number of beef cattle-feeding households	4th
Sweet potatoes	5th
Raw milk production	6th
Wheat and barley	9th
Chinese cabbages	9th
Japanese radishes	10th
Carrots	10th
Rice	14th
Total number of farm households	13th
Farm households in prefecture (per cent)	13th
Total number forest households (67% joint farm-forest households)	16th

Source: Nôrinsuisanshô, Tôkei Jôhôbu, 2003. *Dai-77 Nôrinsuisanshô Tôkeihyô* [*The 77*[th] *Yearbook of Ministry of Agriculture, Forestry and Fisheries, Japan*], 2000-2001, Tokyo, Nôrin Tôkei Kyôkai, pp.4, 160-230, 426.

Matsuoka's sectionalist orientation ensured that he undertook activities designed to deliver policy benefits not only to farm and forest owners in general, but also to specific sub-groupings of producers within this larger occupational category. The production profile of Kumamoto Prefecture as a whole indicated that certain commodities were going to be more important to Matsuoka than others (see Table 2.3). The prefecture was well known for its production of rushes for *tatami* mats, tobacco, corn, horticultural products such as tomatoes, strawberries and ginger root, fruit such as *mikan*, livestock products such as beef and milk, and sweet potatoes. As Table 2.3 shows, Kumamoto Prefecture had the second highest number of tobacco farms of any prefecture in Japan in 2000.

Forests were also a feature of Kumamoto's primary industry landscape. More than two-thirds (67 per cent) of farm households were also forest households (*rinka*),[149] meaning that agricultural landholders also owned some forestland. Of the 24,049 forest households in Kumamoto Prefecture in 2000, 15,221 owned 1-3 hectares.[150] Only 1,223 households were making sales of forestry products, although 7,278 household members were engaged in forestry.[151]

Matsuoka also emphasised forestry policy because he had spent most of his time in the MAFF in the Forestry Agency, where he had been posted to prefectures such as Hokkaido and Akita. However, forestry policy was not restricted to taking care of the interests of small-scale forest owners in his electorate, just as his agricultural policy specialism did not apply simply to farmers in his own constituency. Forestry policy meant looking after the interests of timber companies that logged the forests, including state forests, and companies wanting to convert forestland to other uses such as factories and industrial complexes, residential sites, golf courses and leisure facilities, agricultural land and public land. In addition, forest policy extended to the construction of forest roads (*rindô*) through non-state-owned forests with the assistance of national government (and, to a lesser extent, prefectural government) subsidies. The Kyushu region had more forest roads than any other region except for Tohoku even though its forest area was relatively small, and almost all of these roads were managed by municipal governments. Forestry public works, as with all public works, took the representation of forestry interests out of the realm of sectional interests, and into the sphere of local interests and also into clientelistic politics, where Matsuoka could intercede on behalf of individual clients or small groups of clients.

Clientelistic interests

The search for votes and political funding encouraged Matsuoka to undertake deals for particular clients, usually through direct, personal contacts between himself and those seeking his intermediation on their behalf.[152] In this role, Matsuoka acted as a political broker or private mediator for individual clients, who sought personal, private favours, and as a political 'fixer' for small groups of clients who petitioned for particular policy favours. The key aspect of such clientelistic relations was the personalised connection between Matsuoka and those seeking his mediation, and the delivery of the requested favours as political patronage. Matsuoka's conduct of politics on an individual basis (*kojin honi no seiji*) led inevitably to a political culture of patronage (*onkoshugiteki na seiji bunka*).[153]

Matsuoka's ability to deliver such benefits was critically dependent on Japan's system of discretionary governance by the central government bureaucracy and on the ability of ordinary ruling party backbenchers to maintain direct channels of communication with, and influence over, serving government officials. Bureaucrats had the power to grant the favours; they decided which public works projects should go where (*kashozuke*),[154] and which private companies should undertake these projects. Using their discretionary powers, bureaucrats could even use the promise of subsidies to influence local governments in their selection of which tradespeople they would select to work on projects.[155] Bureaucrats were also responsible for deciding whether, or how much of, a particular subsidy would be provided to a particular group, and for a host of other kinds of administrative decisions that impacted on the lives of individuals and the incomes of companies and producers of various kinds. Government officials could arrange for exemptions for particular individuals, groups or companies from certain administrative rules and regulations.

Within their administrative fiefdoms, bureaucrats were the government, and politicians such as Matsuoka were able to use their position as Diet members to prey on government in this sense. In his role as broker or 'fixer', Matsuoka's job was to solicit, obtain or extract administrative favours from bureaucrats. He maintained a parasitic relationship with them, feeding off the benefits they provided and converting these benefits into political goods for his own purposes.

Central government dispensations and subsidies, including those for particular public works projects such as roads, sports and tourist facilities, served as a basis for Matsuoka to build a range of clientelistic relationships with supporters. He established personal links with individual local politicians, local businessmen and company executives, as well as with the leaders and officials of local groups and sectoral organisations, who were dependent on central government administrative discretion and largesse and who could supply votes and funding in exchange for favours mediated by Matsuoka.

In this context, Matsuoka functioned as a 'benefit and concession-guiding type of politician' (*riken yûdôgata no seijika*).[156] Whilst benefits involved directing public works to particular local regions (*rieki yudo seiji*), 'concessions' (*riken*) usually meant obtaining specific favours for particular clients. A large majority of such favours in Matsuoka's case were for construction companies and other businesses servicing the public works industry.[157] Public works contracts provided these companies with business opportunities they would not otherwise have had. Through his involvement in the public works industry, Matsuoka was able to build personal links with construction company executives and others who then became a major source of political funding in return for the delivery of political patronage in the form of public works contracts.[158] The system worked as follows: Matsuoka would 'drop funds from the agriculture, forestry and fisheries budget into the locality (*jimoto*), Matsuoka would then allow local companies to get involved in the public works projects funded by this budget, and finally, Matsuoka would then receive political funds from these companies'.[159]

Matsuoka's services as a broker were, however, not totally dependent on the provision of central government largesse. In his search for votes and political funds, he was encouraged to field personal requests for favours not only with bureaucrats, but also with other politicians and political leaders, with members of administrative staffs and organisational officials, and even with business leaders.

Some of Matsuoka's key linkages and activities as a mediator or political broker were generated via his *kôenkai*, which served as a communications hub between Matsuoka, the Diet politician, and those local entities who submitted various requests for his patronage. His *kôenkai* provided a mechanism for channelling the particularistic demands of private individuals, local politicians, company executives and organisational officials directly to Matsuoka, the politician. Through his *kôenkai*, Matsuoka undertook direct associations with his supporters and offered them his patronage in the form of various services and policy activities in exchange for votes and political funds. The *kôenkai* thus institutionalised clientelism and its inevitable corollary—patronage politics. Matsuoka personally delivered favours to particular clients in exchange for their personal loyalty and support.

This was essentially a feudal system of politics. Matsuoka was treated like an overlord whom supplicants approached seeking favours in exchange for which they pledged their loyalty (in the form of votes and/or political funds). The behaviour in which it resulted was covert and completely lacking transparency: it easily led to corruption.

NOTES

1   Kumamoto Prefecture is located in the northeast of the large island of Kyushu, which lies at the southwestern end of the main Japanese island of Honshu.
2   Matsuoka Toshikatsu Official Site, 'Matsuoka Toshikatsu no Rirekisho' ['Curriculum Vitae of Matsuoka Toshikatsu'], Available from http://www.matsuokatoshikatsu.org/site002//public/008.html
3   Some of the following details were obtained from 'Matsuoka Toshikatsu no Rirekisho'. Available from http://www.matsuokatoshikatsu.org/site002//public/008.html
4   Nakanishi Akihiko and Journal Reporter Group, 'Matsuoka Toshikatsu to Iu Giwaku Nin' ['The Suspicious Person Called Matsuoka Toshikatsu'], *Bungei Shunjû*, 1 September 2002, pp.184–85.
5   *ibid*, p. 184.
6   *ibid*., p. 185. According to two other reports, Matsuoka spent two years at the National Defence University and then withdrew.
7   *ibid*., p. 185.
8   The Japanese title was Nôrinshô.
9   The National Personnel Authority divides MAFF entrance exams into three levels. The MAFF applies the following categories of levels as follows: Level I is the highest and encompasses both *jimukan* (law, economics and administration), and *gikan* (agronomy, forestry, engineering, human sciences, livestock science etc.). Levels II and III cover all fields. Special exams are also offered for livestock science and veterinary science. In addition, the MAFF has its own interview tests to those who pass these exams. Visit: http://www.maff.go.jp/saiyou/saiyou_top.html
10   Nishikawa Shinichi, 'Tako Tsubo ni Tojikomotte Ôkoku o Gyûjiru Nôgyô Doboku' ['The Agriculture Civil Engineering Technical Bureaucrats Stuck in a Foxhole and Controlling Their Kingdom'], *Shûkan Daiyamondo*, 20 April 2002, p.48.
11   'Matsuoka Toshikatsu: Purofuiru' ['Profile'], in Seisaku Jihôsha, *Seikan Yôran [A Handbook of Politicians and Bureaucrats]*, 1990, First Half Year Edition, Tokyo, Seisaku Jihôsha, March 1990, p.269.
12   The Japanese title is Rinyachô. Along with the Food Agency and the Fisheries Agency, this has been one of the three agencies of the MAFF, although the Food Agency, which, from 1942 onwards bought and sold rice and regulated rice marketing throughout Japan as well as regulating rice production since 1969, has now been disbanded. What was left of its rice-market and production-related functions were taken over a new Food Department of the MAFF.
13   Matsuoka Toshikatsu, 'Shinsan-Kôtoku Chiiki no Shinkihon Keikaku no Gaiyô' ['Outline of the New Industry-Industrial Special Regions New Basic Plan'], *Rinya Jihô*, April 1977, p.4.
14   Matsuoka, 'Shinsan-Kôtoku Chiiki no Shinkihon Keikaku', p.35.
15   Its Japanese title is Nôrinsuisanshô.
16   It became the Ministry of Agriculture, Forestry and Fisheries at the height of international negotiations over the law of the sea, when Nakagawa Ichirô was minister and led the Japanese side in these

negotiations. At the time, the ministry thought it would be a good idea to emphasise the fact that fisheries were part of its administrative domain.
17 The National Land Agency is now incorporated into the Ministry of Land, Infrastructure and Transport as a result of the reorganisation of the central government bureaucracy in January 2001.
18 Nakanishi and Journal Reporter Group, 'Matsuoka Toshikatsu to Iu Giwaku Nin', p. 185.
19 Nakanishi Akihiko and Special Reporting Group, 'Suzuki Muneo, Matsuoka Toshikatsu: Riken no Kyôbô' ['Suzuki Muneo and Matsuoka Toshikatsu: Conspiracy for Concessions'], *Bungei Shunjû*, May 2000, p. 100.
20 'Ni Chaneru Kako Rogu' ['Channel 2, Previous Entries/Log'], Giin Section, available from http://piza.2ch.net/giin/kako/987/987905181.html
21 Nakanishi and Journal Reporter Group, 'Matsuoka Toshikatsu to Iu Giwaku Nin', p. 185.
22 Book review by Brad Glosserman of Ofer Feldman, *The Japanese Political Personality: Analyzing the Motivations and Culture of Freshman Diet Members*, St Martin's Press/Macmillan Press, 2000, *The Japan Times*, 13 July 2000.
23 *ibid.*
24 'Matsuoka Toshikatsu no Rirekisho'. Available from http://www.matsuokatoshikatsu.org/site002//public/008.html
25 Visit: http://www.matsuokatoshikatsu.org/index1.html
26 'Matsuoka Toshikatsu no Rirekisho'. Available from http://www.matsuokatoshikatsu.org/site002//public/008.html
27 Nakanishi and Special Reporting Group, 'Suzuki Muneo, Matsuoka Toshikatsu', p. 100.
28 Nakanishi and Journal Reporter Group, 'Matsuoka Toshikatsu to Iu Giwaku Nin', p. 180.
29 *ibid.*, p. 181.
30 *ibid.*
31 Aum Shinrikyô was a religious group, whose followers carried out a poison gas attack on the Tokyo subways in March 1995 that killed 12 people.
32 Nakanishi and Journal Reporter Group, 'Matsuoka Toshikatsu to Iu Giwaku Nin', p. 181.
33 Nakagawa died in circumstances that remain obscure. Various rumours circulated following his death. One held that Nakagawa had feuded with faction leader Mitsuzuka Hiroshi and/or Suzuki Muneo, who was his secretary at the time. Another rumour held that Nakagawa might have been killed by the CIA or the Soviet's KGB. *Tokyo Shinbun*, 3 August 2005.
34 Nakanishi Akihiko and Special Reporting Group, 'Suzuki Muneo, Matsuoka Toshikatsu', p. 105.
35 Nakanishi and Journal Reporter Group, 'Matsuoka Toshikatsu to Iu Giwaku Nin', p. 185.
36 *ibid.*, p. 185. See also Chapter 6 on 'The Identical Twins of Nagata-cho'.
37 Nakanishi Tomiki, 'Matsuoka Toshikatsu (Jimin Daigishi): Igai na Sugao to Shûkin Ryoku' ['Matsuoka Toshikatsu (LDP Diet Member): An Exceptional "Warts and All" and Money Collecting Power'], *Shûkan Asahi*, February 2002, p.28.
38 Nakanishi and Journal Reporter Group, 'Matsuoka Toshikatsu to Iu Giwaku Nin', p. 185.
39 *ibid.*
40 *ibid.*
41 *Yomiuri Shinbun*, 19 February 1990.
42 'Matsuoka Toshikatsu: Purofuiru', p. 269. Even though this is what Matsuoka is quoted as saying in the Diet handbooks, other sources suggest that he did not want endorsement from the LDP.
43 Machidori, 'The 1990s Reforms Have Transformed Japanese Politics', p. 39.
44 Visit: http://piza.2ch.net/giin/kako/987/987905181.html
45 *ibid.*
46 'Matsuoka Toshikatsu no Rirekisho'. Available from http://www.matsuokatoshikatsu.org/site002//public/008.html
47 Sômuchô, Tôkei Kyoku, 1991. *Heisei 2-nen Kokusei Chôsa Saishû Hôkokusho Nihon no Jinkô (Shiryô Hen)* [*1990 National Census Closing Report Japanese Population (Data Edition)*], Tokyo, Sômuchô, Tôkei Kyoku, 1991, pp. 2-3, 526-7; Sômuchô, Tôkei Kyoku, *Heisei 2-nen Kokusei Chôsa Saishû Hôkoku Dai*

3-kan Dai 2-ji Kihon Shûkei Kekka Sono 2 Todôfuken, Shichôson-hen 43 Kumamoto-ken [*1990 National Census Report, Vol. 2, Second Basic Statistical Results 2 Prefectures and Municipalities, Edition 43 Kumamoto Prefecture*], Sômuchô, Tôkei Kyoku: Tokyo, pp. 288–89 and 294–303; and Vol. 3, pp.2–3.

48   *Heisei 2-nen Kokusei Chôsa Saishû Hôkoku Dai 3-kan Dai 2-ji Kihon Shûkei Kekka*, pp. 288–89 and 294–303; and Vol. 3, pp. 2–3.

49   Calculated from nationwide average figures in Nôrinsuisanshô, Tôkei Jôhôbu, 1992. *Dai-66 Nôrinsuisanshô Tôkeihyô*, [*The 66th Statistical Yearbook of Ministry of Agriculture, Forestry and Fisheries*], 1989–90, Nôrin Tôkei Kyôkai, Tokyo, pp. 5 and 20.

50   For the total cast vote in the district of Kumamoto (1) in the 1990 Lower House election, see the Appendix.

51   Kitamatsu Masahiko *et al.*, 'Matsuoka Toshikatsu Daigishi Tettei Bunseki: Sono Ôsei naru Shûkin Nôryoku no Kiseki' ['An Exhaustive Analysis of Matsuoka Toshikatsu Diet Member: The Tracks of a Vigorous Money-Collecting Ability'], *Zaikai Tenbô*, December 2002, p.47.

52   *ibid.*

53   *ibid.*

54   See Chapter 6 on 'The Identical Twins of Nagata-cho' for an elaboration of Matsuoka's relationship with Suzuki Muneo.

55   Kitamatsu, *et al., op.cit.*, p. 46.

56   '"Nishi no Muneo" Matsuoka Toshikatsu no Sokkin Hisho mo Yukue o Kuramashita' ['"Muneo of the West" Matsuoka Toshikatsu's Close Associate and Secretary Also Disappears'], *Shûkan Bunshun*, 4 July 2002:38.

57   Curtis, *The Logic of Japanese Politics*, p. 143.

58   Tatebayashi, *Giin Kôdô*, p. 4.

59   Nakanishi and Journal Reporter Group, 'Matsuoka Toshikatsu to Iu Giwaku Nin', p. 185.

60   *Yomiuri Shinbun*, 19 February 2005.

61   Personal interview, Ministry of Finance official, January 2003.

62   These are known as civil engineering 'types' (*dokenya*), who form the most powerful subgrouping of *gikan* within the MAFF. Their business is public works including land improvement. Because of their command over a large slice of the MAFF budget, 'they have an air of arrogance and so are easy to spot in the MAFF'. Itô Terî and Editorial Department, 'O'warai Nôrinsuisanshô' ['The Comical MAFF'], *Shûkan Daiyamondo*, 20 April 2002, p. 73.

63   In the 2001 Upper House election, for example, MAFF OB Fukushima Keishirô, who stood for the national (PR) constituency of the Upper House, received such support. His campaign organisation was headed up by Takagi Yûki, who had just retired as MAFF administrative vice-minister. Takagi was subsequently appointed as President of the Agriculture, Forestry and Fisheries Finance Corporation, a top 'descent from heaven' (*amakudari*) post for MAFF OBs.

64   Nakanishi and Journal Reporter Group, 'Matsuoka Toshikatsu to Iu Giwaku Nin', p. 185.

65   *Yomiuri Shinbun*, 15 February 1990.

66   *ibid.*

67   'Karate 4-dan (Shinseitô) ga 2-dan (Jimintô) o Haritao shita Yoru' ['The Night that the Karate 4th Level from the Renewal Party Pushed Over the Karate 2nd Level from the LDP'], *Shûkan Asahi*, 9 December 1994, p. 32.

68   In 2005, at the age of 88, he was advisor to the prime minister, former chairman of the Japan Karate Association and advisor to the Karate no Michi (The Karate Way) World Federation.

69   *Yomiuri Shinbun*, 15 February 1990. Matsuno was ranked fifth amongst vote-winners in the 1986 elections, and although he increased his vote slightly in the 1990 elections, it was still not enough to get him over the line. See below.

70   'Matsuoka Toshikatsu no Rirekisho'. Available from http://www.matsuokatoshikatsu.org/site002//public/008.html

71   *Yomiuri Shinbun*, 15 February 1990.

72   The JSP won 136 seats in this election, a substantial increase on the 83 seats it had at the time of the

Diet's dissolution. The *Asahi* newspaper described the JSP's electoral victory in 1990 as a 'huge onslaught' (*daiyakushin*). It followed the JSP's outstanding performance in the Upper House elections of 1989.
73 'Matsuoka Toshikatsu: Purofuiru', p. 269.
74 Matsuno was not the only senior LDP politician with ties to farm and rural-regional interests to be defeated in this election. So were two other big names in the agricultural scene in Kyushu—Etô Takami from Miyazaki (1) and Yamanaka Sadanori from Kagoshima (3). Etô was Minister of Transport at the time, while Yamanaka was the chairman of the LDP's Tax System Investigation Committee. He was held personally responsible for the introduction of the consumption tax. In fact, he was described as 'the real parent of the consumption tax' (*shôhizei no umi no oya*), despite this, he lost by only 28 votes. Other losers of note were Horinouchi Hisao, former Minister of Agriculture, Forestry and Fisheries, as well as other senior LDP politicians who had held posts such as Minister of Construction, Director-General of the Science and Technology Agency, and the Director-General of the Defence Agency. The election result produced a real clean-out of LDP 'big shot Diet members' (*daibutsu giin*) and caused quite a stir at the time.
75 Asahi Shinbunsha Senkyo Honbu, 1990. *Asahi Senkyo Taikan: Dai 39-kai Shûgiin Sôsenkyo (Heisei 2-nen 2-gatsu), Dai 15-kai Sangiin Tsûjô Senkyo (Heisei Gannen 7-gatsu)* [*Asahi General Survey of Election: The 39th House of Representatives General Election (February 1990), The 15th House of Councillors Regular Election (July 1989)*], Asahi Shinbunsha, Tokyo, p. 194.
76 Asahi Shinbunsha Senkyo Honbu, *Asahi Senkyo Taikan: Dai 39-kai Shûgiin Sôsenkyo*, p. 194.
77 *Yomiuri Shinbun*, 19 February 1990.
78 Asahi Shinbunsha Senkyo Honbu, *Asahi Senkyo Taikan: Dai 39-kai Shûgiin Sôsenkyo*, p. 194.
79 *ibid.*
80 Curtis, *The Logic of Japanese Politics*, p. 59.
81 Asahi Shinbunsha Senkyo Honbu, *op.cit.* p. 194.
82 *Yomiuri Shinbun*, 15 February 1990.
83 'Matsuoka Toshikatsu: Purofuiru', p. 269.
84 Calculated from data in *Seikan Yôran*, 1990, First Half Year Edition, pp. 472–75.
85 *Seikan Yôran*, 1990, First Half Year Edition, pp. 472–75.
86 Of the two former MAFF *gikan* in the Diet in 1990—a graduate from Tokyo University's Faculty of Agriculture, and a graduate from Kyoto University's Faculty of Agriculture—both were agricultural engineering Diet members (*nôgyô doboku giin*).
87 Asahi Shinbunsha Senkyo Honbu, *Asahi Senkyo Taikan: Dai 39-kai Shûgiin Sôsenkyo*, p. 194.
88 I am grateful to Yusaku Horiuchi for calculating the correlation coefficient between the percentage of farm households in each municipality and the percentage of the total vote Matsuoka received. The correlation coefficient was 0.162 (insignificant). The regression coefficient was also insignificant (0.220).
89 Asahi Shinbunsha Senkyo Honbu, *Asahi Senkyo Taikan: Dai 39-kai Shûgiin Sôsenkyo*, p. 194.
90 'Karate 4-dan', p. 34.
91 Nakanishi and Journal Reporter Group, 'Matsuoka Toshikatsu to Iu Giwaku Nin', p. 185.
92 These terms are borrowed from Tatebayashi, *Giin Kôdô*, p. v. He argues that some Lower House electorates were divided up geographically (*chiiki wari*) along the lines of different candidates' *jiban*, while others were partitioned along policy sector (*seisaku bunya wari*) lines. Some were divided up by a mixture of both as in the case of Kumamoto (1). The division along the lines of each candidate's *jiban* amounted to a strategy of mutual respect for each candidate's geographic sphere of influence. In the elections, each candidate committed himself to representing a specific locality. The sectoral (policy field) division cut across the geographic division and was a strategy to fill in particular policy gaps across the entire electorate, which was partitioned according to the policy fields in which the candidates specialised. This often amounted to a division by economic or industry sector, such as agriculture, commerce, construction and small business. *Giin Kôdô*, p. 49. According to Tatebayashi, these two major kinds of vote divisions were designed to mitigate competition amongst candidates from the same

party (i.e. LDP) and to maximise the number of LDP candidates elected from the district. *Giin Kôdô*, p. v.
93 *Yomiuri Shinbun*, 20 February 1990.
94 *ibid.*
95 The factional lineage extends from former Prime Minister Kishi through to former Prime Minister Fukuda and Abe Shintarô. Its most recent leader is former Prime Minister Mori. Matsuoka joined when Mitsuzuka Hiroshi was faction leader after Abe died. Other Diet members representing Kumamoto (1) were from the Watanabe and Takeshita factions, so the Mitsuzuka faction was a logical choice for Matsuoka. Mitsuzuka was also a member of Seirankai to which Nakagawa and Tamaki, Matsuoka's original Diet member patrons belonged.
96 Kitamatsu *et al.*, 'Matsuoka Toshikatsu Daigishi Tettei Bunseki', pp. 46-47. See also below.
97 *ibid.*
98 *ibid.*
99 See also below.
100 In the 1993 elections Noda won just over two-thirds of his total vote in the cities of Kumamoto (1).
101 Its Japanese title was Shinseitô. In December 1992, the Takeshita faction (Keiseikai), which was the largest in the LDP, split into two with the departure of the group led by Hata Tsutomu and Ozawa Ichirô. In June 1993, 44 LDP Hata faction members resigned from the LDP and formed the Renewal Party led by Hata, with Ozawa as secretary-general, and 10 junior more left-wing LDP members left to form the New Party Harbinger (Shintô Sakigake) led by Takemura Masayoshi. In the subsequent (July) Lower House election, the LDP failed to win a majority.
102 This point is generalised by Tatebayashi, *Giin Kôdô*, p. 49. He argues that the different groups that supported the LDP were thus distributed amongst the various Diet members according to a policy division of labor. *Giin Kôdô*, p. 34.
103 See below.
104 There were 1.67 million forest-owner members of the 990 forest associations in Japan. Visit: http://www.zenmori.org/profile/gaiyou1.html. By 2003, their numbers had fallen to 1.64 million individual members of 970 forest associations. Personal communication, General Affairs Department, National Federation of Forest Associations, December 2005. Even though, unlike the agricultural cooperatives, they are not called 'cooperative unions' (*kyôdô kumiai*), they are in fact cooperatives organised along the same structural and functional lines as Nokyo. As the majority of forest owners in Japan are small-scale, like farm owners, they establish forest associations and cooperate for the purpose of forest management, the purchase of production and harvesting inputs, and the sale of timber. Prefectural federations of forest associations operate in every prefecture, and a national federation, the National Federation of Forest Associations (Zenkoku Shinrin Kumiai Rengôkai), or Zenshinren is the national body.
105 Visit: http://www.kumamori.or.jp
106 See Chapter 4 on 'Exercising Power as a *Nôrin Giin*'.
107 Its Japanese title was Nihon Shintô.
108 Asahi Shinbunsha Senkyo Honbu, *Asahi Senkyo Taikan: Dai 40-kai Shûgiin Sôsenkyo (Heisei 5-nen 7-gatsu), Dai 16-kai Sangiin Tsûjô Senkyo (Heisei 4-nen 7-gatsu)* [*Asahi General Survey of Election: The 40th House of Representatives General Election (July 1993), The 16th House of Councillors Regular Election (July 1992)*], Tokyo, Asahi Shinbunsha, 1993, p. 19.
109 Visit: http://piza.2ch.net/giin/kako/987/987905181.html
110 *ibid.*
111 *ibid.*
112 *Yomiuri Shinbun*, 14 July 1993.
113 Kitamatsu, *et al.*, 'Matsuoka Toshikatsu Daigishi Tettei Bunseki', p. 47.
114 *ibid.*
115 *ibid.*
116 'Karate 4-dan', p. 34.

117 Hasegawa Hiroshi, 'Kanjûdanomi no Hazama de Shundô' ['Wriggling Through the Gaps of Bureaucratic Demands and Requests'], *Aera*, 18 February 2002, p. 25.
118 Nakanishi and Journal Reporter Group, 'Matsuoka Toshikatsu to Iu Giwaku Nin', pp. 185–86.
119 'Matsuoka Toshikatsu, Nishi no Muneo no Gyôten Sukyandaru: Mitsukoshi "Nisenmanen Bîruken" Fumitaoshi Kosaku' ['Matsuoka Toshikatsu, Muneo of the West's Astonishing Scandal: The Plot Not to Pay for the Mitsukoshi "¥20 Million Beer Coupons"'], *Shûkan Bunshun*, 5 September 2002, p. 168.
120 'Matsuoka Toshikatsu, Nishi no Muneo no Gyôten Sukyandaru', p. 169.
121 *ibid.*
122 *ibid.*
123 *ibid.*, p. 168.
124 *ibid.*
125 *ibid.*
126 *ibid.*, p. 169.
127 *ibid.*
128 *ibid.*, p. 170.
129 Asahi Shinbun Election Headquarters, *Asahi Senkyo Taikan: Dai 40-kai Shûgiin Sôsenkyo*, p. 19.
130 *ibid.*, p. 153.
131 'Karate 4-dan', p. 32.
132 *ibid.*
133 *ibid.*, p. 34.
134 *ibid.*, p. 35.
135 Nakanishi and Special Reporting Group, 'Suzuki Muneo, Matsuoka Toshikatsu', p. 100.
136 'Matsuoka Toshikatsu Daigishi ni Hisho no Taishokukin & Kyûyo Pinhane Giwaku' ['Suspicion that Matsuoka Toshikatsu is Raking Off His Secretary's Retirement Money & Allowances'], *Flash*, 5 February 2002, p. 15.
137 Visit: http://piza.2ch.net/giin/kako/987/987905181.html
138 Tatebayashi generalises this point. See *Giin Kôdô*, p. 2.
139 *ibid.*, p. 49.
140 *ibid.*
141 Hasegawa, 'Kanjûdanomi no Hazama de Shundô', p. 23.
142 Matsuoka Toshikatsu Official Site, 'Kumamoto-ken kara no Seisaku Teian' ['A Policy Proposal from Kumamoto Prefecture'], in *Katsudô Hôkoku* [*Activity Report*]. Available from http://matsuokatoshikatsu.org/index1.html
143 For a comprehensive analysis of this phenomenon, see Kôno Takeshi and Iwasaki Masahiro, 2004. *Rieki Yûdô Seiji—Kokusai Hikaku to Mekanizumu* [*Politics That Benefit Local Interests—Mechanism and International Comparison*], Ashi Shobô, Tokyo.
144 Hasegawa, 'Kanjûdanomi no Hazama de Shundô', p. 25.
145 Nakano generalises this point. See Nakano Minoru, 1992. *Gendai Nihon no Seisaku Katei* [*Policy-Making Process in Contemporary Japan*], Tôkyô Daigaku Shuppankai, Tokyo, p. 124.
146 Matsuoka was made chairman of the LDP subcommittee disbursing this expenditure. See Chapter 4 on 'Exercising Power as a *Nôrin Giin*'.
147 Nakanishi and Journal Reporter Group, 'Matsuoka Toshikatsu to Iu Giwaku Nin', p. 183.
148 For details, see Chapter 4 on 'Exercising Power as a *Nôrin Giin*'.
149 This meant that they owned one hectare or more of forestry area.
150 Nôrinsuisanshô, Tôkei Jôhôbu, *Dai-77 Nôrinsuisanshô Tôkeihyô*, p. 427.
151 *ibid.*, pp. 428, 430.
152 For the definitive study of clientelism in Japanese politics, see Ethan Scheiner, 2006. *Democracy Without Competition in Japan: Opposition Failure in a One-Party Dominant State*, Cambridge University Press, New York.
153 Tatebayashi generalises this point. See *Giin Kôdô*, p. 11.

154 *Kashozuke* means the designation (by bureaucrats) of an area that will become a place where public works subsidised by central and prefectural governments (usually a combination thereof) will be carried out. As Itô explains, 'politicians, through their introduction of budgets to local areas (constituencies,) which are called the designated places (*kashozuke*), respond to the expectations of companies and voters'. Itô, 'Heisei Jiken Fuairu: Nôrin Jigyô Hojokin o Dokusen Suru Matsuoka Toshikatsu', p. 64.

155 For example, for a long time, in relation to the MAFF's agricultural structure improvement (*nôgyô kôzô kaizen*) projects (*jigyô*), 'a system of "group leader administration" (*hanchô gyôsei*) operated in which the opinions of the assistant divisional chiefs in charge (*tantô kachô hosa*), namely *gikan*, were particularly influential. This was because the standards for authorising (*nintei*) the project district (*chiku*) and for allocating the project cost were not transparent. Such a process created room over a long period for new selections and budget allocation to be made at the discretion of the person in charge, who had greater specialised knowledge and experience.' See the interim report of Watanabe Yoshiaki, who chaired the MAFF's 'Committee for Investigations Relating to Agricultural Structure Improvement Public Works', which was established in late 1999. The report was quoted in Ishii Kôki, 'Nôsuishô Osen: Amakudari Konsarutanto ga Genkyô da' ['MAFF Contamination: The Amakudari Consultants Are the Ringleaders'], *Bungei Shunjû*, May 2000, p. 199.

156 'Za Sankuchuari: Jimintô "Nôrin Zoku"' ['The Sanctuary: LDP "Agriculture and Forestry Tribe"'], *Sentaku*, Vol. 30, No. 2, February 2004, p. 59.

157 In Japanese, the concept of *riken* often involves the idea of businesses colluding with public organisations and politicians.

158 See Chapter 6 on 'The Identical Twins of Nagata-chô'.

159 Nakanishi and Special Reporting Group, 'Suzuki Muneo, Matsuoka Toshikatsu', p. 104.

3

ACCOMMODATING ELECTORAL REFORM

Matsuoka, like every other Lower House Diet member, faced a vastly altered political world and electoral landscape as a result of the overhaul of the Lower House electoral system in 1994. The most important aspect of the changes was the restructuring of the Lower House electoral system into a combination of 300 first-past-the-post single-member districts (SMDs) and 11 regional blocs electing 200 candidates on a proportional representation (PR) basis.[1] The electoral boundaries in Kumamoto Prefecture were redrawn, which changed its electoral composition from two MMDs to five SMDs. Now that Matsuoka was contesting a seat as the only LDP candidate, the altered electoral arrangements directly affected his electoral prospects and required some adjustment to his campaign strategy.

CONSTITUENCY REORGANISATION

The reorganisation of seats in Kumamoto Prefecture transferred Matsuoka from Kumamoto (1) with five seats to Kumamoto (3) with one seat. He was no longer competing against members of his own party, but in order to win the seat, he had to obtain a plurality—the most votes of any candidate standing for the seat. Matsuoka himself was opposed to the small constituency system, possibly fearing that it would make electoral battles even tougher, stating that 'candidates will alter their opinions and behaviour in line with whoever is powerful at the time'.[2]

Kumamoto (3) was located in the middle of Kyushu at the North Eastern part of Kumamoto Prefecture, known as 'fire country' (*kaji no kuni*) because of Mt. Aso. It was relatively large in geographic size in comparison to the other

SMDs in the prefecture, which was indicative of its lower population density. Only Kumamoto (5) at the western end was bigger in area.

The electoral reorganisation not only sought to equalise the value of votes as far as possible across the new SMDs, but also to draw the geographic boundaries of the new districts around the *jiban* of sitting members. While the overall size of electoral districts might have shrunk, electoral restructuring was implemented in such a way that politicians such as Matsuoka were able to maintain their geographically concentrated voting bases. Kumamoto (3) encompassed Matsuoka's hometown (Aso Town) in his home county (Aso County), and it also retained Kamoto County and Kikuchi County as well as two cities, Yamaga City and Kikuchi City.

CHANGING THE CHARACTER OF THE ELECTORATE

The key consequence of electoral reform for Matsuoka was that his constituency became more rural. Kumamoto (3) lost Kumamoto City, with farm households constituting only 1.1 per cent of total households (see Table 2.1). It also lost the highly urbanised areas of Arao City and Tamana City (see Table 2.1). Kumamoto (3) had only two cities, Yamaga City and Kikuchi City, which had higher proportions of farm households (see Table 2.1). Matsuoka could now forget about having to battle it out for votes in the big cities of Kumamoto Prefecture, including Kumamoto City.

At the time of the electoral reorganisation, Kumamoto (3) was classed as semi-rural (*junnôsonteki senkyoku*),[3] which was defined as an electoral district with more than 20 per cent of the population employed in primary industry.[4] Thus, Matsuoka's electorate changed in socio-economic character from semi-urban Kumamoto (1) to semi-rural Kumamoto (3). His constituency became more rural and agricultural because it encompassed mainly rural counties.

Matsuoka went from an electorate with an average of 6.4 per cent farm households out of the total across all municipalities in 1990 to one with an average of 19.9 per cent farm households across all municipalities in 2000 (see Table 2.2 and Table 3.1). The electorate's average figure of 19.9 per cent farm households compared with the national average of only 2.75 per cent.[5] Similarly, population density in Kumamoto (3) at 160 persons per square km was less than half the population density across the whole country at 340 persons per square km in 2000.[6]

Matsuoka's constituency was a good example of an electorate that became more homogeneous in socio-economic composition and occupational character as a result of the electoral reorganisation. His smaller constituency of Kumamoto (3) was much less diverse than the larger one of Kumamoto (1). The redrawing of electoral boundaries meant that agricultural and forestry interests became more concentrated in his electorate.

The practical effects of electoral reorganisation were, therefore, to divest Matsuoka's constituency of a large number of urban voters. This enabled him to concentrate on representing rural regions and farm and forestry interests. Instead of turning Matsuoka into a something-for-everyone kind of politician, electoral reform, by changing the composition of his constituency, actually enhanced Matsuoka's position as a rural-regional representative and *nôrin giin*.

This outcome ran directly counter to the conventional wisdom about the impact of the 1994 electoral reform. The former MMD system, in which different candidates from the same party (the LDP) could offer specialised representation of particular interests, had changed to an SMD system in which the party candidate needed a plurality to win the seat. This was expected to force party candidates to develop wide appeal that would attract a range of voters and their interests. In theory, the changeover to the new system should have meant that Diet members representing SMDs, including Matsuoka, could no longer afford to rely so heavily on farm and rural votes. Because their special-interest supporters could not deliver a plurality they would have to broaden their appeal to a wider cross-section of voters.

In Matsuoka's case, however, electoral reform had the reverse effect. The composition of his new electorate supported even stronger sectoral specialisation. He was able to project himself more starkly as a representative of agricultural and forestry interests. The practical effects of the new system were to make the electoral battle easier for Matsuoka because he had a logical appeal for the large proportion of naturally conservative voters (many with agricultural and forestry interests) in his newly constructed electorate. Whereas in the past he had to fight hard for city votes (especially against candidates such as Noda, Hosokawa, Uozumi and the JSP and Kômeitô candidates), his prospects were now for an easier time appealing largely to rural county voters.

The impact of electoral reform for Matsuoka demonstrated that the new electoral system did not turn all SMD candidates from policy specialists into policy generalists. In some cases, the electoral reforms supported an even stronger policy specialism, and the incentives for Matsuoka to represent sectional interests were reinforced. Moreover, Matsuoka retained his *jiban* in the new electorate, to which he was encouraged to supply regionally concentrated policy services. So the incentives for localism also remained.

Thus, being a *nôrin giin* and being a politician who had a strong geographically focused *jiban* served Matsuoka well in the changed system. He was able to continue to base his electoral appeal on what he could personally deliver to his constituents in the way of pork-barrel benefits as well as agricultural and forestry policy concessions, and not cleaving to some general manifesto of the LDP as a whole.[7] Although Matsuoka was the only candidate now standing for the LDP in Kumamoto (3), his election campaign remained centred on his personal vote-seeking style and was conducted primarily on an individual basis (*kojin honi*). Such an orientation was further encouraged by the fact that his main opposition was now his old rival Uozumi, ex-LDP, standing for the New Frontier Party,[8] which effectively split the conservative vote in Kumamoto (3).

ORGANISING THE VOTE

For Matsuoka, as for all LDP election candidates following electoral reform, his *kôenkai* remained the principal means by which he mobilised votes and organised campaign activities.[9] Not only did he maintain his *kôenkai*, but he also strengthened it. The LDP's Kumamoto Prefecture No.3 Electoral District Branch (Jiyûminshutô Kumamoto-ken Daisan Senkyoku Shibu) became the Kumamoto branch of Matsuoka's *kôenkai*, located in Kikuyo Town in Kikuchi County. There were two other branches, the Johoku branch located in Yamaga City and the Aso Office in Minami Aso Village, Aso County. The *kôenkai* had both a Youth Division (Seinenbu) and a Women's Division (Fujinbu), which Matsuoka addressed from time to time to rally support. He also arranged for delegations of both groups to visit Tokyo from time to time for study tours of the Diet and LDP headquarters.

The *kôenkai* contained a very tight network in Aso Town, which was the core of Matsuoka's *jiban*. The town mayor, Kawasaki Atsuo, was chairman of Matsuoka's *kôenkai* in that town. Kawasaki's father became the first town mayor when Aso Town was created out of the amalgamation of five towns and villages in 1954, and he occupied that position for four terms. Extending over a period of 36 years (although not continuously) Aso Town politics was firmly within the grasp of father and son.[10]

Moreover, the local administrative set-up within Aso Town became virtually synonymous with Matsuoka's electoral organisation, extending right down to the grass roots and forming the core of his personal supporters' organisation. A branch of the Aso Town Associates' Group (Dôshikai)—previously called the Aso Town Construction Associates Group (Kensetsu Dôshikai)—entrenched itself in each of the town's 52 wards. Each branch had a head separate from the head of the ward (*kuchô*), which was an administrative position. Members of the group filled most of the important positions in the town office. The group was formed in the current mayor's father's generation, and according to its treasurer, about half the voters were members. It functioned as a mechanism for Matsuoka to organise voting support. The Dôshikai's president in 2003 was the president of a local construction company who had also served as chairman of the local assembly, and who was the vice-president of the local branch of Matsuoka's *kôenkai*.[11]

Over a long period, the Dôshikai was considered to be synonymous with a political control regime linking Matsuoka to affiliated prefectural assembly members and town mayor Kawasaki. The section head of the group reputedly had more power than the ward head, while its president, along with mayor Kawasaki, headed up Matsuoka's election countermeasures organisation (*senkyo taisaku soshiki*) within his *kôenkai*. The organisational chart of the Dôshikai corresponded exactly to Matsuoka's election organisation.[12]

The Dôshikai also allegedly controlled the way people voted in Aso Town. According to an influential figure privy to the internal affairs of the Matsuoka political control regime, 'at election time, a "trustworthy" person was sent with a "dangerous" person to a polling station. The "dangerous" person had to show the "trustworthy" person how they voted. If they hadn't done the right thing, they would be ostracised in the village. It's not like it is in the cities'.[13]

A former member of the Dôshikai, who was previously involved in election campaigns, recounted a similar story, describing how, in some localities, members would go to vote in groups and show their ballot papers to each other.[14] As Hasegawa observes, 'this behavior is reflective of a closed society that puts a priority on regional and blood relations. As long as it continues, the Matsuoka-affiliated prefectural assembly member-mayor Kawasaki regime will be supported by its bedrock.'[15]

Such a system not only operated in national elections, but also in Aso Town elections, helping to entrench the Matsuoka political control regime even more deeply. The Dôshikai decided in detail how many votes would go to which candidate in the town elections. As a result, 16–17 members of the group were always elected to the 18-member town assembly. Such behaviour represented a form of political vote-rigging (*dango*). The town assembly members, who were elected through this kind of vote distribution, were unable to say anything against Matsuoka's regime, and so it became even more embedded in the local political scene. What is more, town office officials (most of whom were members of the Dôshikai) carried out vote counting in elections. When a young, would-be lawyer and leader of a citizens' movement, Izeri Seigo, stood for the mayoralty against Kawasaki in April 2002, he was defeated by only nine votes. A number of locals expressed their suspicions about unfair counting in the election.[16] Iseri was heard to comment that 'the rule of law was merely something that I studied in law school; it did not exist in reality'.[17]

This was a vote-gathering regime that centred on Matsuoka alone, not his party. It was designed to attract a personal vote to Matsuoka himself, not to the LDP, with the glue being the pork-barrel benefits Matsuoka could deliver to his *jiban* and the patronage that Matsuoka could provide to his cabal of personal supporters. He provided a crucial link for town locals to the centre of power in Tokyo, and to the prefectural political world.

ORGANISING THE FUNDS[18]

Electoral reform occasioned a major restructuring of Matsuoka's fund-gathering arrangements. The big difference between the pre and post reform periods was the rationalisation of Matsuoka's political funding groups and his acquisition of government subsidies through the party branch. This all took place in 1996, the first year in which a national election was fought under the new system. The previous year (1995) was effectively a transitional year, the first year in which a formal report was made of political funding (¥10 million) flowing through the local party branch, the LDP Kumamoto Prefecture No.3 Electoral District Branch to Matsuoka.[19] In 1996, the party branch became a much bigger source of funding at ¥55 million.[20] Matsuoka, as chairman of the local LDP branch, was legally qualified to use funds from the public subsidy paid to the branch by LDP headquarters. He could use the money to support his *kôenkai* activities. In fact, the address of the local party branch of the LDP was the same as Matsuoka's *kôenkai* address in Kikuyo Town, Kikuchi County.

In 1996, Matsuoka's political funding groups were all rolled into one: the Matsuoka Toshikatsu New Century Politics and Economics Discussion Association, which collected the largest amount of any of his funding sources in that year—¥128 million.²¹ Matsuoka's *kôenkai* also generated a substantial amount—¥97 million.²² This three-fold structure remained thereafter, with varying proportions of the total official funding gathered from the three sources each year. Generally speaking, the largest amounts came from Matsuoka's political funding group, except for 2000 when the biggest quantity was sourced from the LDP party branch (¥155 million). ²³ In that year, Matsuoka's *kôenkai* ranked second with ¥132 million and his political funding group ranked third with ¥120 million.²⁴ Matsuoka also received money from his faction, the Mitsuzuka faction.

Matsuoka remained in the Mitsuzuka faction until 1998, when he moved to the Etô-Kamei faction (Shisuikai) led by Kamei Shizuka and Etô Takami. The faction was composed of members of the Kamei group, which had spun off from the faction headed by former Prime Minister Mori, and former Nakasone faction members. Etô was a prominent member of the *nôrin zoku* who became joint leader of the faction. Kamei was known to hand out 'pocket money' to get his faction members to vote in the way he wanted. In 1998 when Matsuoka joined, he received a ¥2 million contribution from Kamei's *kôenkai*.²⁵

Suzuki Muneo also remained an important source of financial backing for Matsuoka. Suzuki's 21ˢᵗ Century Policy Research Association donated a total of ¥6 million to Matsuoka's New Century Politics and Economics Discussion Association over four years: ¥2 million in 1996, ¥1.5 million in 1997, ¥2 million in 1998 and ¥500,000 in 1999.²⁶ In addition, Matsuoka received ¥500,000 from the LDP's Hokkaido House of Representatives Proportional Representation District No. 1 Branch in 1999. The LDP Kumamoto Prefecture Electoral District No. 3 Branch also received ¥6 million in 2000.²⁷ This made the total amount ¥12.5 million from Muneo over this period.²⁸

Matsuoka's political funding group did not record the exact amount received in several lots from Suzuki's 21ˢᵗ Century Policy Research Association in 1995: ¥2 million in August 1995, ¥3.5 million in September 1995 and ¥3 million in December 1995—for a total of ¥8.5 million. It only recorded a total of ¥2 million.²⁹ Taking the additional amounts into account produced the grand sum of ¥19 million.

ELECTORAL PERFORMANCE AFTER THE REFORM
The 1996 Election

In the intervening period since the previous general election in which the Japan New Party boom overhang continued to be felt, Matsuoka, who was re-elected in 1996—the first poll held after the electoral reorganisation—increased his voice in the Federation of Kumamoto Prefecture Liberal Democratic Party Branches (Jiyû Minshutô Kumamoto-ken Shibu Rengôkai).[30] He achieved a closely fought victory over Uozumi standing for the New Frontier Party (see Appendix). In the final vote count, Matsuoka beat Uozumi only by a whisker (1,275 votes). It was Matsuoka's continuing support in Aso County, where Uozumi could only obtain around a third of the votes secured by Matsuoka, which once again stood him in good stead and saved him from defeat. It was fortunate for Matsuoka that the ever-popular Hosokawa stood for re-election in Kumamoto (1) rather than in Kumamoto (3).

The most outstanding feature of the 1996 election was that it once again underscored Matsuoka's dependence on the voters of Aso County. Overall, Matsuoka won 64.9 per cent of the total vote cast in Aso County, supplying just under 40 per cent of his total vote. He won first place in all the towns and villages in the county (see Table 3.1), with the proportion of his vote ranging from 56.7 per cent in Nishihara Village to 75.9 per cent in Kugino Village. His vote tally in Aso Town was higher than ever at 31,081 votes, which was 72.2 per cent of the total vote. The figures pointed to Matsuoka's successful consolidation of his *jiban* in Aso County and his increasing reliance on this region for electoral support. Clearly Matsuoka had successfully made the transition from Kumamoto (1) to Kumamoto (3), transferring the hard vote based on his hometown and home county into the new electorate. He retained the part of the electorate that had always supported him strongly and it became an even bigger rock on which his electoral performance rested.

Matsuoka gained uniformly second place in all the towns and villages of Kikuchi County, with Uozumi beating him into 1st place. Nevertheless, Matsuoka's vote tally in Kikuchi County was nearly three times what it had been three years earlier in 1993 (compare Table 2.2 and Table 3.1). Matsuoka was even more popular in Kamoto County where he won first place in all the towns except for Ueki Town (see Table 3.1).

Table 3.1 Farm household composition/votes cast for Matsuoka by municipality in Kumamoto (3) in 1996 Lower House election

Name of municipality	No. of farm[a] households	Farm households as % of total in municipality/ies	Votes cast for Matsuoka	% of total cast vote	% of Matsuoka's total vote	Placing among 4 candidates
Cities						
Yamaga City	3,224	16.1	11,637	33.3	14.4	2nd
Kikuchi City	1,490	13.2	7,365	41.2	9.1	2nd
	1,734	20.0	4,272	25.0	5.3	2nd
Counties						
Kamoto County	17,293	20.7	69,088	47.9	85.6	1st
Kahoku Town*	5,052	29.5	14,233	43.1	17.6	1st
Kikuka Town*	795	54.3	1,798	49.5	2.2	1st
Kamoto Town*	1,158	55.6	2,632	52.9	3.3	1st
Kao Town*	603	22.8	2,464	47.7	3.0	1st
Ueki Town	695	47.2	1,821	49.3	2.3	1st
Kikuchi County	1,801	19.1	5,518	35.5	6.8	2nd
Shichijo Town	4,570	11.1	23,774	37.6	29.5	2nd
Kyokushi Village	618	39.4	1,438	49.5	2.2	2nd
Ozu Town	533	36.8	1,477	52.9	3.3	2nd
Kikuyo Town	1,148	12.6	5,681	47.7	3.1	2nd
Koshi Town	729	8.0	4,710	49.3	2.3	2nd
Shisui Town	505	7.4	3,672	35.5	6.8	2nd
Nishigoshi Town	620	14.8	2,663	49.5	2.2	2nd
	417	4.7	4,133	52.9	3.3	2nd

Aso County	7,671	31.1	31,081	64.9	38.5	1st
Ichinomiya Town	727	22.4	3,369	57.6	4.2	1st
Aso Town	1,643	27.6	8,867	72.2	11.0	1st
Minamioguni Town	578	38.3	1,783	59.4	2.2	1st
Oguni Town	854	28.5	3,701	62.0	4.6	1st
Ubuyama Village	286	47.2	786	65.1	1.0	1st
Namino Village	264	50.9	770	63.2	1.0	1st
Soyo Town	733	49.8	2,304	74.3	2.9	1st
Takamori Town	616	25.1	2,806	60.0	3.5	1st
Hakusui Village	584	44.6	1,883	67.0	2.3	1st
Kugino Village	420	57.1	1,412	75.9	1.7	1st
Choyo Village	401	18.7	1,690	57.6	2.1	1st
Nishihara Village	565	33.0	1,710	56.7	2.1	1st
Total/average	20,517	19.9	80,725	45.1	100.0	1st

Notes: [a] Farm household data are for 2000.

Sources: Sômuchô, Tôkei Kyoku, 2001. *Heisei 12-nen Kokusei Chôsa Hôkoku Dai 2-kan Dai 1-ji Kihon Shûkei Kekka Sono 2 Todôfuken, Shichôson Hen–43 Kumamoto-ken* [Year 2000 National Census Report Vol. 2 Primary Basic Statistical Results 2 Prefectures and Municipalities Edition 43 Kumamoto Prefecture], Tokyo, Sômuchô, Tôkei Kyoku, pp: 288–91 and 294–303; Asahi Shinbunsha Senkyo Honbu, 1997. *Asahi Senkyo Taikan: Dai 41-kai Shûgiin Sôsenkyo (Heisei 8-nen 10-gatsu), Dai 17-kai Sangiin Tsûjô Senkyo (Heisei 7-nen 7-gatsu)* [Asahi General Survey of Election: The 41st House of Representatives General Election (October 1996), The 17th House of Councillors Regular Election (July 1995)], Tokyo, Asahi Shimbunsha, p. 278.

Matsuoka's total county vote shot up considerably in the 1996 election, rising by more than 20,000 votes. It supplied 85.6 per cent of Matsuoka's supporting votes (see Table 3.1). The most substantial increases occurred in Kamoto County and Kikuchi County. The smaller proportional rise in Aso County suggested that Matsuoka's level of support there was just about at saturation point. The general picture of Matsuoka's electoral performance in the counties underlined his position as the farm and rural-regional representative.

Amongst voters in urban areas, Matsuoka gained second place (to Uozumi) in Yamaga City and Kikuchi City, winning 41.2 per cent of the vote in Yamaga City but only 25.0 per cent in Kikuchi City (see Table 3.1), where Uozumi was more popular. Overall, city votes shrank to 14.4 per cent of Matsuoka's total vote (see Table 3.1), underscoring once again Matsuoka's unequivocal conversion to farm and rural-regional representative of the district.

Based on these figures, the composition of Matsuoka's voting support was clear: strong backing from rural counties, and especially strong support from his *jiban*, which was a reliable and continuing source of votes. Not surprisingly, in Kumamoto Prefecture Matsuoka came to be known as 'Aso's Matsuoka'. His slim margin of victory also served to underscore the regionally concentrated nature of his support base.

In the campaign itself, name recognition was no longer an issue. Matsuoka was a third-time candidate who had already made a name for himself in both policy activities and in *rieki yûdô seiji*.[31] His public promises (*kôyaku*) prior to the election contained the usual all-embracing goals of 'establishing stable politics', 'implementing social welfare policies', 'reviving agriculture, forestry and fishery industries and agricultural, mountain and fishing villages', 'promoting administrative reform, including educational reform', and 'maintaining basic transport and information networks'.[32]

As for money, figures reveal that financial support skyrocketed in 1996 compared with earlier years. However, the election was marred by violation of the Public Office Election Law by one of Matsuoka's local secretaries, and the issuing of a subsequent search warrant against his name. When interviewed about it, Matsuoka said, 'I'm very disappointed. So that it never happens again, we're making sure that everyone in the [electoral] office knows [the rules], and we've put up notices around the office'.[33]

Matsuoka was recommended by the Nokyo National Council in the 1996 election, but did not appear on a list of prefectural *nôseiren*-endorsed candidates.[34]

A member of a Kumamoto agricultural cooperative posted a comment about Matsuoka on the Internet, which said, 'Matsuoka *sensei* is really terrible. When the votes for the LDP in the Kumamoto PR electorate were low, he put incredible pressure on each unit agricultural cooperative. Most people were voting for Hosokawa *sensei* [who was standing for the New Frontier Party]. We really had lots of trouble in the election when Hosokawa *sensei* gained a lot of votes…'[35]

For Matsuoka to receive electoral support from the prefectural *nôseiren*, he had to demonstrate sympathy for, and understanding of, the organisation's agricultural policy campaigns (*nôsei undô*), and to make a public promise of adherence to a position that would reflect the intentions of Nokyo along with farmers in politics. In exchange for recommendation and authorisation (*kônin mo suishin mo*), he would have to sign a policy agreement with the organisation and become a staunch friend (*meiyû*) of the league.[36]

After the election, Matsuoka, along with 139 other successful Diet members who had been recommended by the National Council, fronted up at a 'Gathering to Talk About the Future of Japanese Agriculture' hosted by the National Council and the National Central Union of Agricultural Cooperatives (Zenkoku Nôgyô Kyôdô Kumiai Chûokai, or Zenchû). The management of each prefecture's *nôseiren* was also in attendance. Because Matsuoka was chairman of the LDP's Agriculture and Forestry Division (Nôrin Bukai) at the time, he featured prominently in the speechmaking.

He also attended a 'Meeting to Talk with Diet Members' hosted by the Kumamoto Prefecture *nôseiren* and the Kumamoto Prefecture Nokyo Central Union (Kumamoto-ken Nôgyô Kyôdô Kumiai Chûokai) in September 1997. The four LDP Diet members officially endorsed and recommended by the prefectural *nôseiren* in the 1996 elections reportedly attended this meeting. Each of the Diet members who participated issued their national policy reports and exchanged opinions with Nokyo officials on topics and issues relating to farmers in Kumamoto including rice production adjustment (*gentan*), the quantity of rice for government purchase (*seifumai*), the new basic law for agriculture, and the management of farms producing buckwheat noodles (*soba*) and *igusa* as well as livestock products.[37]

THE 2000 ELECTION

The Lower House election in 2000 represented the peak of Matsuoka's electoral performance in his entire Diet career. He was up against a bunch of new

Table 3.2 Farm household composition/votes cast for Matsuoka by municipality in Kumamoto (3) in 2000

Name of municipality	No. of farm households	Farm households as % of total in municipality/ies	Votes cast for Matsuoka	% of total cast vote	% of Matsuoka's total vote	Placing among 4 candidates
Cities	3,224	16.1	17,832	56.4	16.3	1st
Yamaga City	1,490	13.2	10,427	62.0	9.6	1st
Kikuchi City	1,734	20.0	7,405	50.0	6.8	1st
Counties	17,293	20.7	91,295	65.2	83.7	1st
Kamoto County	5,052	29.5	21,919	69.2	20.1	1st
Kahoku Town*	795	54.3	2,657	76.2	2.4	1st
Kikuka Town	1,158	55.6	3,753	77.8	3.4	1st
Kamoto Town	603	22.8	3,532	71.4	3.2	1st
Kao Town	695	47.2	2,499	71.7	2.3	1st
Ueki Town	1,801	19.1	9,478	63.5	8.7	1st
Kikuchi County	4,570	11.1	35,339	56.2	32.4	1st
Shichijo Town	618	39.4	2,279	66.4	2.1	1st
Kyokushi Village	533	36.8	2,239	72.3	2.1	1st
Ozu Town	1,148	12.6	7,594	59.0	7.0	1st
Kikuyo Town	729	8.0	7,295	56.4	6.7	1st
Koshi Town	505	7.4	5,475	51.3	5.0	1st
Shisui Town	620	14.8	4,147	58.1	3.8	1st
Nishigoshi Town	417	4.7	6,310	49.7	5.8	1st
Aso County	7,671	31.1	34,037	74.7	31.2	1st
Ichinomiya Town	727	22.4	3,314	63.0	3.0	1st
Aso Town	1,643	27.6	9,275	79.2	8.5	1st
Minamioguni Town	578	38.3	2,156	73.3	2.0	1st

Oguni Town	854	28.5	4,161	74.1	3.8	1st
Ubuyama Village	286	47.2	906	78.3	0.8	1st
Namino Village	264	50.9	876	77.4	0.8	1st
Soyo Town	733	49.8	2,257	77.3	2.1	1st
Takamori Town	616	25.1	3,104	70.5	2.8	1st
Hakusui Village	584	44.6	2,082	78.1	1.9	1st
Kugino Village	420	57.1	1,390	82.0	1.3	1st
Choyo Village	401	18.7	1,999	73.8	1.8	1st
Nishihara Village	565	33.0	2,517	75.7	2.3	1st
Total	20,517	19.9	109,127	63.6	100.0	

Note: [a] Farm household data are for 2000.

Sources: Sômuchô, Tôkei Kyoku, *Heisei 12-nen Kokusei Chôsa Hôkoku Dai 2-kan Dai 1-ji Kihon Shûkei Kekka Sono 2 Todôfuken, Shichôson Hen-43 Kumamoto-ken*, pp: 288–91 and 294–303; Kumamoto-ken Hômu Pêji/Senkyo Kanri Iinkai, *(Dai 42-kai) Shûgiingiin Sôsenkyo (Shôsenkyoku) Kaihyô Kekka: Heisei 12-nen 6-gatsu 25-nichi* [*(The 42nd) House of Representatives General Election (Single-Member Districts), The Results of the Vote Count: 25 June 2000*], p. 3; <http://www.pref.kumamoto.jp/gyousei/senkan/osirase/no42/pdf/04.pdf>.

candidates, all relative unknowns (see Appendix) against whom he was a clear favourite. For the first and only time, Matsuoka won first place in all the towns and cities of Kumamoto (3) (see Table 3.2).

Because Uozumi (having lost the last election to Matsuoka by a whisker) had moved to the Upper House, a split in the conservative vote was avoided. The absence of Uozumi as an electoral rival meant that Matsuoka won close to 80 per cent of the LDP vote. This did not stop him, however, from simultaneously standing on the LDP party list in the Kyushu bloc.

In addition, Matsuoka benefited from being a jointly endorsed LDP-Kômeitô candidate (the average increase in support for LDP candidates across electorates from this arrangement was reportedly 20,000-30,000 votes). These were not personal votes, but party-influenced votes. Even so, the distribution of Matsuoka's support remained relatively the same across the electorate. As Table 3.2 indicates, Aso Town in Aso County remained Matsuoka's most reliable source of support, providing the highest number of votes ever for Matsuoka (34,037) and almost a third of his total vote with 74.7 per cent of votes in that county going his way. Clearly Matsuoka's *jiban* in Aso County remained absolutely unshakeable and unassailable. This was despite a deliberate effort by the DPJ candidate, Hamaguchi Kazuhisa, to try and pick up the anti-Matsuoka vote by holding large-scale gatherings in Matsuoka's home district of Aso Town.[38]

The key differences between the 1996 and 2000 elections were the rise in Matsuoka's support in Yamaga City and Kikuchi City (no doubt partly due to Kômeitô's endorsement) and the massive increase in support for him across the counties (by more than 20,000 votes in total) (compare Table 3.1 and 3.2). The latter could be attributed to the consolidation of Matsuoka's power as a *nôrin giin* and his attainment, by 2000, of agriculture and forestry tribe Diet member (*nôrin zoku*) status.[39] In the interim, Matsuoka had played a key role in guiding benefits to the localities of Kumamoto (3) and in influencing agriculture and forestry policymaking in LDP and Diet committees. Matsuoka's own public election promises, with echoes of 1996, contained the usual mix of bland generalities and motherhood statements, such as 'reviving the market and economy', 'implementing a social welfare policy', 'implementing educational reform and public safety countermeasures', 'rejuvenating regional areas', and 'strengthening international and diplomatic undertakings for resolving population, food and environmental problems'.[40]

Matsuoka was the only candidate standing in Kumamoto (3) to receive the official recommendation of the Nokyo National Council and the Kumamoto Prefecture *nôseiren* in the 2000 election. In order to receive the recommendation, Matsuoka had to pass through a comprehensive vetting process by the Nokyo organisation. Several steps were involved: approval of his answers to a public questionnaire set by the organisation and the signing of a policy agreement relating to political topics, which demonstrated Matsuoka's real understanding of the concerns of the organisation, and an application by the prefectural *nôseiren* to the National Council for recommendation by the entire body—both national and prefectural—operating in a unified fashion.[41]

Less obvious to the public view was Matsuoka's ever-tighter network of contacts with construction industry executives in his electorate. According to the son of one such executive, these 'recruits' to Matsuoka's cause were not always willing. During the election, the president of a concrete company found his name on a list of Matsuoka promoters. When he went to Matsuoka's campaign-launching ceremony as instructed by his secretary, he was asked to take a position on the podium. The president got fairly angry at this kind of treatment and consequently voted for the DPJ.[42]

THE QUINTESSENTIAL SPECIAL-INTEREST POLITICIAN

In winning and retaining the seat of Kumamoto (3) in the 1996 and 2000 elections, Matsuoka's political and policy behaviour were once again predominantly shaped by local and sectional interests as well as by the personal interests of certain clients. These interests were significantly but not substantially affected by electoral reform.

Local interests

Winning the seat of Kumamoto (3) meant that Matsuoka had to work really hard as the representative of that electoral district. The SMD system intensified the electoral competition that Matsuoka faced while concurrently shrinking the geographic size of his electorate, which encouraged an even stronger predisposition towards localism.[43] The new electoral system thus entrenched rather than curbed Matsuoka's 'constituency-service-oriented politics'.[44] Matsuoka was a good example of how electoral reform in Japan had the contra-indicated effect of unleashing the unrestrained forces of localism.

Another significant difference was that the new electoral system encouraged Matsuoka into a whole constituency-service orientation rather than simply a predominant focus on his *jiban*. While the latter remained extremely important, the need for a plurality meant that Matsuoka had to direct his appeals to the entire electorate and not just to a specific part of it. This meant doing whatever he could to direct public resources across the whole electorate. In this way, Matsuoka attempted to demonstrate that he was the most effective representative of that district.

The changeover to the new electoral system thus strengthened the incentives for Matsuoka to engage in pork-barrelling, to make promises about what he was going to do for his constituency and to broaden regionally concentrated policy services beyond his *jiban*. Although Matsuoka's primary electoral payback was to the voters of Aso County, he could not afford to neglect the rest of the electoral district because of his need for a plurality.

One of the most significant impacts on Matsuoka's representation of interests was, therefore, the incentive to engage even more intensively in *rieki yûdô seiji*. To differentiate himself from his rivals (from different parties) under such a system, he had to demonstrate the advantages he had as an incumbent. Guiding benefits to the whole district was the best way to do this. Moreover, because he was now the only member of the LDP elected from his Kumamoto constituency, his power strengthened in the prefecture.[44]

Through his successful acquisition of public works projects for a number of areas in his constituency, but particularly for the towns and villages of Aso County, Matsuoka consolidated his reputation as a politician who guided benefits to local areas (*rieki yûdôgata no seijika*).[45] Matsuoka was described as 'very useful' in alerting the central agencies to local interests and in securing budgets and projects.[46] In fact, in his constituency, Matsuoka's record of obtaining funds for various public works projects was soon unsurpassed. As far as the residents in the deserted rural and mountainous villages of Kumamoto were concerned, where agriculture was in decline and where young people had all left for the cities, public works were indispensable as the only industry in town. For them, it was said, 'Matsuoka was a necessary evil'.[47] Matsuoka commented during a general meeting for party reform at LDP headquarters in November 1997: 'If you want to call me a "civil engineering Diet member" (*doboku giin*), then do that. There are no Diet members who aren't thinking about elections'.[48] In this sense, Matsuoka 'did not hide the fact that he was a "concessions king"'.[49]

Matsuoka became a great believer in getting down to the grass roots and conducting on-the-spot investigations of particular issues of concern to locals. His website exhibited photographs of numerous visits to this place or that, discussing matters with farmers and others. Matsuoka always returned to his locality at the end of each week and met up with local people, talking to them and getting to know what they wanted.

Besides bringing public works back to his electorate, representing local interests was crucial to Matsuoka's electoral fate in other ways. Localism was more than just *rieki yûdô seiji*. It required him to exert influence on behalf of local politicians in his constituency over particular matters of concern to them, such as budget allocations to particular municipalities, local government amalgamations, the impact of the central government's decentralisation policies and the distribution of fiscal powers between central and local governments. Matsuoka was often visited by delegations of local leaders and politicians from his electorate, wanting him to intercede with the central government on issues affecting local government in their area. For example, some municipalities in Kumamoto (3) were alarmed about the potential impact of local government mergers on government spending in their localities, such as cuts in public works that could undermine the regional economy. In February 2003, LDP Diet members representing Kumamoto Prefecture, including M atsuoka, met to exchange opinions with municipal mayors. A majority of the local mayors felt that consideration should be given to the distinctive situation in each district in the local government merger process. The LDP Diet members' group confirmed that they would consider the earnest opinions of regional representatives and strive to reflect them in policies.[51]

Matsuoka also regularly hosted study tours of the Diet by his local constituents. He was happy to show groups of visitors from his *jiban* various aspects of Diet and party operations in Tokyo. The Aso branch of the association of ward heads, for example, visited Matsuoka in Tokyo and asked to be shown around the Diet. Matsuoka was able to say that he hoped everyone had gained some idea of where he worked and how important his job was as a Diet member.[52] Another such tour included 18 people from Oguni Town, Kikuchi City and Omori Town. Matsuoka showed them around the Diet and introduced them to various Diet members.

Sectional interests

Because of the predominantly rural nature of his support base, Matsuoka was concerned with conditions in rural-regional industries. After electoral reform, his support rate in individual municipalities of Kumamoto (3) became highly correlated with the percentage of farm households in that municipality. For example, a significant correlation in the 2000 poll results could be observed between the percentage of farm households in a municipality and the percentage of the total vote Matsuoka received.[53]

The switch to a single-member electorate did not, therefore, force Matsuoka to sacrifice his agricultural and forestry policy specialism in order to broaden his appeal to a wider cross-section of voters. He retained and indeed consolidated his representation of farm and forestry interests. If anything, an agricultural and forestry policy niche beckoned him even more strongly. There were many farm votes to be retained as well as new ones to be won in his new constituency. Indeed, he had to cater to farm households and forest owners even more assiduously in Kumamoto (3) than in Kumamoto (1).

As a *nôrin giin*, Matsuoka combined two predominant types of representation. The first was representation of large aggregated interests such as agriculture and forestry in terms of advancing particular macro-policy objectives, often defined in cooperation with large integrative interest groups such as Nokyo or the forest associations. Matsuoka made a point of attending 'request roundtables' of the Kumamoto Nokyo organisation at which he listened to what locals were saying about particular problems.

The second was representation of special interests through the application of micro-policies and/or allocations of specific-purpose subsidies to particular groups of beneficiaries. Agricultural and forestry policy generated a lot of pork-barrel benefits including subsidies for agriculture and rural infrastructure that could benefit particular localities, or agricultural cooperatives or other groups of farmers. So it doubled as policy that could also serve local interests in the electorate.

On the other hand, as the sole Diet representative from Kumamoto (3), Matsuoka had to take care of all aspects of the lives of his supporters in those municipalities where he won a plurality of votes—their welfare, economic, social and livelihood needs and so on. When his margin of victory was so small, as his generally was, every vote counted. He could not afford to concentrate exclusively on regional areas and agricultural and forestry issues and ignore the cities. After all, supporters in Yamaga City and Kikuchi City delivered 14.4 per cent of his total vote in the 1996 elections and 16.3 per cent in 2000. So Matsuoka had to be concerned with city businesses and other economic and social issues of concern to urban voters.

Such policies were, however, a sideshow to the Matsuoka's main orientation towards localism and sectionalism. These were the primary strategies by which he sought electoral success. The strongly bifurcated rural-urban split in his electoral strategy adopted during his days as Diet representative for Kumamoto (3) was abandoned. Once again, Matsuoka's changed electoral circumstances challenged predictions that the new electoral system would exclusively produce policy generalists rather than policy specialists, and median vote-seekers rather than special-interest vote-seekers. Quite the reverse, Matsuoka's specialism became more pronounced in keeping with the more homogeneous nature of his constituency in rural counties, which provided the large majority of his supporting votes.

Every New Year Matsuoka delivered a National Policy Report to the inner circle of his *kôenkai* in the Aso Town Gymnasium.[54] The report referred to Matsuoka's achievements in all areas of public policy, but particularly to the agriculture and forestry policies that he had influenced and to his achievements in terms of bringing pork-barrel benefits back to his electorate.

Clientelistic interests

Because the new electoral system intensified competition amongst candidates for a plurality, Matsuoka had an even stronger incentive to offer his services as a mediator to those seeking personal favours in order to secure votes and political funds. Matsuoka faced an environment of heightened competition for bribery and 'financial influence corruption'.[55] Electoral reform failed to convert Matsuoka into a new style of politician, primarily concerned with programmatic policies rather than with special interests and individually brokered deals.

Such activity involved engineering benefits not only for the leaders of particular interest groups and other public, semi-public and private organisations, but also for businessmen as well as for local government politicians and officials. Matsuoka constantly received petitioners in his Diet office seeking his patronage in the form of favours regarding various matters. Matsuoka's natural expectation was to deliver benefits in exchange for money or votes. Once when an executive of a public interest corporation (*shadan hôjin*) visited Matsuoka to petition him for a favour, Matsuoka shouted at him, 'I am not doing this job as Diet member for a hobby. If you don't bring money, bring votes'.[56] Such a comment revealed the depth of Matsuoka's orientation towards clientelistic interests and his propensity to act as a political broker. As one member of the Japanese public averred, 'he is a conscientious person who will do anything if you give him money'.[57] When the Mediation for Profit Prohibition Law (*Assen Ritoku Kinshihô*), which made it a crime for politicians to receive compensation for mediating with bureaucrats and others over matters such as public works contracts, was debated in the Diet in 2000, Matsuoka opposed the bill. He remarked that 'denying mediation results in the denying of the politicians themselves. It is erroneous to assert that mediation is evil'.[58] He also commented on the subject of collecting contributions from construction companies, saying that these were effectively 'a repayment for services rendered'.[59]

NOTES

1. The figure of 200 was reduced to 180 prior to the 2000 election.
2. Nakanishi, 'Matsuoka Toshikatsu', p. 28.
3. Shigeki Nishihira, 1995. 'Shosenkyoku Bunrui Kijun no Teian' ['Proposals for a Classification Standard for the Single-Member Electorate System'], *Chûô Chôsahô*, No. 449, March , p. 5; electoral data kindly supplied by author, worksheet, p. 2.
4. A total of 26 SMDs out of the total of 300 were categorised in this most rural of electoral categories. The percentage of population employed in primary industries in Kumamoto (3) at the time was 25.0 per cent. *ibid.*
5. This figure was calculated from data available at http://www.pref.kumamoto.jp/statistics/siryo/h15nenkan/xl_data/nenkan_data/nenkan-SB1.xls>. The figures are based on the 2000 census.
6. Data was obtained from: http://www.pref.kumamoto.jp/statistics/siryo/h15nenkan/xl_data/nenkan_data/nenkan-SB1.xls, and http://www.stat.go.jp/data/kokusei/2000/kihon1/00/zuhyou/a001.xls
7. As Curtis has aptly commented, where candidates in the new Lower House electoral districts 'differ tends to depend more on the kind of constituency they are running in than the party they belong to. Elections take place in separate districts. They are rarely nationwide referenda on broad policy issues. In a country such as Japan…where parties are loosely structured, what candidates say their policies are depends on what they think will get them elected in their particular districts.' Refer to Curtis, *The Logic of Japanese Politics*, p. 164.
8. Its Japanese title was Shinshintô.
9. Ellis S. Krauss and Robert Pekkanen, 'Explaining Party Adaptation to Electoral Reform: The Discreet Charm of the LDP', *Journal of Japanese Studies*, Vol. 30, No.1, Winter 2004, pp.10–13.
10. Hasegawa Hiroshi, 'Jimin "Gajô" no Chikaku Hendô' ['A Tectonic Shift in an LDP "Stronghold"'], *Aera*, 24 November 2003, p. 26.
11. *ibid.*
12. *ibid.*
13. *ibid.*
14. *ibid.*
15. *ibid.*
16. *ibid.*
17. *ibid.*
18. See also Chapter 6 on 'The Identical Twins of Nagata-chô'.

19 Kitamatsu, et al., 'Matsuoka Toshikatsu Daigishi Tettei Bunseki', p. 47.
20 ibid. See also below.
21 ibid.
22 ibid. All these figures were confirmed by the nationwide analysis of political funding conducted by the Asahi Shinbun in 1998. See 'Seiji Shikin Zenkoku Chôsa Kekka: 96-nen Sôsenkyô, Shôsenkyoku Kanren Tôsensha 384 Ninbun o Kôkai' ['Results of the National Investigation of Political Funding': Disclosure of 384 Successfully Elected Persons in the 1996 General Election'], available from: http://www.asahi.com/paper/special/shikin/
23 Kitamatsu, et al., 'Matsuoka Toshikatsu Daigishi Tettei Bunseki', p. 47.
24 ibid.
25 Kitamatsu, et al., 'Matsuoka Toshikatsu Daigishi Tettei Bunseki', p. 48.
26 ibid., pp. 48–9.
27 These are the years in which they were reported; they were actually received in the preceding year
28 Kitamatsu, et al., 'Matsuoka Toshikatsu Daigishi Tettei Bunseki', pp. 48–9.
29 ibid. p. 49.
30 Itô, 'Heisei Jiken Fuairu: Nôrin Jigyô Hojokin o Dokusen Suru Matsuoka Toshikatsu', p. 65.
31 See Chapter 4 on 'Exercising Power as a Nôrin Giin'.
32 Visit: http://www.kumanichi.co.jp/senkyo/senkyo-33.html
33 Visit: http://www.kumanichi.co.jp/senkyo/senkyo-33.html
34 Dai 41-kai Shûgiin Giin Sôsenkyo: Fuken Nôsei Undô Soshiki Suishin, Shiji Tôsen Giin Ichiran' ['The 41st House of Representatives General Election: A Summary of Elected Diet Members Recommended and Supported by Prefectural Agricultural Policy Campaign Organizations'], Nôsei Undô Jyânaru, No. 10, November 1996, p.22.
35 Visit: http://piza.2ch.net/giin/kako/987/987905181.html/
36 'Genchi Rupo—Kumamoto ken' ['On the Spot Report—Kumamoto Prefecture'], Nôsei Undô Jyânaru, No. 24, March 1999, p. 29.
37 'Suishin Giin to Meiyû no Paipu o Futoku' ['Fattening the Staunch Friend Pipe with Recommended Diet Members'], Nôsei Undô Jyânaru, No. 15, September 1996, p. 23.
38 Yomiuri Shinbun, 20 June 2000.
39 See also below, Chapter 4 on 'Exercising Power as a Nôrin Giin', and Chapter 5 on 'Exercising Power as a Nôrin Zoku'.
40 Visit: http://www.kumanichi.co.jp/senkyo/senkyo2000/kouho/kouho31.html
41 'Dai 42-kai Shûgiin Sôsenkyo ni mukete' ['With a View to the 42nd House of Representatives Election'], Nôsei Undô Jyânaru, No. 30, April 2000, p. 3.
42 Visit: http://piza.2ch.net/giin/kako/987/987905181.html
43 As Curtis observes, '[a]s a general matter, the smaller the district, the greater is the tendency for candidates and voters to be concerned with local issues'. The Logic of Japanese Politics, p. 163. As he points out, 'the U.S. House of Representatives—with single-member districts—is better known for pork-barrel politics than for sober consideration of how best to design policies to serve the broad national interest'. The Logic of Japanese Politics, pp. 163–4.
44 ibid., p. 164.
45 Itô, 'Heisei Jiken Fuairu: Nôrin Jigyô Hojokin o Dokusen Suru Matsuoka Toshikatsu', p. 65.
46 '"Muneo no Bôrei"'ni Maketa Meiyû "Matsuoka Toshikatsu"' ['The Sworn Friend "Matsuoka Toshikatsu" Who Lost to "Muneo's Ghost"'], Shûkan Shinchô, 20 November 2003, p.28.
47 Hasegawa, 'Kanjûdanomi no Hazama de Shundô', p. 24.
48 Nakanishi and Special Reporting Group, 'Suzuki Muneo, Matsuoka Toshikatsu', p. 105.
49 Mainichi Shinbun, 18 November 1997.
50 Visit: http://piza.2ch.net/giin/kako/987/987905181.html
51 Matsuoka Toshikatsu Official Site, 'Sessoku na Shichôson Gappei no Saikô o' ['Reconsidering the Hasty Municipality Mergers'], in Katsudô Hôkoku [Activity Report]. Available from http://matsuokatoshikatsu.org/site002//public/003.html

52 Matsuoka Toshikatsu Official Site, 'Kokkai Kengaku' ['Diet Study Tour'], in *Katsudô Hôkoku* [*Activity Report*]. Available from http://matsuokatoshikatsu.org/index1.html
53 The correlation was r=0.629. The regression coefficient between percentage of population in farm households and the percentage of the total vote received by Matsuoka was 0.535, which is also highly significant. This data was obtained from Horiuchi Yusaku who kindly did the calculations for the author.
54 Hasegawa, 'Kanjûdanomi no Hazama de Shundô', p. 25.
55 'Karate 4-dan', p. 35.
56 Nakanishi and Journal Reporter Group, 'Matsuoka Toshikatsu to Iu Giwaku Nin', p. 184.
57 Visit: http://piza.2ch.net/giin/kako/987/987905181.html
58 Nakanishi and Journal Reporter Group, 'Matsuoka Toshikatsu to Iu Giwaku Nin', p. 182.
59 Kitamatsu, *et al.*, 'Matsuoka Toshikatsu Daigishi Tettei Bunseki', p. 46.

4
EXERCISING POWER AS A *NÔRIN GIIN*

Where and how Matsuoka would exercise power in Japanese politics was to some extent predetermined. His career background, political connections and electoral support dictated both the policy interests that he represented and the policy activities that he pursued. These factors led him inexorably to his role as a *nôrin giin*, a representative of agricultural and forestry interests in the party and in the Diet.

PARTY COMMITTEE

As soon as Matsuoka entered the Diet in 1990 (see Table 4.1), he joined the LDP's Comprehensive Agricultural Policy Investigation Committee (CAPIC) (Sôgô Nôsei Chôsakai), one of the investigation committees of the PARC.[1] CAPIC was formed in 1968 to discuss medium and long-term policy issues for agriculture. Since that time, it had remained one of the two most important agricultural policy committees of the PARC (the other was the Agriculture and Forestry Division). CAPIC was concerned with the larger questions of agricultural policy such as the structure of agriculture, the future of Japanese agriculture and agricultural policies as well as rice production and pricing in the context of these larger, sector-wide issues. For that reason, CAPIC was generally considered to be the 'strategy division' of agricultural policy, whilst the Agriculture and Forestry Division was the 'tactics division'.[2] Unlike the division, whose membership was capped, LDP Diet politicians could freely register to join CAPIC, which had a very large membership as a result (around 245 in 1990).

Matsuoka joined CAPIC for several very important reasons. First, he wanted to demonstrate his credentials as a politician representing agricultural interests.

Membership was a good indicator of his intended policy direction and activities. It showed the strength of his interest in a specific policy domain.[3] By representing his farming constituents in the party Matsuoka also helped to secure his re-election. Only by winning successive elections could he build seniority in the party, thereby fulfilling one of the most important qualifications for appointment to higher office, both in the party and in government.

Second, membership of CAPIC provided a means by which Matsuoka could take positions on particular policies in which his constituents and supporting organisations had an interest.[4] These standpoints were vital in allowing others to grasp his 'revealed policy preference'.[5]

Third, because Matsuoka had spent 19 years in the MAFF, he considered himself well versed in agricultural policy, so it was natural for him to gravitate toward a committee that considered government measures for agriculture. Besides exhibiting the characteristics of a 'status incentive politician', Matsuoka also demonstrated the features of 'single-issue incentive politician'. As Glosserman explains

> …members of this group focus on a single policy issue…Many were dissatisfied with their previous lifestyle; all of them took up politics out of a desire to be involved in policy on issue in which they had a long-standing interest.[6]

Fourth, CAPIC was where Matsuoka could refine his expertise and skills in the domain of agricultural policy.[7] Developing his agricultural policy niche would furnish additional means for career progression and thus increase his power and influence in the party. Regular membership of a committee would qualify him for an executive position in that committee, which in turn would provide a ladder to higher office including sub-cabinet posts and ultimately ministerial positions. Membership was also proof to party executives and faction leaders of his actual activities in policy domains.[8] It allowed Matsuoka to demonstrate to party leaders that he had policy ability, which would be linked to future re-election and a successful career.[9]

Fifth, becoming a member of CAPIC was a vital step in putting Matsuoka into a position where he could influence party policy on agriculture and forestry, and thus government policy. If an ordinary LDP backbencher such as Matsuoka wanted to shape government policy, he had to join an LDP policy committee. Materialising policy influence for LDP backbenchers took place primarily in the committees of the PARC, which was the party's deliberative organ for policy decisions.¹⁰ The party policy committees performed the crucial functions of 'advance scrutiny' (*yotô shinsa*) and 'prior approval' (*jizen shônin*) of government policies and bills. Being on a committee enabled Matsuoka, individually, to exert influence on government policy. The exercise of such influence was vital in enabling him to claim credit for particular agricultural policy measures. CAPIC discussed both government bills and policies and amended them to take account of the interests of the special-interest members, such as Matsuoka. Even as a first-term legislator, Matsuoka would be free to participate in the debates in the committee and thus influence policy outcomes. He could even exercise denial rights (*hitei riken*) over a government-proposed policy.¹¹ Because decisions in PARC committees were taken on the basis of a consensus, a single individual such as Matsuoka, or a handful of like-minded politicians could hold up the business of government.

Sixth, being on a PARC committee enabled Matsuoka to directly communicate with and influence bureaucrats, who monopolised key steps in the policymaking process, such as policy formulation and bill-drafting. In particular, CAPIC would be a key locus of interaction between Matsuoka and MAFF officials, who often attended committee sessions and provided input into the decisions taken by the committee.

Finally, getting a start in an agricultural policy committee of the PARC was mandatory if Matsuoka were ever to take the additional step from *nôrin giin* to *nôrin zoku*. As an agricultural and forestry *zoku*, Matsuoka could exercise unparalleled influence both within party circles over agricultural policies and over bureaucrats in the allocation of subsidies and public works to his constituency. This was vital if Matsuoka were to guide benefits back to his *jiban* and his wider electorate, as well as to provide favours to key backers as a broker.

In 1991, Matsuoka made a logical progression in his memberships of key PARC committees relating to agriculture. He gained entry into the Agriculture and Forestry Division (see Table 4.1). LDP members of the Lower House Committee on Agriculture, Forestry and Fisheries (AFF), to which Matsuoka was appointed in that year, automatically became members of this division. Together with his continuing membership of CAPIC, joining the division gave him coverage of the two most important PARC committees on agriculture. The division was also concerned with forestry policy matters: an area in which Matsuoka could claim a great deal of expertise and career experience.

DIET COMMITTEES

Besides the PARC agricultural committees, Matsuoka's ambition was to become a member of the Lower House AFF Committee. Diet committees were formal bodies without decision-making power, and a place for opposition party, rather than ruling party, policy activity. In practice, the PARC committees were a more significant locale for LDP politicians to exercise policy influence because they provided an opportunity for them to amend government bills and proposed measures. Nevertheless, being a member of the AFF Committee would reinforce Matsuoka's policy specialism and represent an important step towards becoming a *nôrin zoku*.

Because of the popularity of the AFF Committee amongst LDP Lower House Diet members, Matsuoka was not able to join right away. As his former LDP colleague from Fukushima (2) explains

> [t]here is no way a first-term Diet member can join the agricultural and forestry committee or the construction committee. The older Diet members monopolise posts where there is a possibility of links to concessions (*riken*) and which are advantageous for elections. New Diet members have to wait their turn.[12]

Such norms meant that new members and young members of the ruling party such as Matsuoka had no right of choice. The distribution of memberships across the various Diet committees reflected the will of senior Diet members who had won a number of elections.[13] Membership of PARC divisions and investigation committees were much more a matter of individual choice.

When he entered the Diet, Matsuoka was first allocated to the Lower House Regional Administration Committee (see Table 4.1), which was concerned with policies relating to regional development (public works projects such as road construction and airports), as well as regional industries such as agriculture and forestry. This policy domain, along with that of the Construction Committee, was closely linked to rural areas, which had relatively higher proportions of agricultural, forestry and fisheries population.[14] For Matsuoka, the Regional Administration Committee was a stepping-stone to the AFF Committee.

In 1990, Matsuoka was also appointed to the Diet's Special Committee Relating to Land Problems etc. (see Table 4.1), which was also indirectly concerned with agriculture and forestry because these were land-based industries. Also, given his professional career experience in the National Land Agency, Matsuoka could put his expertise to good use in this special committee.

In 1991, Matsuoka achieved his ambition of AFF Committee membership (see Table 4.1). Getting on to this committee in only his second year as a Diet member was a significant coup for Matsuoka. It complemented his membership of the equivalent PARC committees, it helped to build his agriculture and forestry policy specialism further, and it was a necessary condition for his later accession to membership of the *nôrin zoku*.

MAFF PARLIAMENTARY VICE-MINISTER

In August 1995, in his second term, Matsuoka was appointed MAFF parliamentary vice-minister (see Table 4.1) in the reshuffled coalition cabinet of former Prime Minister Murayama Tomiichi, who headed a coalition government of the LDP and former JSP until January 1996. From Matsuoka's perspective, a parliamentary vice-ministership was another vital step on the ladder of political advancement.

Formally speaking, the position accorded Matsuoka considerable power over the MAFF. According to Clause 3, Article 17 of the State Administration Organisation Law (*Kokka Gyôsei Soshikihô*), the post of parliamentary vice-minister had more power and authority than a ministry's own administrative vice-minister, which was the top position in a ministry. The relevant clause stated that the parliamentary vice-minister's role was to assist the minister, to participate in the planning of policies and plans, to manage affairs of state, as well as to receive orders from the minister and to undertake the duties of the minister in his or her absence. The parliamentary vice-minister could stand in for the minister in undertaking ministerial duties, while the administrative vice-minister could not.[15]

In practice, however, the council of parliamentary vice-ministers, which was a sub-committee of the cabinet, met only one or twice a month and did not decide anything. It just listened to explanations from bureaucrats and was more of an arena to exchange opinions.[16] A parliamentary vice-ministership was considered a junior learning position within a ministry, a post reserved for second or third-term Diet members. Because it was a post that normally went to relatively junior politicians, parliamentary vice-ministers could not do important business.[17]

Nevertheless, being appointed to such a position in his second term was a tribute to Matsuoka's standing in agricultural and forestry policymaking circles and his demonstrated expertise on the various relevant committees. The position enabled him to hone his policy skills and to develop closer personal links with serving MAFF officials, as well as to consolidate his ties to all the relevant interest groups operating in the sector. In this respect, for Matsuoka as for other politicians, a parliamentary vice-ministership was a crucial step in breaking into the structure of concessions (*riken kikô*) in his chosen policy sector. In that respect, it gave him a leg-up to becoming a *zoku giin*.[18] In fact, the LDP reportedly used the parliamentary vice-minister's post as a mechanism for cultivating *zoku giin*, by linking politicians with specific ministries in this way.[19] The parliamentary vice-ministership thus served as a pointer to Matsuoka's political career and his political ambitions in the agriculture and forestry sector.

Moreover, it was common for the parliamentary vice-minister to become a director of the corresponding Diet committee, facilitating the passage of draft bills that the ministry had submitted for party perusal, and conducting negotiations with the opposition parties. In exchange, the ministry provided various benefits for their parliamentary vice-minister's electorate and for the industry world with which they had connections.[20] Both of these advantages suited Matsuoka's own ambitions and interests.

Accordingly, in 1995, Matsuoka became one of the directors of the AFF Committee to match his appointment as MAFF parliamentary vice-minister (see Table 4.1). He held this position until 1999—well beyond the end of his parliamentary vice-ministership. Formally, becoming a director was a matter of election by the members of the committee, but Matsuoka was actually nominated by the chairman according to his factional affiliation. Selection on this basis ensured a factional balance amongst the directors from the LDP. There were usually four LDP directors of the AFF Committee with the balance coming from the opposition parties. The directors were like vice-chairmen and a stepping-stone to the chairmanship. The directors played an important role in managing the conduct of committee business, meeting both before and after committee discussions in order to draft the agenda, to draw up the consensus of the meeting and to undertake crucial coordination functions.

Becoming MAFF parliamentary vice-minister in 1995 was serendipitous for Matsuoka because it was the interim period between the passage of the New Food Law (Law for Stabilisation of Supply-Demand and Price of Staple Food, or *Shuyô Shokuryô no Jukyû oyobi Kakaku no Anteihô*) in November 1994 and its implementation a year later in November 1995. The new law engineered the most radical change in the nation's Food Control system governing rice pricing and distribution in the post-war period. Under the law, the Food Agency devolved some of its controls over rice marketing to non-government players, and so Matsuoka was parliamentary vice-minister at a crucial time. When various questions were put to Matsuoka about rice under the new regime, he dutifully became a mouthpiece for the MAFF, commenting

> [w]e have to make [rice] production adjustment a success and keep a balance between demand and supply [in order to prevent producer rice price falls in a more liberalised market]. The government and the ruling parties have decided on some assistance for production adjustment, including compensation measures. As the government, we need to secure the budget to be able to do these sorts of things, and that's what I'll be endeavouring to do from now on.[21]

Like the MAFF spokesman that he was, Matsuoka opposed the idea of giving government assistance to all farmers participating in the planned distribution system for rice (*keikaku ryûtsûmai*), which was the distribution route that remained under government management. In Matsuoka's view, only those producers undertaking production adjustment should get assistance. He also pointed out that if imported rice affected the consumption of domestic rice, it would be necessary to think about developing new kinds of demand for processed rice.[22]

In 2000, Matsuoka became chairman of the AFF Committee (see Table 4.1). He had to be elected to the position in the plenary session of the Diet, but his party (effectively his faction) put his name forward after an internal discussion, and he received a formal nomination by the chairman of the Diet (*gichô*). As AFF Committee chairman, it was Matsuoka's job to report back to the plenary session on the committee's investigations of various aspects of the legislation submitted to it by the cabinet and by individual Diet members. Each party then made its final decision on the legislation based on this report.

LDP COMMITTEE EXECUTIVE 1995–2000

In 1995, Matsuoka became chairman of the Agricultural Basic Policy Subcommittee (Nôgyô Kihon Seisaku Shôiinkai) of CAPIC (see Table 4.1). It was his first executive position on an LDP agricultural policy committee. The subcommittee handled all matters relating to agricultural basic laws (*kihonhô*) and basic plans (*kihon keikaku*), as well as broader policy issues relating to agricultural production policy, technical development of farming, and rice policy, including rice production adjustment. Matsuoka remained chairman of the subcommittee almost without interruption until 2003, a long time in which to serve in the same executive position (see Table 4.1). Over this period, Matsuoka fashioned the Agricultural Basic Policy Subcommittee into his own policy kingdom.

In 1998 and 1999, as chairman, Matsuoka played a leading role in the formulation of LDP policy on the new agricultural basic law[23] to replace the existing Agricultural Basic Law (*Nôgyô Kihonhô*) of 1961, as well as the forerunners to the new law, the 'Agricultural Policy Reform Outline' (*Nôsei Kaikaku Taikô*) and the 'Policy Program' (*Seisaku Puroguramu*). In interviews with Nokyo's National Council on these issues, Matsuoka called, amongst other things, for mutual understanding amongst the LDP, MAFF and farmers, for the need to entrust prices to markets but to protect farm incomes through policy measures, to maintain rural communities and to promote concrete policies leading to the establishment of new income policies for farmers, particularly for farmers operating under disadvantageous conditions in mountainous areas.[24] Matsuoka assiduously attended national gatherings of Nokyo representatives focusing on these policy issues, where the views of farmers and farm households could be directly transmitted to LDP agricultural committee executives. Matsuoka directly invited farmers and agricultural organisations to make input into the new basic law.

> The LDP considers this [law] to be the most important issue [in agricultural policy], and continues to discuss it in the Agricultural Basic Policy Subcommittee. In order to realise the policy that we are aiming for, I would like to request that farmers and agricultural groups tackle it with us in order to complete the basic law outline.[25]

Complementing Matsuoka's rising power as a *nôrin giin* was his accession to the chairmanship of the Uruguay Round-Related Countermeasures Implementation Subcommittee (UR Kanren Taisaku Jisshi Shôiinkai) in 1995 (see Table 4.1). The UR committee was a subcommittee of the Nôrin Bukai. Its main task was to decide the allocation of ¥6.01 trillion on projects and other policy measures for farmers and rural dwellers under the UR countermeasures policy and to make sure that all the funds were spent.[26]

The chairmanship of the subcommittee put Matsuoka in charge of subsidies for agricultural and rural development projects funded by the UR countermeasures expenditure.[27] Because of public criticism of the lavish amount of government subsidies being scattered (*baramaki*) in rural areas, Matsuoka made a very defensive speech about the countermeasures policy in front of 30 young men from the local agricultural cooperatives in Kumamoto, saying, '[a]griculture is always victimised as a "rogue" and bad people say nasty things about it. City dwellers do not understand anything. I will not allow even one yen to be cut from the ¥6.01 trillion'.[28] He claimed to have 'defended the package 100 per cent'.[29]

When the money was being distributed, Matsuoka and his close political associate, Suzuki Muneo (later indicted and convicted on political corruption charges),[30] ran the show pretty much as they liked. The farmers' organisation of the JCP (Nôminren),[31] complained on its website that, even though Muneo and Matsuoka obtained around ¥6 trillion in subsidies to compensate farmers for the liberalisation of rice imports, Matsuoka used part of the subsidies for construction work on building spas, some of which had closed in the red, and other facilities that were using up the budget of Aso Town, his hometown. In Nôminren's view, the UR agricultural countermeasures expenditure had been turned into engineering works.[32]

The spas referred to by Nôminren were hot spring resorts called 'Refresh Villages',[33] which were built in various places across rural areas of Japan, including Kumamoto.[34] In Matsuoka's hometown, a theme park called 'Hana Aso Bi' was constructed at a cost of ¥920 million, with ¥460 million coming from the UR countermeasures package. According to one report, the structure was excellent, but stepping inside, some people said that it looked no different from a 'drive-in' souvenir store on a highway.[35] Another facility built with UR countermeasures expenditure was a 'Tofu Museum', which, according to some, was on a par with a junior high school laboratory. It was questionable what, if any, benefits those employed in agriculture actually gained from facilities such as these.[36]

In addition to these projects, total expenditure on a hot spring resort called 'Mizube Plaza Kamato' amounted to ¥1 billion, with approximately ¥500,000 allocated from the UR countermeasures package.[37] Another resort, or 'general exchange terminal',[38] which included hot spring facilities, a direct selling market and restaurants etc., called 'Sanfurea' was built in Kikuyo Town, Kikuchi County. Budgeted as an 'agricultural improvement project', which would bring rural and city residents together, it cost ¥1.2 billion with more than ¥600 million coming from the UR countermeasures budget.[39] The town office sang its praises as the 'Kikuyo Hot Spring'.[40]

Both Sanfurea and Mizube Plaza Kamato were located in Matsuoka's electorate.

> The UR countermeasures package was distributed most heavily to *zoku* Diet members who say what the MAFF wants. Local people involved in agriculture commented sarcastically: 'the only people who were strengthened by the UR budget were *zoku* Diet members and civil engineering and construction types. We haven't heard anything about agriculture in Kumamoto being strengthened'.[41]

Others commented that although the UR countermeasures expenditure was officially funding for 'agricultural' measures, it was just a bonus to general contractors (*zenekon*).⁴²

The construction works for the Kikuyo Hot Spring were successfully bid for by a *zenekon* with a head office in the heart of Tokyo (Toyo Construction, which was said to be on friendly terms with Matsuoka).⁴³ According to company executives who were connected to Matsuoka's electorate, the Matsuoka office in Kumamoto City intervened in the choice of the sub-contractors in the works, and, indeed, in the actual orders from the *zenekon*. Although these facts were denied by Matsuoka's office and by the Kikuyo Town office, they were verified by Araki Katsutoshi, a construction company executive, Kumamoto prefectural assembly member, and one of Matsuoka's most important political followers.⁴⁴

Many rural prefectural assembly members like Araki ran construction companies that relied on public works orders from both the central and local governments for their business and profitability. When Araki asked Tominaga Kiyotsugu, mayor of Kikuyo Town, whether they would use local businesses as sub-contractors, the mayor replied: 'Matsuoka's office deals with those sorts of issues'.⁴⁵ Araki then spoke to a secretary in Matsuoka's Kumamoto City branch office about the matter. He was also advised by a Kumamoto Prefecture Agricultural Department official to go and pay his respects to Matsuoka if he wanted to participate in the project.⁴⁶ As another person in the construction industry elaborated about Matsuoka and the role he played in the allocation of construction contracts.

> Being a hardliner with a big voice, a considerable part of the public construction in Kumamoto Prefecture now 'consults' Matsuoka's office. With respect to construction in Kumamoto (3), as in Aso Town, 'consultation' must be close to 100 per cent. He has become that influential.⁴⁷

Matsuoka also fiercely defended the UR countermeasures expenditure against budget cuts. In February 1997, when Prime Minister Hashimoto, in answer to a question in the Lower House Budget Committee, said that not only the UR countermeasures expenditure but also agriculture, forestry and fisheries-related expenditure would not be treated as a 'sacred area' in the government's fiscal reconstruction program, the LDP set up another subcommittee chaired by Matsuoka. This Uruguay Round-Related Works Implementation Promotion Subcommittee (UR Kanren Jigyô Jisshi Suishin Shôiinkai) was established by a joint council (*gôdô kaigi*) of CAPIC and the Nôrin Bukai (Matsuoka was also chairman of this committee at the time). The new subcommittee conferred on the conditions for implementing the countermeasures (that is, ensuring that expenditure targets could be found) and reviewed the contents of the works funded by the UR package. It came to a number of resolutions, including that 'the full amount of ¥6.01 trillion in expenditure should be preserved and special measures should be taken to secure the budget in the future'.[48] The group then lobbied the government's Fiscal Structural Reform Council (Zaisei Kôzô Kaikaku Kaigi) as well as the party's executive to get its objectives met.

When later interviewed by the National Council about whether agriculture and forestry-related public works should be excluded from public works, Matsuoka responded as follows

> [e]ver since the Hosokawa Cabinet, fiscal reform has been discussed under the principle of 'economy for economy's sake' and discussion has been led by the financial world (*zaikai*). This was the background against which the idea that agriculture-related public works should be excluded from public works originated. Because the role that agriculture and forestry plays is indispensable to the lives of the people, nothing is more closely related to the public benefit than agriculture and forestry. Therefore, I strongly believe that agriculture-related public works have to be included in works for public benefit. The Uruguay Round was an international treaty that Japan agreed to for the benefit of the entire nation's trading interests, but that means we must take measures for agriculture. At the time, the Hosokawa Cabinet promised to undertake assistance measures for agriculture, and after that, I and others made similar promises now that we're back in power. Excluding agriculture and forestry-related public works from public works is the last thing we can give in to...[Finally] the important thing is to demonstrate the position of agriculture in farm households. In other words, it is necessary to show clearly how to ensure farm household income. It is necessary at least to show that you can get this much if you produce this much.[49]

In 1996, Matsuoka became acting chairman then chairman of the LDP's Agriculture and Forestry Division, a position he held until 1997 (see Table 4.1). This was an appointment made by the LDP's Executive Council, as were all the top executive appointments in the PARC, including the chairmanships of other agricultural committees, such as CAPIC, the Forestry Policy Investigation Committee (Rinsei Chôsakai), and the Agriculture, Forestry and Fishery Products Trade Countermeasures Special Committee (Nôrinsuisanbutsu Bôeki Taisaku Tokubetsu Iinkai, or Bôtaii).

Customarily, junior and middle-ranking Diet members were appointed as division chairmen. The divisional chairs were distributed according to faction but factors such as how many times they had been elected, their contribution to the division and whether they had shown 'presence' (*sonzaikan*) were also taken into account.[50]

When someone asked Matsuoka, 'what are the divisions of the LDP all about?', he answered

> [a]s the ruling party, each division in the LDP drafts (*ritsuan*) numerous policies. Large numbers attend the divisional meetings and active debate takes place, especially in the agriculture and forestry-related divisional meetings. Although society might misunderstand the role of the divisions, in reality, it is quite obvious that policies are formed (*seisakuka*) as intense debate takes place and accumulates.[51]

Immediately after the 1996 Lower House election the LDP set up a new executive regime relating to agriculture and forestry, to which Matsuoka, as chairman of the Agriculture and Forestry Division, was appointed along with three other LDP agriculture committee chairmen, including the chairman of CAPIC. The purpose of the new executive was to push various agricultural policy issues rapidly to a conclusion. For Matsuoka, his accession to the divisional chairman's position was a trigger for his elevation to higher status in the party's agricultural and forestry policymaking machinery. According to one MAFF OB, 'in 1995, at the time that Matsuoka became parliamentary vice-minister of the MAFF, he didn't have that much power, but in the following year (1996) when he became the party's Agriculture and Forestry Division chairman, he suddenly became powerful'.[52]

Electoral reform appeared to have no impact whatsoever on Matsuoka's policy specialism. In fact, he retained and strengthened it, following the same career track that he would have without electoral reform and remaining a *nôrin giin*. It was at this time in 1996, when the first Lower House election was held under that new system, that Matsuoka's seniority in a range of committees enabled him to exert wide-ranging powers over all major agricultural policies. He participated in the joint council (*gôdô kaigi*) of the Agriculture and Forestry Division and CAPIC, which played a vital role in the final stages of agricultural budget formulation. Participating in the joint council provided a means whereby the LDP agricultural policy executives, who were also Diet members pressured by Nokyo and its National Council, could directly influence the MAFF minister on the verge of cabinet negotiations on the final budget draft.

Matsuoka also secured membership of the LDP's general agriculture and forestry executive (*nôrin kanbu*), consisting of the chairmen of all the important PARC committees on agriculture and forestry. The executive was in charge, for example, of deciding the LDP's producer rice price in the ultimate stage of decision-making within the party on the issue. In 1996, it was active in realising

Nokyo's producer rice price demand in defiance of the government's (MAFF's) plan to lower the basic rice price amidst a severe over-supply situation. The government-LDP negotiations ground on to the very last minute, producing an additional package of ¥10 billion for 'special countermeasures works', which met producers' expectations.

While serving as acting chairman of the Agriculture and Forestry Division in 1996, Matsuoka also became chairman of the Livestock Commodity Prices Etc. Subcommittee (Chikusanbutsu Kakakutô Shôiinkai). It was normal for *nôrin giin* to become chairman of an Agriculture and Forestry Division subcommittee first, and then move on to become chairman of the division itself, if they proved successful in their subcommittee post.

This subcommittee traditionally formulated party policy on the price stabilisation bands for beef, the indicative stabilisation price for dairy products and the guaranteed price for raw milk for processing. Like CAPIC's Rice Price Committee (Beika Iinkai), the livestock price subcommittee played a key role in determining the LDP's position on support prices for these products. Provision was made at its meetings for the submission of producer requests from Nokyo representatives.

When, in March 1996, Matsuoka was asked by the National Council what his views were on livestock prices, he commented that the beef liberalisation in 1991 and the URAA of 1994 was a 'double punch' as far as livestock and dairy farmers were concerned. He trotted out the usual homilies about the most important policy issues being how to promote motivated farm households, to modernise the dairy and livestock industries, and to expand production. He thought greater consideration should be given to the fact that the dramatic reduction in production costs (which were driving down the administrative prices for livestock commodities) could be attributed to the rise in the value of the yen. On the cost side for farmers, Matsuoka noted the expense of disposing of animal waste, which he thought should also be taken into account in determining the administrative prices. In calculating livestock prices for that year, Matsuoka thought that farmers' feelings were the most important factor. He undertook to apply himself to the livestock price decision whilst giving consideration to concrete problems.[53]

In May 1996, when the National Council again provided a vehicle for the publication of his views on the livestock price issue, Matsuoka commented that 'we have to put our best efforts into obtaining a price decision that doesn't weaken the motivation of livestock farmers'.[54] He added that it was necessary to find a solution for dairy beef farmers so that they could cope with the liberalisation of beef. Fluctuations in the cost of feed also had to be taken into account in determining prices for livestock commodities. Another factor was farmers' debt levels. These had been declining, not because management was prospering, but because farmers had stopped investing in facilities owing to the uncertain business conditions. In order to give farmers certainty in the future, Matsuoka commented that

> [w]e have to make livestock price decisions that don't weaken the motivation of livestock farmers. Because of this, I, as LDP Livestock Commodity Prices Etc. Subcommittee chairman, will decide to maintain the current prices, and will also undertake a radical review of the formula for calculating prices and the way in which production cost investigation is done, which up until now, has been extraordinarily disadvantageous to farmers.[55]

In 1997, Matsuoka took over as chairman of the Rice Price Committee, which was concerned with the producer rice price and production issues such as rice production diversion programs (*gentan*). One of his main tasks in that committee was to establish a New Rice Policy (*Arata na Kome Seisaku*) designed to compensate farmers for falls in rice prices. In this capacity, Matsuoka attended a 'National Gathering of Representatives for the Establishment of a Rice Policy and the Stabilisation of Rice Crop Management', organised by the National Council and Zenchû in October 1997. Approximately 1,200 representatives attended from local agricultural cooperatives nationwide, and they made a direct request to the participating LDP Diet politicians for a New Rice Policy that would include income compensation for rice farmers. Because of the sense of crisis in national rice policy caused by falls in prices for rice farmers, a large number of LDP Diet members took part in the meeting. Matsuoka attended as chairman of the Agricultural Basic Policy Subcommittee and gave a speech. In it, he stated that 'expanding production adjustment is the only way to deal with the problem of excess rice. We must collect as big a budget as possible in order to do this'.[56]

At a similar meeting organised in November by Nokyo groups to demand the necessary funding for a New Rice Policy, Matsuoka again emphasised the need for production adjustment.[57] The National Council followed up with a direct approach to the *nôrin kanbu*, in which council representatives sat down with LDP politicians at a roundtable conference in the LDP headquarters. Matsuoka attended as the chairman of the Rice Price Committee along with the chairmen of the other main LDP agricultural policy committees, including the CAPIC chairman, and the chairman of the Agriculture and Forestry Division, a position that Matsuoka had relinquished by November 1997.

A week later a much larger rally of Nokyo representatives was held at LDP headquarters. Matsuoka, as chairman of the Rice Price Committee, delivered some of the main greetings. He spoke about his resolve and the political judgement that it was necessary to get ¥40 billion as a countermeasures policy to compensate farmers' income for falls in the price of rice for the current year's crop. However, there were insufficient funds to cover this expenditure.[58]

Matsuoka was interviewed by the National Council on the 25 November 1997, six days after the new policy was announced. He began by pointing out that

> [b]ecause of bumper harvests, the government's rice stocks have risen to more than 3.5 million tonnes and as a result, market prices for rice have plummeted. It seems that everything goes against rice farmers, and the main reason for establishing a New Rice Policy is to how to break through this situation. In order to reduce the amount of rice in stock (where the balance between supply and demand has not recovered), only three choices are possible: a) rice should be exported overseas, b) demand and consumption should be increased in other areas, and c) production should be controlled. Because the first two options are problematic, emphasis should be placed on production control (*gentan*). We asked for a large number of opinions from various fields and established a framework that guaranteed farmers' income. I strongly demanded that the MAFF raise the necessary funds. This was done by pulling money from various sources: by getting ¥25 billion from the Ministry of Home Affairs as their contribution, by the Food Agency making efforts to cut its expenditure by 5 per cent, and by getting ¥45 billion in new sources of revenue from various places. Putting all these funds together including those from the agriculture and forestry budget produced a total of ¥610.1 billion over two years.[59]

The National Council issued a special 'thank you' to the three agricultural and forestry executives (*nôrin sanyaku*), including Matsuoka, for their great efforts in finalising the New Rice Policy. Importantly, the UR agricultural countermeasures expenditure was left untouched.

From 1997 to 1999, Matsuoka served as acting chairman of the Agriculture, Forestry and Fishery Products Trade Countermeasures Special Committee. Its task was to tackle agriculture, forestry and fisheries trade-related issues for the LDP. In August 1998, the trade committee set up a Study Team (Nôrinsuisanbutsu Bôeki Chôsakai Sutadei Chîmu) initially to analyse and investigate in detail the contents of the URAA and report back to the larger committee. It also set about constructing a strategy for the next round of agricultural trade negotiations in close consultation with the government (MAFF), LDP and Nokyo organisations. This was the World Trade

Organization (WTO) tripartite council (*WTO sansha kaigi*), a consultative council established by the MAFF in late 1998 to facilitate the formation of a consensus amongst agricultural bureaucrats, the LDP's *nôrin kanbu* and representatives of agricultural, forestry and fisheries groups on trade-related issues. The executive leadership of the trade countermeasures special committee was put in charge of positively advancing 'Diet members' diplomacy' (*giin gaikô*)[60] on trade issues, by sending delegations to Asia, the European Union and other countries about the upcoming negotiations on agricultural trade under WTO auspices.

At one time in 2000 Matsuoka served in seven executive posts in LDP agriculture[61] and forestry[62] policy committees, as well as being AFF Committee chairman and, once again, chairman of the Rice Price Committee (see Table 4.1). In 2000 and 2001 he also became acting chairman of the Mountain Village Development Countermeasures Special Committee (Sanson Shinkô Taisaku Tokubetsu Iinkai) of the PARC (see Table 4.1). This committee was concerned with matters relating principally to special support for agriculture and other industries in mountainous regions as well as with the provision of public works and community facilities for farmers and rural dwellers.

In December 2000, Matsuoka became chairman of the Management Income Study Meeting (Keiei Shotoku Sutadei Kaigô) (see Table 4.1), a group within the LDP's Agricultural Basic Policy Subcommittee, which Matsuoka also chaired at the time. The group was set up to tackle a new policy that would provide direct income subsidies to farmers. These would be called 'agricultural management income stabilisation countermeasures' (*nôgyô keiei shotoku antei taisaku*) and they would be in line with similar systems already introduced in the United States, Canada, and the European Union.[63] Other proposals discussed by the group called for encouraging farmers to expand the size of their land holdings and to boost per capita productivity.

At the time he took up the chairmanship of the Management Income Study Meeting, Matsuoka had just been appointed MAFF deputy minister in the second Mori Cabinet (see Table 4.1). On the very day he was appointed, he led a meeting of a newly formed group, which came up with a set of proposals to 'raise farmers' incomes by channelling agricultural subsidies directly to them instead of by buying produce at government-set prices'.[64] The policy 'targeted 400,000 farmers earning their living solely through agriculture or who had made farming their main source of income'.[65]

The development of this proposal into a formal government policy was overtaken by the advent of the Koizumi administration four months later in April 2001, and Koizumi's commitment to structural reform. Koizumi campaigned first for the presidency of the LDP and then in the 2001 Upper House election on a platform of dismantling the old order, principally the political and bureaucratic institutions blocking economic reform. Thereafter, direct income subsidies had to be instrumental in achieving structural reform of the farm sector.

Being an executive of particular LDP agriculture and forestry committees did not prevent Matsuoka from attending the meetings of other committees in which he did not play an executive role. The most crucial issue for Matsuoka's farm vote-gathering strategy was policy that affected farm household incomes from agriculture. He had to demonstrate a commitment to maximising incomes for farmers in order to retain high levels of voting support from farm households. This meant poking his head into the proceedings of any committee that was deliberating on matters relating to farm incomes. For example, he attended discussions of the LDP Farmers' Pension Subcommittee (Nôgyôsha Nenkin Shôiinkai) and was vitally concerned with the prices of all agricultural commodities. For this reason, he regularly participated in all the major agricultural and forestry committees concerned with pricing issues. For example, when the Vegetable, Fruit Tree and Upland Field Crops etc Countermeasures Subcommittee (Yasai, Kaju, Hatake Sakubutsutô Taisaku Shôiinkai) met to decide prices for upland field crops, and to discuss production countermeasures for wheat, soybeans, sugar beets and sugar cane, Matsuoka was there making his contribution to the decision.

Matsuoka was also an especially prominent figure in both rice and leaf tobacco price decisions; the producer rice price because it captured the largest number of agricultural producers, and tobacco because it was especially important in Kumamoto. By 1995 the prices of farm products such as rice were being increasingly marketised. Leaf tobacco was the last fortress for the LDP to exercise decision-making authority over the price. The normal procedure was for Japan Tobacco Inc. (JT) (Nihon Tabako Sangyô Kabushiki Kaisha)—which was the monopoly buyer of domestically produced leaf tobacco—to submit its proposed purchasing price to the government's Leaf Tobacco Advisory Council (Hatabako Shingikai), which would recommend a certain purchase price. It was customary, however, for 'tobacco-related Diet members such as Matsuoka to decide [the tobacco price] by *nemawashi* [preparing the groundwork] before the council. Some would say that this 'constituted not only "prior examination" (*jizen shinsa*) as Prime Minister Koizumi described it, but also "prior decision" (*jizen ketchaku*)'.[66] When the leaf tobacco price was lowered substantially in 1995, Matsuoka hurled an empty can of juice at a JT executive, yelling, 'what is the salary of the president? Say it!'[67]

Matsuoka was also opposed to an increase in the tobacco tax in 1999. He made a special visit to the Prime Minister's Official Residence (Kantei) to talk to Aoki Mikio who was chief cabinet secretary at the time. Matsuoka suggested to Aoki that a pachinko tax should be introduced instead of a tobacco tax in order to avoid a decline in the consumption of domestically produced tobacco leaves. A Matsuoka critic lamented, '[w]hile this guy is around, there is no way Japan will establish an anti-smoking right or a policy to lower the smoking rate. It seems that the concession for the tobacco farmers is more important than the health of the citizens'.[68]

MAFF DEPUTY MINISTER

In December 2000, in former Prime Minister Mori's second cabinet, Matsuoka was appointed to the newly created position of MAFF deputy minister,[69] taking up the position on 6 January 2001, when the restructuring of government ministries and agencies took effect. In keeping with tradition, Matsuoka was nominated to the post by his faction boss, Kamei.[70]

Matsuoka's accession to the deputy minister's post was a pointer to how important he had become as one of the LDP's leading politicians on matters relating to agriculture and forestry. As deputy minister, Matsuoka had the power to act on behalf of the MAFF minister. He could explain draft bills to the Diet from the state minister's gallery and also in the Lower House AFF Committee. He could answer questions in the Diet on behalf of the minister, substituting for the practice of bureaucrats answering these questions in lieu of the minister (*seifuiin*). He could also attend deputy ministers' councils in the Kantei.

As one of the two new deputy ministers in the MAFF (the other was Tanaka Naoki), Matsuoka was assisted by two parliamentary secretaries (Kaneda Hideyuki and Kunii Masayuki): all LDP Diet members. They were all expected to 'exert political weight on the bureaucracy'[71] by providing additional expertise and support for the MAFF minister, thus bolstering his position against the bureaucratic weight of his own ministry.

Matsuoka, however, was very keen to exert his own power over the MAFF.[72] Two days after the government reorganisation took effect, Matsuoka told *Kyodo News* that the new MAFF deputy ministers and parliamentary secretaries would set up a council at the ministry to resolve key policy issues and problems. The council would meet on a weekly basis. There were no such fora in any ministry or agency before the reorganisation. Matsuoka's plan was to lessen the party's reliance on bureaucrats in policy formation and reduce their power.

Some MAFF bureaucrats called the initiative an attempt to create an extra organ, since there already was a top meeting for ministry officials.[73] Matsuoka pressed on regardless. As he explained, '[d]epending on the agenda, the council will call for participation of directors-general from the Food, Forestry and Fisheries agencies and may include Agricultural Minister Yoshio Yatsu'.[74] Matsuoka remarked at the time that politicians 'have knowledge and ability in dealing with the ministry's administrative affairs and are competent enough to equal bureaucrats in handling policy matters'.[75]

Leading the way for the other central ministries and agencies, a meeting of MAFF deputy ministers and parliamentary secretaries was held in the MAFF on 9 January 2001, three days after bureaucratic reorganisation came into effect. At the meeting

> [e]ach MAFF bureau director and directors-general of the MAFF's agencies reported respectively on important policy issues relating to their areas of administration. On the basis of their reports, the deputy ministers and parliamentary secretaries gave the necessary directions (*shiji*) and executed the required coordination (*chôsei*), thus putting into practice policy planning (*ritsuan*) under political leadership.[76]

Matsuoka led the meeting along with Deputy Minister Tanaka, as well as the two new parliamentary secretaries. He was jubilant after the meeting, claiming '[w]e politicians now directly engage in the task of formulating policies…We are here to do the job of working out important policies. No policy can be decided on without being discussed at our meetings'.[77]

In terms of actual policies, Matsuoka's biggest impact as deputy minister was felt in the area of agricultural trade. He travelled with MAFF Minister Yatsu for the purpose of conducting foreign (agricultural trade) policy activities (*gaikô katsudô*) overseas. On his return to Tokyo, he reported back to the LDP committees concerned with agricultural trade issues (the WTO *sansha kaigi*, the Agriculture, Forestry and Fishery Products Trade Investigation Committee Study Team, and the committee itself).

His greatest coup, however, was his pivotal role in the government's decision to invoke safeguards on imports of rushes for tatami mats (*igusa*), raw *shiitake* mushrooms and leeks from China. Matsuoka exerted direct influence on the MAFF to agree to provisional safeguard measures (emergency import restriction measures) being invoked for 200 days under WTO rules against imports of these commodities from China. The decision was officially justified as a response to rapidly expanding import volumes of these products between 1997 and 1999. Matsuoka told the press, 'I want to invoke the [safeguard] measure'.[78] Even prior to his becoming deputy minister, Matsuoka had pushed this option strongly in the Agricultural Basic Policy Subcommittee, presenting a report on the provisional invocation of safeguards.[79]

Matsuoka's constituency was in Kumamoto Prefecture where 95 per cent of domestic rushes were grown. Reports suggest that he was under severe pressure from *igusa* farmers in Kumamoto Prefecture as well as the prefectural Nokyo organisation and the National Federation of Igusa Production Groups (Zenkoku Iseisan Dantai Rengôkai).[80] There was even a rumour that the Kumamoto Nokyo threatened not to support him in the next election 'unless he made significant efforts'.[81] In 2000, when safeguards were becoming an issue, the LDP's Kumamoto Prefecture No. 3 Electoral District Branch (effectively a branch of Matsuoka's *kôenkai*) received a political donation of ¥1.8 million from the prefectural *nôseiren*. Previously Matsuoka had received only ¥100,000 from the *nôseiren*, obtaining most of his donations from the construction industry.[82] The Kumamoto Prefecture No. 3 Electoral District Branch also received ¥1 million from the *nôseiren* branch in Yatsushiro region, where most of the *igusa* was produced. It was the first time that Matsuoka had obtained donations from this region, which was in fact located in Kumamoto (5).[83] According to a former leader of Yatsushiro Nokyo, 'the donation was made with getting Matsuoka to make efforts for agriculture in general in mind, but there was some anticipation in regards to the safeguards. While we hadn't made any [donations] before, we decided that this was an opportunity, and the donation was made on the decision of the league head'.[84]

Another Diet member commented that Matsuoka's 'standpoint was as if his single-handed efforts led to the invoking of the safeguards'.[85] Matsuoka lobbied the Ministry of Economy, Trade and Industry (METI) and the Ministry of Foreign Affairs (MoFA) hard, saying 'what can you do?' in relation to invoking the safeguards.[86] At one point he was reprimanded by Minister Yatsu, who warned, 'you are the deputy minister so this attitude will not do. If you made any mistakes, it would be disastrous'.[87] However, raw *shiitake* mushrooms were of primary concern to MAFF Minister Yatsu, and leeks were added to the list in order to symbolise action against the influx of Chinese vegetable imports. Also in the minister's and deputy minister's minds was the fact that an Upper House election was looming in July 2001.

On 23 April 2001, the Japanese government imposed emergency import restrictions for 200 days (until 8 November) on the three products. It was the first time that the Japanese government had imposed ordinary safeguards under WTO rules. The measures imposed punitive tariffs on imports above a certain volume, which was designed to bring the prices of the three products up to levels in Japan.

When the Chinese government hinted at retaliation, Matsuoka was despatched to China to explain the Japanese decision and to try and find a compromise. The Japanese government was hoping that negotiations might induce the Chinese side to voluntarily restrict exports. Matsuoka urged China's vice minister of foreign trade and economic cooperation to restrict exports of the products. He also conferred on matters relating to tree-planting cooperation as part of a greening project in China. Matsuoka told the Chinese that the rapid increase in imports of leeks, *shiitake* mushrooms and tatami rushes into Japan had had a bad influence on Japanese farmers. He also explained that the Japanese government had conducted the requisite investigation and examination in order to increase tariffs immediately.[88]

The Chinese government retaliated against the safeguards by imposing 100 per cent tariffs on Japanese motor vehicles, mobile phones and air-conditioners, exports of which virtually stopped. The loss to the Japanese car industry amounted to ¥51.2 billion, which, when added to the countermeasures budget for the three safeguard categories, came to ¥85.5 billion.[89] Such considerations made it impossible for the Japanese government to institute full safeguard measures in December 2001 after the expiry of the provisional safeguard measures in November. The Koizumi government backed down in the face of the Chinese action, which was illegal under WTO rules, but, at the time, China was not a member of the WTO.

This did not stop Matsuoka doing his best to pressure his own government to institute full safeguard measures.

> Matsuoka and others repeatedly demanded that the government present a firm attitude. Matsuoka argued that 'there were no cases where full safeguards were not implemented after the invocation of provisional safeguards. It is a national disgrace. Implement the full safeguards'.[90]

This opinion was not necessarily shared by other leading *nôrin zoku*, such as Yatsu and Nakagawa Shôichi. They reasoned with Matsuoka that 'if the regular invocation is implemented, Japan will lose at the WTO panel. If Japan loses, the three farm products can enter from China at a stroke'.[91]

Matsuoka's position as deputy minister lasted only from January 2001 to April 2001 because Koizumi became LDP president and prime minister in April 2001 and appointed a new cabinet. This meant that Matsuoka was only in the position for just over three months, which limited the extent to which he could exercise his new-found power.

EXECUTIVE COMMITTEE POSITIONS FROM 2001

After stepping down from his post as deputy minister, Matsuoka resumed his executive posts in LDP agricultural and forestry committees: as chairman of the Agricultural Basic Policy Subcommittee as well as acting chairman of the Mountain Village Development Countermeasures Special Committee and chairman of the State-Owned Forests Problems Subcommittee (see Table 4.1).

Rice policy reform

Matsuoka continued to lead the Agricultural Basic Policy Subcommittee through 2002 (see Table 4.1). In that role, he was right at the centre of party deliberations on rice policy reform. He was made chief of the survey group conducting investigations relevant to the reform in Hokkaido. At the local level, the survey group listened to explanations from agricultural cooperative representatives, and exchanged opinions with them.[92] Matsuoka's Hokkaido group was one of several local survey groups reporting back to the subcommittee, which called on the government to listen to voices at the grassroots level and which also indicated the directions of the plan for rice policy reform.[93]

In July 2003, Matsuoka attended the 'National Nokyo Representatives Convention on the Rice Policy Reform Countermeasures' along with Horinouchi Hisao, chairman of CAPIC. In the final stages of formulating rice policy reform, Matsuoka resolved to guarantee about ¥300 billion to fund the new policy, with the subcommittee leaving the final decision to the top three agricultural and forestry executives (*nôrin sanyaku*)[94] and the *nôrin kanbu*.[95]

After the final decision was made on concrete policy for the so-called New Rice Policy Reform (*Arata na Kome Seisaku Kaikaku*), a roundtable of 'three related parties' was held. The three parties were Matsuoka, Horinouchi and the chairman of Zenchû. Horinouchi and Matsuoka presented a report to the Zenchû chairman, outlining the contents of the decision. Because the ¥300 billion to fund the policy, which had been requested by Nokyo, had been secured, the chairman expressed his thanks for Matsuoka's efforts.[96] The next day the Agricultural Basic Policy Subcommittee acknowledged the final decision on the policy. Horinouchi and Matsuoka then reported back to a combined meeting of CAPIC and the Agriculture and Forestry Division, which also acknowledged the new policy.[97]

In November 2003, Matsuoka was instrumental as chairman of the Agricultural Basic Policy Subcommittee in determining the amount of acreage to be taken out of rice production. This was a political issue because it influenced the price of rice for both producers and consumers by impacting on the amount of rice circulating in the domestic market. In 2003, the committee decided to leave the *gentan* acreage at the same level as in the previous year in order to maintain, as the subcommittee described it, a stable 'consumer' price.[98]

Trade policy

In 2003, Matsuoka established an executive connection with the LDP's Agriculture, Forestry and Fishery Products Trade Investigation Committee in the key post of secretary-general (see Table 4.1). From the time of its inception in 2001,[99] the trade investigation committee took over as the main PARC committee dealing with Japan's agricultural trade negotiating position at the WTO and on bilateral Free Trade Agreements (FTAs). Both sets of international trade negotiations put pressure on Japan's agricultural sector for market-opening concessions. In the position of secretary-general, Matsuoka became one of the group of executives (*kanbukai*) of the committee, and thus played a pivotal role in its proceedings. In this position, he was also assiduous in attending gatherings of Nokyo representatives on agricultural trade matters, especially those organised by Zenchû.

In 2004, when Matsuoka was still secretary-general of the committee, officials from the Ozu branch of Kumamoto Prefecture *nôseiren* visited him in Tokyo and he allowed them to attend one of the committee's meetings. Afterwards, the officials commented very favourably on how Matsuoka had conducted the committee proceedings as the organiser of the debate and how Matsuoka had made the relevant ministries and agencies come up with countermeasures on the spot. One of the officials said: 'I now understand very well how a policy will be realised in that way. I did not know how expert Diet representative Matsuoka was in policy until I actually saw it with my own eyes'.[100]

In March 2004, the committee approved the Japan–Mexico Free Trade Agreement. This was Japan's first free trade agreement encompassing the agricultural, forestry and fisheries sector. The judgement of the committee was that the bilateral agreement was a well-balanced settlement, which protected areas that should be protected and included some acceptable market opening in other areas. Because further liberalisation of oranges and orange juice was permitted under the agreement, the committee verified that all possible domestic countermeasures would be taken.[101]

Tobacco price

In 2003, in recognition of his interest and role in LDP committee proceedings on the producer price of tobacco, Matsuoka became chairman of the Leaf Tobacco Price Investigation Subcommittee (Hatabako Kakaku Kentô Shôiinkai) (see Table 4.1). This subcommittee examined the Leaf Tobacco Advisory Council's report on what the producer price of leaf tobacco should be. The price was decided annually each November. After receiving the report, the subcommittee held hearings at which it received submissions from organisations of tobacco farmers and Japan Tobacco about matters such as its (JT's) buying price of tobacco and so-called production countermeasures (*seisan taisaku*) for farmers.

In November 2003, the committee undertook a comprehensive examination of tobacco farm management. This was considered necessary in light of the drastic fall in tobacco farmers' income because of fire damage, the rapid change in the situation surrounding tobacco in recent years and other factors. The subcommittee decided to leave the cultivated area of tobacco and the purchase price of leaf tobacco produced in 2004 where they were. This decision was in line with the report of the Leaf Tobacco Advisory Council.

Avian flu countermeasures

In March 2004, Matsuoka was made secretary-general of the newly established LDP Avian Influenza Countermeasures Headquarters (Tori Infuruenza Taisaku Honbu) (see Table 4.1). It was set up to alleviate the anxiety of both consumers and producers about the outbreak of avian flu in Oita and Yamaguchi prefectures. The headquarters deliberated on matters such as amending the 1951 Livestock Infectious Diseases Prevention Law (*Kachiku Densenbyô Yobôhô*). Matsuoka had earlier visited Oita Prefecture in February 2004 with a group of LDP Diet representatives and delegates from the party's Oita and Kumamoto federation of branches. Based on the results of the investigation, Matsuoka felt a strong need for a response from the whole party. The headquarters was thus created on the basis of his proposal, enabling Matsuoka to claim the credit.[102]

After Matsuoka was appointed secretary-general of the committee, he asked questions on two occasions in the Budget Committee of the Lower House about what was going to be done to stop the spread of avian flu. His questions followed a new outbreak in Kyoto Prefecture, and there were fears that the infection could spread. Matsuoka demanded that the headquarters and the government face the problem cooperatively to the best of their ability.[103]

The headquarters later organised emergency measures for avian influenza and, on the basis of these countermeasures, a part of the Livestock Infectious Diseases Prevention Law was revised. The amendment institutionalised subsidies to livestock farmers who cooperated with restriction orders on the movement of birds. It also increased the fines and penal restrictions on traders and others who neglected to notify the government of infected birds.[104] However, this did not stop the spread of avian flu to Ibaraki Prefecture in 2005.

In June 2005, Matsuoka received a direct delegation from the Japan Egg Producers Association (Nihon Keiran Seisan Kyôkai). They were concerned about the impact on egg production and egg-producing farm households of the recent outbreak of avian flu in Ibaraki Prefecture, saying it would cause trouble for their management (that is, their income). They wanted government 'countermeasures'—financial compensation of some sort. Matsuoka commented that measures for consumers were also in order to allay any concerns they might have.[105]

Mountain village development

In 2004, Matsuoka became acting chairman of the Mountain Village Development Countermeasures Committee (see Table 4.1) because Uesugi, the former chairman, lost his seat in the November 2003 election. In 2005, the committee was retitled the Mountain Village Development Committee (Sanson Shinkô Iinkai). The committee dealt with matters relating to the administration of the Mountain Village Promotion Law (*Sanson Shinkôhô*), which was originally passed in 1965 by bipartisan consent with a 10-year period of application. It had already been extended three times and was due to expire again at the end of 2004. It had been principally concerned with the provision of infrastructure (*kiban seibi*) in mountain areas where there were farmers. Such infrastructure included agricultural and forestry roads, as well as 'livelihood facilities' in rural villages in mountainous areas under the rubric of 'the development of the rural living environment'. This referred to the provision of mobile phone access, electricity, and medical, nursing and welfare facilities for mainly old people left in these villages. The committee also concerned itself with expenditure for mountain village development in the MAFF budget.

Matsuoka, as acting chairman, engineered a visit by key members of the committee to Kumamoto at the end of January 2005. He chaired a meeting of the committee in Oguni Town, Aso County, in the electorate of Kumamoto (3) (which, along with Minami Oguni Town, were the only places in Aso County that failed to elect Matsuoka in first place in the 2003 election, as shown in Table 7.1). The idea behind holding the meeting in Oguni Town was to get a sense of what the locals wanted from their political representatives and the relevant government ministries and local public officials.[106] The meeting was officially to hold an on-the-spot investigation and to exchange opinions. It lasted for two and a half hours in a hotel with about 90 people present, including the chairman of the Kumamoto Prefecture branch office of the National Mountain Village Promotion League (Zenkoku Sanson Shinkô Renmei), as well as Nokyo and forest association officials, and local municipal mayors. Matsuoka presided at the meeting, which was also attended by relevant officials of the MAFF, Ministry of Land, Transport and Infrastructure (MLIT), and the Ministry of Internal Affairs and Communications (MIAC), and the prefectural government. Uozumi attended as vice-chairman of the committee. Most of the discussion was about public works projects, including a transport centre, the Oguni Dome and other public buildings and amenities.[107] Local representatives demanded an extension and revision of the Mountain Village Development Law and the establishment of a cellular phone communication base since there were many areas where it was not possible to use cellular phones. Women attending the meeting demanded better medical treatment, nursing, and welfare facilities.[108] The meeting provided Matsuoka with an opportunity to engage in some surrogate election campaigning. Voters could see him as an influential LDP Diet politician who was not only committed to his constituents and local issues but also effective in bringing subsidies and projects back to the local area.

A week after the meeting was held, an outline of amendments to the law was discussed and approved by the Mountain Village Development Committee. It was then presented to the Diet as Diet members' legislation requiring cooperation from the opposition parties. Successful passage of the amended law extended it for another 10 years. It allotted a greater role to municipalities in the implementation of development plans for mountain villages as a hallmark of greater decentralisation. Under the amendment, the power to formulate these plans was passed from prefectures to municipalities. In addition, the amended law strengthened countermeasures against damage done by birds and animals and also extended to installation of information and communications infrastructure.

Forestry policy

In 2004, Matsuoka became chairman of the LDP's Forestry Policy Basic Problems Subcommittee (Rinsei Kihon Mondai Shôiinkai) (see Table 4.1), a subcommittee of the Forestry Policy Investigation Committee. The subcommittee had the task of holding annual hearings on how much timber should be used each year by government bodies.[109] In October each year the subcommittee also met with the Forestry Policy Investigation Committee in a joint council (*gôdô kaigi*) for purposes of discussing the contents of budget demands for the following year.

Working on this committee was advantageous for Matsuoka because it put him in direct contact with the leadership of the forest associations and its national body, the National Federation of Forestry Associations (Zenkoku Shinrin Rengôkai, or Zenshinren). Matsuoka's position as chairman continued into 2005 when the committee reviewed the issue of reform of the forest associations. It listened to explanations from Forestry Agency officials on the issue as well as representations from the head of Zenshinren. The subcommittee also took up the issue of the development of the forest and timber industry in Japan. Matsuoka used the committee as a venue from which to push his environmental message about the value of forests in protecting the environment.[110]

Direct income subsidies

In 2004, Matsuoka became chairman of the LDP's Management Countermeasures Project Team (see Table 4.1), a subcommittee of CAPIC (of which Matsuoka was also acting chairman at the time). The project team was established to deliberate on the policy proposed by the MAFF under the 'New Food, Agriculture and Rural Areas Basic Plan' (*Arata na Shokuryô, Nôgyô, Nôson Kihon Keikaku*), or New Basic Plan (*Shin Kihon Keikaku*). This policy would replace price-related subsidies with direct income subsidies to farmers as the main method of supporting farmers' incomes—a so-called Japanese style direct payments system. Matsuoka was made chairman of the project team because he was one of the prime movers in the December 2000 LDP proposal to provide direct income support to farmers facing declines in agricultural prices led by the rice price.[111]

The project team's job was to discuss various proposals and views advanced by the MAFF, which were presented to the committee, investigate how similar policies were implemented in other countries, particularly in the United States and in the European Union, review the implications of such a system for Japan's food self-sufficiency and come up with its own views about what form the new policy should take and then present them to CAPIC. The MAFF wanted to restrict the payments to 'core farms'—farms of larger size—with the idea of encouraging the amalgamation of farm plots and the structural reform of agriculture. However, many LDP farm politicians in Matsuoka's committee saw the policy as potentially 'destroying' small-scale farmers.

As the new scheme represented a radical departure from the government's past policy of assisting all agricultural producers regardless of farm size, farmers were also pressing for the widest possible eligibility for the new subsidy program. In April 2004, an explanatory meeting (*setsumeikai*) concerning the details of the LDP version of the New Basic Plan was presented by the agricultural and forestry executive (including Matsuoka as chairman of the project team) to representatives of agricultural, forestry and fishery groups. About 100 of these representatives attended.

The government delayed the decision about which farms would be eligible for the new form of state support while Matsuoka's project team tried to ensure that small-scale farmers would not be left out. The team came up with a series of farm management income stabilisation countermeasures, with Matsuoka making proposals to the team. It was only prepared to consider policies that would not exclude small-scale farm households. It wanted to make sure that the new policy for direct income subsidies did not destroy this type of farming.

The project team took its proposals into the larger Agricultural Basic Policy Subcommittee, which also received representations from Nokyo spokesmen about how the management stabilisation countermeasures should be applied as part of the revised basic plan. The subcommittee was in charge of deciding the policy for the LDP. In July 2005, a meeting of the subcommittee was held at LDP headquarters in which there were explanations about the formulation of the New Basic Plan, which had begun in March. Matsuoka emphasised that he wanted to 'establish a farm industry with strong legs and loins and positively promote high quality agricultural products that could be exported'.[112]

Matsuoka was reappointed as chairman of the Agricultural Basic Policy Subcommittee after the 11 September 2005 election (see Table 4.1). He acknowledged that it was a post that carried heavy responsibilities, but he said that he had to listen to the opinions of many people.[113] As chairman of the subcommittee, Matsuoka presided over the final agreement within the party on the direct payments system, and between the party and the MAFF.

The subcommittee's immediate task (until the end of October) was to wrap up discussion on the conditions for 'bearers' of agriculture to receive direct income subsidies from the government. In a late September meeting, the subcommittee decided to thrash out the bearer issue.[114] In the same month, Matsuoka also conferred directly with executives from the National Council of Nokyo Youth Organisations (Zenkoku Nôkyô Seinen Soshiki Kyôgikai) about which farmers should be the target of the new direct payments system. The board of directors of the organisation proposed that the government should consider the targets of the New Basic Plan with some flexibility so that hard-working farmers would not miss out.[115]

The following month the subcommittee held hearings on the policy for the purpose of eliciting the opinions of key agricultural organisations on the issue. Appearing at the hearings were representatives from Zenchû, the Japan Chamber of Agriculture (Nôgyô Kaigishô), the Nokyo Youth Council (Zenseikyô) and agricultural production corporations in Niigata and Aomori, two prominent rice-producing prefectures.

In early October, the subcommittee firmed up a concrete plan for the direct payments system, while the LDP's *nôrin kanbu* deliberated on the essentials of the draft.[116] In late October the subcommittee convened a meeting to conclude the party's final draft of the 'Japanese edition direct payments system' (*Nihonhan chokusetsu shiharai*). In the final days of deliberation on this draft, the subcommittee met every day at 8.30am. Matsuoka noted that, as chairman of the subcommittee, it was his responsibility to decide the essentials of the draft plan after discussion in the subcommittee.

The subcommittee drafted a final report and presented it for approval to the combined council of the Agriculture and Forestry Division and CAPIC. The subcommittee then held a press conference in the Diet building about the 'Management Income Stabilisation Countermeasures Outline' (*Keiei Shotoku Antei Taisaku Taikô*) and explained its contents.[117]

After the government and LDP agreed on the new policy in November, Matsuoka, as head of the LDP's subcommittee, explained that the new policy would target a 50 per cent integration of farm households in terms of cultivated land area by fiscal 2007.[118] He was referring to those farm households that would be forced to amalgamate (either through conversion to community farms or farm corporations) in order to be eligible for direct farm subsidies under the new policy.

SECOND-STRING INTERESTS

Matsuoka's committee memberships reveal an unerring commitment to agriculture and forestry policy, and reflect a policy specialism gained over a number of decades. His chosen policy field was the escalator that took him to the top executive positions in the relevant Diet and party committees. He used these to build his political reputation, standing and influence as a policymaker. However, like any successful politician, Matsuoka developed second-string and third-string interests that complemented his major policy focus on agriculture and forestry. Significantly, his secondary interests were in disaster and environmental policy—areas that were closely related to agriculture and forestry.

In his second year in the Diet, Matsuoka joined the Lower House Special Committee on Disaster Countermeasures (Saigai Taisaku Tokubetsu Iinkai), rising to be a director in 1992 and staying on the committee in this executive position until 1999 (see Table 4.1). Disaster policy was attractive to Matsuoka because it was lucrative in terms of bringing subsidies (including for public works) back to his electorate in order to rectify the damage caused by earthquakes, storms and typhoons. Typhoons, torrential downpours and earthquakes did a lot of damage to farms and forests. Nine typhoons struck Japan between June and October 2003 causing a total of ¥7.4 billion worth of damage to agricultural, forestry and fisheries, including crops, agricultural land and agricultural, forestry and fishery facilities. In the same year, a large earthquake in Niigata caused ¥96.8 billion worth of damage to agricultural, forestry and fisheries industries, the most since the war.[119] Under government policy, those affected in these industries had to be compensated in addition to the reconstruction involved, so disasters became an important rationale for public works and subsidy outlays.

Whenever a natural calamity hit his local electorate, Matsuoka made a point of conducting on-the-spot investigations of the damage, liaising with local government politicians and officials, and working hard to get funds from the disaster restoration budget directed to areas that had suffered damage. In July 2005, for example, Matsuoka visited Oguni Town in his electorate following a torrential downpour that did substantial damage in the town. He listened to a description of the damage at the town office, and then visited each of the wards in the town to see for himself.[120] In September of the same year, he visited Oguni Town again to hear requests from the town assembly members about restoration of the damage done by the heavy rain, along with representatives of the MAFF, the MIAC, the Cabinet Office and MLIT.[121]

After the meeting, he attended the ordinary general meeting of the Japan Flood Control and Riparian Works Association (Nihon Chisan Chisui Kyôkai). As he liked to preach about the importance of industries he was inclined to support for political reasons, he stated on his website that

> [f]lood control and riparian works are extremely important for creating safe land that is not damaged. The idea that looking after mountains and water is looking after the country has been around since the feudal times of the Sengoku era...It is part of the heritage that we carry on today.[122]

Besides being formally a part of the membership and executive of the Lower House committee on disasters, Matsuoka was active in attending the relevant LDP divisions and committees of the PARC when disaster struck in his area of Japan in order to show that he was responding to the needs of his constituents suffering disaster damage. In July 2003, following torrential downpours that did terrible damage to areas in the mid-western part of Kyushu, Matsuoka attended a joint council (*gôdô kaigi*) of the LDP's Special Committee on Disaster Countermeasures (Saigai Taisaku Tokubetsu Iinkai) and its Cabinet Division (Naikaku Bukai). The joint council received a report from the ministries concerned in the presence of the PARC Chairman Asô Tarô. The council confirmed the need to tackle countermeasures in a unified fashion with the government after discussing matters such as the disaster prevention radio system in the disaster-stricken area and to what degree forecasts of precipitation were possible in advance.[123]

Matsuoka also entered the realm of environment policy by joining the Diet's Environment Committee, rising to be a director in 1994 and chairman of the LDP's Environment Division in the same year (see Table 4.1). A divisional chairmanship was a senior position for a second-term Diet member, but the division was not very popular because it was recognised as 'high politics' and not directly connected to votes.[124] As a result, competition for executive positions in this policy sector was not high. The environment was also a policy set generally recognised as representative of urban interests.[125]

Matsuoka was interested in it, however, because it was connected, indirectly, to both the agricultural and forestry industries. Environmental policy related to agriculture because one of the main arguments used by Matsuoka and others for continued support and protection of the Japanese farming sector was that it preserved the environment. The same could be said for forests, which had well-acknowledged environmental functions. Their positive environmental value was one of the main rationales for maintaining them. Moreover, environmental arguments resonated amongst the broader public, including urban dwellers, and thus offered powerful national-interest grounds on which to defend both Japan's agriculture and its forests. In this way, environmental policy could be used as a bridge between city and urban areas.

THIRD-STRING INTERESTS

Matsuoka acquired not only a second but also a third string to his policy bow. One area of interest from relatively early in his Diet career was communications. He became a member of the Lower House Communications Committee (Tsûshin Iinkai) and the corresponding LDP Communications Division (Tsûshin Bukai) in 1992 and became vice-chairman of the LDP's Communications Division in 1993 (see Table 4.1). The Communications Division (prior to the amalgamation of the Ministry of Home Affairs with the Ministry of Posts and Telecommunications in January 2001) handled matters relating to posts and telecommunications as well as broadcasting issues—everything under the jurisdiction of the Ministry of Posts and Telecommunications.[126]

Given his family connections to the military and his youthful ambition to join the National Defence Academy, Matsuoka also showed an interest in defence policy in his second term by joining the National Defence Division (Kokubô Bukai). He became the vice-chairman of the division in 1994 (see Table 4.1). Defence put some balance into his policy focus by broadening his scope to encompass issues outside Japan, but given his background and family connections in this area, Matsuoka preferred defence policy to foreign affairs. He only began to attend meetings of the Foreign Affairs Division (Gaikô Bukai) later in his career in order to provide support for his main LDP backer, Suzuki Muneo.

Matsuoka also served on Diet and party policy committees on accounts and administration, ethics in public elections and others (see Table 4.1), including the Special Committee on Okinawa and the Northern Territories (Okinawa oyobi Hoppô Mondai ni kansuru Tokubetsu Iinkai). Here he could again provide backup to Suzuki,[127] who was very influential in this area and who, at one time, served as director-general of the Hokkaido Development Agency.

In 1996, Matsuoka rose to be vice-chairman of the LDP's Diet Policy Committee, which was concerned with advancing the parliamentary legislative process. Its task was chiefly one of coordination: with bureaucrats, who wanted to get their legislation passed,[128] and with opposition parties, whose cooperation was needed for the smooth passage of legislation through the Diet.

In 2003, Matsuoka served as vice-chairman of the LDP's Special Committee on Aviation Problems (Kôkû Mondai Tokubetsu Iinkai),[129] which later set up a subcommittee called the Aviation Industry Countermeasures Subcommittee (Kôkû Jigyô Taisaku Shoiinkai) in order to come up with solutions to the downturn in the aviation industry. At one of the subcommittee meetings, Matsuoka, as a market interventionist, proposed that special financing should be made available to each (aviation) company via the policy investment bank. In his view, support for the aviation industry in Japan was important 'so that globalisation of the country would not stagnate'.[130]

In 2004, Matsuoka became vice-chairman of the LDP's Medical Treatment Basic Problems Investigation Committee (Iryô Kihon Mondai Chôsakai) (see Table 4.1). He claimed to have some knowledge and interest in this area, particularly as it could be applied to his local constituency. In the past, Matsuoka had held meetings on medical issues in Aso County Medical Hall so that voters could evaluate him as a political representative on medical issues. He discussed initiatives such as the introduction of an emergency helicopter, his involvement in reviewing medical laws, the entry of joint-stock companies into medical fields (as a deregulation measure) and problems of Japan's low birth rate and aging population.[131]

Matsuoka laid claim to some of the success in achieving the implementation of a national paediatric emergency telephone consultation service in 2004, which he asserted had been one of his pet projects for some time. He had tackled the problem for the first time 20 years previously when he was assistant director of the National Land Agency's Regional Development Division. During that period, he said he had 'keenly felt the need for an emergency medical system specialising in paediatrics'.[132] Since then, the fulfilment of an entire emergency medical system had been one of his policy objectives including the introduction of an emergency helicopter and local medical treatment in areas such as remote islands and mountain village regions. Matsuoka claimed credit for the fruits of his 'inconspicuous but energetic and patient efforts in all directions',[133] which had finally started to produce results.

In another social welfare policy area, Matsuoka stepped into the lead position on an LDP panel on Minamata disease (see Table 4.1), the Minamata Problem Subcommittee (Minamata Mondai Shoiinkai). Minamata disease was a form of mercury poisoning that broke out in the 1950s and Minamata Bay was located in Kumamoto Prefecture. As chairman of the subcommittee, Matsuoka claimed credit in 2005 for breaking the deadlock between the national government and Kumamoto prefectural government on the issue of who should foot the bill for medical costs for patients with Minamata disease. He instructed the Environment Ministry and other ministries and agencies concerned to work out a compromise under which the national government would bear more of the costs than local government.[134]

Matsuoka also stepped into important positions in the Diet committee system. In November 2003, in the early days of his fifth term, Matsuoka was elected director of the Lower House Budget Committee (Yosan Iinkai) (see Table 4.1). The Budget Committee's purview is all-encompassing. It not only deliberates on the government's budget bill, but also on all other important policies with fiscal implications, such as the dispatch of Self-Defence Force (SDF) troops to Iraq, pension issues, the economy, postal privatisation, foreign policy and so on. The committee also serves as the main arena for question time between government and opposition party leaders. This function put Matsuoka, as vice-chairman, in the thick of government business in the Lower House. In this role, Matsuoka attended meetings of Budget Committee directors, which were held in order to organise the committee agenda as well as to discuss items for examination undertaken after the close of the Diet session.

At the very first meeting of the Budget Committee of the special Diet in September 2005, Matsuoka was reappointed as a director of the committee (see Table 4.1). At the same time he was reappointed as chief of the LDP's Information Research Bureau, a position he first took up in 2004 (see Table 4.1). This made him one of five LDP bureau chiefs. The bureau provided data to determine the basic policy of the party through the collection and arrangement of all sorts of information necessary for LDP activities.[135]

Within the executive ranks of the ruling party, Matsuoka rose to be deputy secretary-general of the LDP in 2000 and held other leading positions in the LDP's organisation (see Table 4.1). The highest general policy position he held was as a member of the Executive Council in 2001–03, reflecting his seniority. The council was the supreme decision-making body of the LDP and the clearing-house for all PARC policy decisions. It had 31 members, all senior members of the party appointed mainly through coordination amongst the party factions.[136] Article 38 of the party rules stipulates that the role of the Executive Council is to 'deliberate on important bills relating to party management and Diet activities and determine whether to support the bills or not.' No policy could become party policy without the approval of the Executive Council, which gave Matsuoka the opportunity to act as a supreme veto point on policy within the party.

ACTIVITIES IN DIET MEMBERS' LEAGUES

Matsuoka's public profile as an agitator on behalf of agricultural interests was further boosted by his high-profile activities in more informal organisations known as Diet members' leagues. From the time he entered the Diet in 1990, Matsuoka remained resolutely resistant to agricultural trade liberalisation and was not open to persuasion. He was very strongly protective of his constituents, and became the leader of a movement (within the LDP) to block the opening of the rice market during the UR negotiations. In 1992, he mobilised the Special Action Diet Members' League to Protect Japanese Agriculture (Nihon no Nôgyô o Mamoru Tokubetsu Kôdô Giin Renmei), a group of 32 politicians who had only been elected once. The group spearheaded the anti-rice liberalisation lobby within the LDP. Matsuoka was its representative organiser (*daihyô sewanin*). The group went into action whenever the government showed a softening attitude towards liberalising imports of rice. In late 1992, it threatened to request the immediate resignation of Prime Minister Miyazawa if the government changed its approach to the rice import issue. It also proposed sending its own mission to Geneva to appeal to the Japanese GATT negotiators to prevent comprehensive tariffication of agricultural import barriers.[137] As the leader of the group, Matsuoka, in his own words, 'made a signed pact with fellow Diet members and conducted a sit-in protest in front of the Diet'.[138] He sat down in front of the Diet building and refused to move.

Matsuoka also demonstrated a similar commitment to defending the interests of his farm supporters in 1996. While holding the position of chairman of the Agriculture and Forestry Division, he simultaneously participated in the LDP's 'action corps', which aligned with agricultural groups to secure the agriculture and forestry budget and which included a number of senior LDP *nôrin giin*.

These are both examples of informal, but organised, policy activities characteristic of special-interest politicians such as Matsuoka. The most transient groups are called 'action corps', but the more substantial groupings in which Matsuoka has been involved, some with an almost semi-permanent existence, are Diet members' leagues (*giin renmei*).

The leagues engage in very public activities that enable Matsuoka to promote particular policy causes and to lobby the party leadership, the ministries and the government leadership. Such activities are essentially a form of public relations. They present a good image to voters, and are designed to demonstrate Matsuoka's experience, knowledge, sense of duty and commitment, as well as his policy interests and credibility. Activities in the leagues raise Matsuoka's public profile and visibility, particularly in his electorate, showing how energetically he is working in *giin katsudô* on behalf of his supporters, how he is engaging with other like-minded legislators on issues in which they are in common agreement and how they are working for common objectives in a way that cuts across both factional and sometimes party membership. The leagues enable Matsuoka to operate as part of an internal pressure group within the Diet and ruling party.

Some of the leagues in which Matsuoka has participated are related to agriculture and forestry, directly or indirectly. Matsuoka became acting chairman of the LDP's Dairy Policy Association (Jimintô Rakuseikai), a group of LDP farm politicians who represented constituencies where dairy farming was important, and who had close links to the dairy farmers' political leagues.

In the role of acting chairman, Matsuoka attended the permanent and central combined council of the dairy farmers' political leagues and made his greetings to the attendees as acting chairman of the LDP group. He saw this as necessary order to demonstrate his continuing support for dairy farming.[139]

Matsuoka also assumed the position of acting chairman of the Forests, Forestry and Forestry Industry Activisation Promotion Diet Members' League (Shinrin, Ringyô, Rinsangyô Kasseika Sokushin Giin Renmei),[140] which maintained links to the Forests, Forestry and Forestry Industry Activisation Promotion Assembly Members' League (Shinrin, Ringyô, Rinsangyô Kasseika Sokushin Giin Renmei) organised by regional prefectural and municipal assemblies. In his executive role, Matsuoka liaised between the two leagues and attended roundtable conferences of the prefectural group.[141] These activities provided him with direct links to prefectural politicians concerned with regional forestry issues.

In an agriculture-related role, Matsuoka became chairman of the Diet members' league called the Association for Researching the Food Labelling Problem for Consumer Protection (Shôhisha Hogo no tame no Shokuhin Hyôji Mondai Kenkyûkai), which focused on the issue of labelling food with the regional district in which it was produced. It also dealt with the lack of precise regulations on food-producing district labelling for supermarkets and others. Meat companies were able to label imported beef as domestic beef during the domestic bovine spongiform encephalopathy (BSE) scare in order to obtain subsidies from the government, a sham in which Matsuoka was indirectly involved.[142] Matsuoka claimed that 'through the enthusiastic action of the Diet members' league the government has started to adopt strict criteria for producing district labelling and penal regulations for offenders'.[143]

Matsuoka also established the Diet Members' League to Promote the Export of Farm Products Etc. (Nôsanbutsutô Yûshutsu Sokushin Giin Renmei) in December 2003, with the catchphrase 'agricultural policy on the offensive' (*seme no nôsei*). He later became leader of the LDP's Agricultural Products Etc. Export Promotion Research Association (Nôsanbutsutô Yûshutsu Sokushin Kenkyûkai), which was formed in February 2004 with approximately 40 members from the Upper and Lower Houses of the Diet. The association was established with the objective of encouraging people in other countries to taste authentic Japanese food using genuine Japanese foodstuffs, to understand and like Japan's culture more, and to make Japanese agriculture into an export industry. The foundation general meeting of the association expressed the view that the amount of exported farm products should rise from the existing level of ¥270 billion to ¥1 trillion over five years in cooperation with the Japan External Trade Organisation (JETRO) and other groups.[144]

Through his membership of Diet members' leagues, Matsuoka managed to span the range of other core LDP interests as well, with small business, traditional Japanese culture, consumer and welfare interests, and health and aviation policy figuring in his membership. For example, through league activities, Matsuoka became something of a small business advocate. He became chairman of the LDP's Small and Medium Enterprise Area Coordination Law Subcommittee (Chûshô Kigyô Bunya Chôseihô Bunkakai), which was a subcommittee of a larger Diet members' league, the Association to Foster Small and Medium Enterprise to Revive the Japanese Economy (Nihon no Keizai o Kasseika shi Chûshô Kigyô o Sodateru Kai),[145] commonly known by the sobriquet 'Association to Reconsider Deregulation' (Kisei Kanwa o Minaosu Kai).

Matsuoka also served as chairman of another subcommittee of this larger Diet members' league, the Coexistence with Large-Scale Stores Problem Subcommittee (Daikibo Tenpo to Kyôson Mondai Bunkakai). On these committees, Matsuoka went out of his way to speak for the owners of the old-fashioned Japanese 'mom-and-pop' stores, the traditional small business owners who sheltered behind a welter of regulations preserving their profits, and who formed a very important bailiwick for the LDP throughout Japan. Because the push for deregulation could affect small businesses in regional areas, Matsuoka railed against the iniquities of deregulation on his website. He cited the example of Germany where, he argued, unregulated development had not been allowed to take place in regional cities, and where a large-scale supermarket opening in a rural district was forced by regulation to deal in products other than those supplied by regional shops.

Thus agriculture and forestry have not been Matsuoka's exclusive interest or zone of political activity. He has had other policy concerns and became a member and executive of committees that were completely unrelated to his primary specialism. In fact, on his website, he claimed 'to be active in a wide range of areas and to be a rarely gifted person…with an extraordinary ability to execute actions'. Furthermore, as Matsuoka gained seniority in the Diet and in the party, he spread his wings further in preparation for, as he saw it, higher office. This required him to gain knowledge and expertise in a wider range of committees, and to demonstrate that he was not simply a narrowly focused, special-interest politician. His agricultural and forestry specialism was the dominant but not the sole dimension of his policy activities. No politician in Japan could afford to be one-dimensional in his representation of agricultural and forestry interests because of the number of votes connected to primary industries was declining all the time.

However, the key difference between Matsuoka's memberships of agriculture and forestry committees and all the rest (apart from the Lower House Special Committee on Disasters where he demonstrated rather more dedicated attachment because of its connection to public works in regional areas) was that his participation was not continuous. It was Matsuoka's persistent attachment to agriculture and forestry committees and the variety of committees in this policy sector on which he served and directed, which pointed to his policy specialism and primary area of interest representation.

Table 4.1   Matsuoka's Committee Memberships et cetera

Lower House committee memberships	
Regional Administration Committee	1990
Special Committee Relating to Land Problems	1990
Agriculture, Forestry and Fisheries Committee	1991–94
Member/Director	1995–96
Director	1997–99
Chairman	2000
Special Committee on Disasters	1991, 1995–97, 1999
Director	1992–1994, 2003, 2005
Communications Committee	1992–93
Environment Committee	
Member/Director	1994
Director	1995
Land, Infrastructure and Transport Committee	2001–2002
Special Committee on Okinawa and Northern Territories	2001
Director	2002
Special Committee on Ethics in Public Elections	2002–2003
Budget Committee	2003, 2005
Director	2004–05
Accounts and Administration Committee	2004
Special Committee on Postal Privatisation	
Director	2005

LDP (PARC) policy committee memberships

Comprehensive Agricultural Policy Investigation Committee	1990–94
Agriculture and Forestry Division	1991–95
Chairman	1996–97
Communications Division	1992
Vice-Chairman	1993–94
Defence Division	1993
Environment Division	1993
Chairman	1994
National Defence Division	
Vice-Chairman	1994
Agricultural Basic Policy Subcommittee	
Chairman	1995–2003, 2005
Uruguay Round-Related Countermeasures Implementation Subcommittee	
Chairman	1995
State-Owned Forests Problems Subcommittee	
Chairman	1995, 1997, 2000–01
Livestock Product Price Sub-Committee	
Chairman	1996
Diet Policy Committee	
Vice-Chairman	1996
Rice Price Committee	
Chairman	1997
Agriculture, Forestry and Fishery Products Trade Countermeasures Special Committee	
Acting Chairman	1997–99
UR-Related Works Implementation Promotion Subcommittee	
Chairman	1997
Comprehensive Agricultural Policy Investigation Committee	1997
Acting Chairman	2004–05
Committee Concerned with the Rice Price	
Chairman	2000
Mountain Village Development Countermeasures Special Committee	
Acting Chairman	2000–01, 2004
UR-Related Countermeasures Implementation Promotion Subcommittee	
Chairman	2000
Agriculture, Forestry and Fisheries Administrative Reform Investigation Team	
Chairman	2000
Management Income Study Meeting	
Chairman	2000

Agriculture, Forestry and Fishery Products Trade
Investigation Committee
    Secretary-General   2003–05
    Chairman   2006
Leaf Tobacco Price Investigation Subcommittee
    Chairman   2003
Forestry Illegal and Unlawful Logging Countermeasures
Investigation Team
    Chairman   2003
Special Committee on Aviation Problems
    Vice-Chairman   2003
LDP Avian Influenza Countermeasures Headquarters
    Secretary-General   2004
Management Countermeasures Project Team
    Chairman   2004–05
Forestry Policy Basic Problems Subcommittee
    Chairman   2004–05
Illegal Logging Countermeasures Investigation Team to
Protect the Global Environment
    Chairman   2004–05
Agriculture and Forestry Executive
    Acting Chairman   2005
Tobacco and Salt Industry Special Committee
    Chairman   2005
Forestry Management Activation Council
    Chairman   2005
Forestry Policy Investigation Committee
    Vice-Chairman   2005
Mountain Village Development Committee
    Acting Chairman   2005
Roads Investigation Committee
    Vice-Chairman   2005
Minamata Problem Subcommittee
    Chairman   2005
Social Welfare System Investigation Committee
    Vice-Chairman   2005
Special Committee on Aviation Countermeasures
    Vice-Chairman   2005
Human Rights Problem Investigation Committee
    Vice-Chairman   2005
Committee to Rapidly Promote Exports of   2006
Japanese Agricultural Products etc.

Cabinet and sub-cabinet positions
Parliamentary Vice-Minister for Agriculture, Forestry and Fisheries: Murayama Cabinet 1995
Deputy Minister for Agriculture, Forestry and Fisheries: Second Mori Cabinet 2001
Minister of Agriculture, Forestry and Fisheries 2006

Party organisation/ executive posts
Agriculture, Forestry and Fisheries Bureau
    Assistant Director 1993
Communications and Information Bureau
    Assistant Director 1993
Construction Bureau
    Chief 1994
National Organisation Committee
    Vice-Chairman 1994
Construction Bureau
    Chief 1995
National Land and Construction-Related Groups Committee
    Chairman 1997
Deputy Secretary-General 2000
Executive Council 2001–03, 2005
Information and Investigation Bureau
    Chairman 2004
PARC Deliberation Commission 2005
Organisation Headquarters
    Vice-Chairman 2005

Sources: MAFF mimeo, Jiyû Minshutô Seimu Chôsakai (ed.) February 1992. *Jiyû Minshutô Seimu Chôsakai Meibo, Heisei 4-nen, 2-gatsu, 3-nichi Genzai* [*Liberal Democratic Party Policy Affairs Research Council Membership List 3 February 1992 to the Present*], pp: 63, 70, 90; *Seikan Yôran*, various issues, *Kokkai Benran*, Tokyo, Nihon Seikei Shinbunsha, various issues; Matsuoka Toshikatsu Website, *Nôsei Undô Jyânaru*, various issues. <http://www.matsuokatoshikatsu.org/index1.html>; <http://www.jimin.jp/jimin/giindata/matsuoka-to.html>.

NOTES

1. These investigation committees (*chôsakai*) are also called 'research commissions'.
2. Tachibana Takashi, *Nôkyô: Kyodai na Chôsen* [*Nokyo: The Enormous Challenge*], Tokyo, Asahi Shinbunsha, 1980, p. 337.
3. Tatebayashi generalises this point. See *Giin Kôdô*, p. 125.
4. *ibid.*, p. 63.
5. Tatebayashi generalises this point. See *Giin Kôdô*, p. 62.
6. Book review by Brad Glosserman, *The Japan Times*, 13 July 2000.
7. See the discussion in Machidori, 'The 1990s Reforms Have Transformed Japanese Politics', p. 39.
8. Tatebayashi generalises this point. See *Giin Kôdô*, p. 2.
9. *ibid.*
10. Tatebayashi, *Giin Kôdô*, p. 65.
11. Tatebayashi generalises this point. See *Giin Kôdô*, p. 71.
12. Quoted in Tatebayashi, *Giin Kôdô*, p. 68.
13. Tatebayashi, *Giin Kôdô*, p. 69.
14. *ibid.*, p. 132.
15. Kan Naoto, 1998. *Daijin* [*Ministers*], Iwanami Shôten, Tokyo, p. 160.
16. *ibid.*, p. 35.
17. *ibid.*, pp. 162–63.
18. Kan generalises this point. See *Daijin*, p. 162.
19. *ibid.*, p. 163.
20. *ibid.*, pp. 162–63.
21. 'Shinshokuryôhô Ketchaku' ["The New Food Law Launched'], *Nôsei Undô Jyânaru*, No. 5, January 1996, p. 12.
22. *ibid.*, p. 12.
23. This was the New Food, Agriculture and Rural Areas Basic Law (*Shokuryô, Nôgyô, Nôson Kihonhô*) of July 1999.
24. 'Matsuoka Toshikatsu, Jimintô Nôgyô Kihon Seisaku Iinkai Iinchô ni Kiku' ['Listening to Matsuoka Toshikatsu, Chairman of the Agricultural Basic Policy Subcommittee'], *Nôsei Undô Jyânaru*, No. 23, February 1999, pp. 12–3.
25. 'Seidoteki na Shikumi wa Dekita. Nôgyô Genba no Jikkô ni Kitai Shitai' ['A Systematic Framework has Emerged. I Expect that This Will be Executed at the Agricultural Grass Roots'], *Nôsei Undô Jyânaru*, No. 16, November 1997, p. 18.
26. See also the discussion in Chapter 5 on 'Exercising Power as a Nôrin *Zoku*'.
27. Nakanishi and Journal Reporter Group, 'Matsuoka Toshikatsu to Iu Giwaku Nin', p. 183.
28. Nakanishi and Special Reporting Group, 'Suzuki Muneo, Matsuoka Toshikatsu', p. 103, citing the *Asahi Shinbun*, 2 August 1997.
29. *ibid.*
30. See also Chapter 6 on 'The Identical Twins of Nagata-cho'.
31. This is short for Nômin Undô Zenkoku Rengôkai (National Federation of Farmers' Campaigns). It claims that it does not support any particular political party, but it is generally regarded as affiliated with the Japan Communist Party. Visit: http://www.nouminren.ne.jp/aboutus/soshiki/gaiyo.htm
32. Visit: http://www.nouminren.ne.jp/dat/200208/2002081202.htm
33. These were 'health resorts' equipped with hot springs, places for doing holistic chi-kung and other places for people to stand together in large numbers. Hasegawa Hiroshi, 'Nôsuishô o Haishi seyo' ['Abolish the Ministry of Agriculture, Forestry, and Fisheries'], *Aera*, 1 April 2002, p. 37.
34. Nakanishi and Journal Reporter Group, 'Matsuoka Toshikatsu to Iu Giwaku Nin', p. 183.
35. Nakanishi and Special Reporting Group, 'Suzuki Muneo, Matsuoka Toshikatsu', p. 103.
36. *ibid.*

37 'Onsen de Kokusai Kyôsôryoku Kyôka Nôgyô Yosan Muda Tsukai no Kôzu' ['Internationalisation Strengthened Through Hot Springs—The Composition of Wasteful Spending in the Agricultural Budget'], *Shûkan Daiyamondo*, 20 April 2002, p. 52. The resort provides a wide range of leisure facilities in addition to a hot spring, including restaurants, shops, hotels. Visit: http://www.mizube-plaza.co.jp/
38 The idea behind this terminology is to 'promote interaction between urban and rural areas, while at the same time securing employment opportunities for farm villages. 'Onsen de Kokusai Kyôsôryoku Kyôka', p. 52.
39 Nakanishi and Special Reporting Group, 'Suzuki Muneo, Matsuoka Toshikatsu', p. 103.
40 Hasegawa, 'Kanjûdanomi no Hazama de Shundô', p. 23.
41 'Onsen de Kokusai Kyôsôryoku Kyôka', p. 52.
42 Nakanishi and Journal Reporter Group, 'Matsuoka Toshikatsu to Iu Giwaku Nin', p. 183. *Zenekon* is an abbreviation of *zeneraru kontorakutâ*, (general contractor), which directly contract construction works from clients and conduct all aspects of construction work. *Zenekon* consist of the top construction companies in Japan such as Takenaka Corporation, Obayashi Corporation, Shimizu Corporation, Kajima Corporation and Taisei Corporation, which are sometimes called Super Zenekon. Even though the core business of these five corporations is the execution of construction works, the corporations also have design, engineering, and research and development sections and possess extensive technological resources for construction. In this respect they are different from ordinary (small-scale) construction companies (*kensetsu kaisha*).
43 Itô, 'Heisei Jiken Fuairu: Nôrin Jigyô Hojokin o Dokusen Suru Matsuoka Toshikatsu', p. 67. Toyo Construction also secured the contract for a cultural exchange centre in Kikuyo Town. Originally a different general contractor involved in the planning stages was going to take the job, but the bid went to Toyo Construction.
44 Hasegawa, 'Kanjûdanomi no Hazama de Shundô', p. 23. See also Chapter 6 on 'The Identical Twins of Nagata-cho'.
45 *ibid.*
46 *ibid.*
47 Itô, 'Heisei Jiken Fuairu: Nôrin Jigyô Hojokin o Dokusen Suru Matsuoka Toshikatsu', p. 65.
48 'UR Taisaku 6 Chô 100 Oku En tô Nôgyô Kankei Yosan Kakuho e Jimintô Sôgô Nôsei Chôsakai, Nôrin Bukai Zenryoku' ['The LDP's Comprehensive Agricultural Policy Investigation Committee and Agriculture and Forestry Division Put All Power Into Securing the UR Countermeasures ¥6.01 Trillion Agriculture-Related Budget'], *Nôsei Undô Jyânaru*, No. 13, May 1997, p. 4.
49 This is an abridged version of Matsuoka's response, reported in 'Zaisei Kôzô Kaikaku ni kakawaru Nôgyô Kankei Yosan Kakuho Taisaku no Torikumi ni tsuite' ['About Grappling with Countermeasures to Secure the Agriculture-Related Budget Endangered by Fiscal Structural Reform'], *Nôsei Undô Jyânaru*, No. 14, August 1997, p.5.
50 Tatebayashi, *Giin Kôdô*, p. 70.
51 Matsuoka Toshikatsu Official Site, 'Jimintô no "Bukai" tte??' ['What are the LDP "Divisions??"'], in *Katsudô Hôkoku* [*Activity Report*]. Available from http://matsuokatoshikatsu.org/sit002//public/053.html
52 'Kinkyû Nyûin shita "Suzuki Muneo" no Funkei no Tomo: Matsuoka Toshikatsu Daigishi no Taiho Jôhô' ['"Suzuki Muneo's" Eternal Friend is Admitted to Hospital in an Emergency: A Report of Matsuoka's Arrest'], *Shûkan Shinchô*, 4 July 2002, p. 26.
53 'Dai 136-kai Kokkai Nôrinsuisan, Yosan Iinkai Shitsumon Ôtô' ['The 136th Diet Agriculture, Forestry and Fisheries and Budget Committees' Responses to Questions'], *Nôsei Undô Jyânaru*, No. 7, May 1996, pp. 28–9.
54 'Jimintô Chikusanbutsu Kakakutô Shôiinchô, Matsuoka Toshikatsu, "Chikusan Nôka no Iyoku o so ga nai Kakaku Kettei ni Zenryoku"' ['LDP Livestock Commodity Prices Etc. Subcommittee Chairman, Matsuoka Toshikatsu, "Full Power for a Price Decision that Does Not Weaken the Motivation of Livestock Farmers"'], *Nôsei Undô Jyânaru*, No. 7, May 1996, p. 11.

55 'Jimintô Chikusanbutsu', p. 11.
56 'Inasaku Keiei Antei, Kome Seisaku no Kakuritsu e' ['Towards the Establishment of a Rice Policy and Stabilisation of Rice Crop management'], *Nôsei Undô Jyânaru*, No. 16, November 1997, p. 20.
57 'Kome Seisaku Kakuritsu e Zaigen Kakuho Motomeru 1300 nin Kesshû shi Zenkoku Daihyôsha Shûkai' ['A National Gathering of Representatives Uniting 1300 People Demand the Securing of a Source of Revenue for the Establishment of a Rice Policy'], *Nôsei Undô Jyânaru*, No. 16, November 1997, p. 21.
58 'Jimintô ni Girigiri made Yôsei, Tôhonbu ni Kesshû shi Sôkekki Shûkai' ['Taking Requests to the LDP, A General Uprising Meeting Concentrating on the Party Heaquarters'], *Nôsei Undô Jyânaru*, No. 16, November 1997, p. 24.
59 'Seidoteki na Shikumi wa Dekita', p. 18.
60 See Chapter 5 on 'Exercising Power as a *Nôrin Zoku*'.
61 In 2000 Matsuoka served as chairman of the LDP's Agriculture, Forestry and Fisheries Administrative Reform Investigation Team (Nôrinsuisan Gyôkaku Kentô Chîmu).
62 In 1995, 1997, 2000 and 2001 Matsuoka served as chairman of the State-Owned Forests Problems Subcommittee (Kokuyû Rinya Mondai Shôiinkai), which was a subcommittee of the Forestry Policy Investigation Committee. The subcommittee was, for example, concerned with the Forestry Policy Reform Outline (*Rinsei Kaikaku Taikô*). It was under Matsuoka's chairmanship in 2000 when the plan for the Forestry Policy Reform Outline was advanced. 'Nôgyô Kankei Seisaku Kettei no Ashidori' ['The Steps of Agriculture-Related Policy Decisions'], *Nôsei Undô Jyânaru*, No. 35, February 2001, p. 29.
63 'Nôgyô Kankei Seisaku Kettei no Ashidori', *Nôsei Undô Jyânaru*, No. 35, February 2001, p. 29.
64 *The Japan Times*, 6 December 2000.
65 *ibid*.
66 Nakanishi and Special Reporting Group, 'Suzuki Muneo, Matsuoka Toshikatsu', p. 103.
67 *ibid.*, p. 102.
68 Visit: http://piza.2ch.net/giin/kako/987/987905181.html
69 This position is also described as 'senior vice-minister'.
70 This was also a position that came to be reserved for Diet members serving their fourth term.
71 *Nikkei Weekly*, 14 August 2000.
72 See also Chapter 5 on 'Exercising Power as a *Nôrin Zoku*'.
73 *The Japan Times*, 8 January 2001.
74 *ibid*.
75 *ibid*.
76 'Chûô Shôchô Saihen Sutâto', ['Reorganisation of the Central Ministries and Agencies Starts'], *Nôsei Undô Jyânaru*, No. 35, February 2001, p. 13.
77 *Yomiuri Shinbun*, 17 January 2001.
78 *The Japan Times*, 21 March 2003.
79 'Nôgyô Kankei Seisaku Kettei no Ashidori', *Nôsei Undô Jyânaru*, No. 37, June 2001, p. 30.
80 Wada Yoshitaka, 'Kenshô: Sêfugâdo wa Naze Hatsudô sareta ka?' ['Investigation: Why Were the Safeguards Invoked?'], *Ekonomisuto*, 23 April 2002, p. 91.
81 Wada, 'Kenshô: Sêfugâdo wa Naze Hatsudô sareta ka?', p. 91.
82 *ibid*. See also Chapter 6 on 'The Identical Twins of Nagata-chô'.
83 *ibid*.
84 *ibid*.
85 *ibid.*, p. 92.
86 *ibid.*, p. 91.
87 *ibid*.
88 Embassy of Japan in China, 'Nihon to Chûkoku no Kankei: Saikin no Ugoki – Matsuoka Nôrinsuisan Fukudaijin ga Hôchû, Yunyû Yasai ya Ryokuka Kyôryoku ni Tsuite Kaidan' ['Japan-China Relations: Recent Activities – Deputy Agriculture, Forestry and Fisheries Minister Matsuoka on a Visit to China,

 	Talks concerning Vegetable Imports and Cooperation in Tree-Planting']. Visit: http://www.cn.emb-japan.go.jp/jp/2nd%20tier/05jckankei/j-c010320jjj.htm
89	Wada, 'Kenshô: Sefugâdo wa Naze Hatsudô sareta ka?', p. 93.
90	Ayukawa Saiji, 'Jimintô de mo Shinkô suru "Matsuoka Hazushi"'['"Removal of Matsuoka" is Even in Progress in the LDP'], *Fôsaito*, May 2002, p. 20.
91	ibid.
92	'Nôgyô Kankei Seisaku Kettei no Ashidori', *Nôsei Undô Jyânaru*, No. 46, December 2002, p. 29.
93	ibid.
94	They were the chairmen of CAPIC, the Agriculture and Forestry Division, and Agricultural Basic Policy Subcommittee.
95	'Nôgyô Kankei Seisaku Kettei no Ashidori', *Nôsei Undô Jyânaru*, No. 51, October 2003, p. 29.
96	ibid..
97	ibid.
98	Matsuoka Toshikatsu Official Site, 'Kome Seisan Chosei Menseki Sueoki o Kettei' ['The Decision to Leave the Area for Rice Production Adjustment As It Is'], in *Katsudô Hôkoku* [*Activity Report*]. Available from http://matsuokatoshikatsu.org/site002//public/048.html
99	It replaced the Agriculture, Forestry and Fishery Products Trade Countermeasures Special Committee.
100	'Jimintô no "Bukai" tte??'. Visit: http://matsuokatoshikatsu.org/sit002//public/053.html
101	Matsuoka Toshikatsu Official Site, 'Nihon, Mekishiko FTA (Jijû Bôeki Kyôtei) Gôi' ['The Japan-Mexico FTA (Free Trade Agreement) Reached'], in *Katsudô Hôkoku* [*Activity Report*]. Available from http://matsuokatoshikatsu.org/sit002//public/053.html
102	Matsuoka Toshikatsu Official Site, '"Jimintô Tori Infuruenza Taisaku Honbu" Tachiageru' ['"LDP Headquarters for Avian Influenza" Established'], in *Katsudô Hôkoku* [*Activity Report*]. Available from http://matsuokatoshikatsu.org/sit002//public/051.html
103	Matsuoka Toshikatsu Official Site, 'Tori Infuruenza Yosan Iinkai de Shitsumon' ['Interpellating on Avian Influenza in the Budget Committee'], in *Katsudô Hôkoku* [*Activity Report*]. Available from http://matsuokatoshikatsu.org/sit002//public/051.html
104	Matsuoka Toshikatsu Official Site, '"Tô Tori Infuruenza Taisaku Honbu" Kachiku Densenbyôo Yôbôhô Kaiseian o Ryôshô' ['"The LDP Avian Influenza Countermeasures Headquarters" Approves the Revised Bill for the Livestock Infectious Diseases Prevention Law'], in *Katsudô Hôkoku* [*Activity Report*]. Available from http://matsuokatoshikatsu.org/sit002//public/053.html
105	Matsuoka Toshikatsu Official Site, 'Tori Infuruenza wa Banzen no Taisaku de' ['All Possible Countermeasures Against Bird Flu'], in *Katsudô Hôkoku* [*Activity Report*]. Available from http://matsuokatoshikatsu.org/index1.html
106	Matsuoka Toshikatsu Official Site, 'Oguni Machi de Genchi Chôsa' ['On-the-Spot Investigation' in Oguni Town'], in *Katsudô Hôkoku* [*Activity Report*]. Available from http://matsuokatoshikatsu.org/site002//public/060.html
107	Visit: http://www.sanson.or.jp/sokuhou/no_901/901-3.html
108	'Oguni Machi de Genchi Chôsa'. Available from http://matsuokatoshikatsu.org/site002//public/060.html
109	For more discussion of Matsuoka's activities on forestry-related committees, see Chapter 7 on 'Electoral Vicissitudes'.
110	See Chapter 7 on 'Electoral Vicissitudes'.
111	'"Keiei Seisaku Taikô" o Matomaru—"Kôzô Kaikaku" no Gutaika Sutâto' ['"Management Policy Outline" Decided—The Start of Concrete Measures for "Structural Reform"'], *Nôsei Undô Jyânaru*, No. 39, October 2001, p. 20.
112	Matsuoka Toshikatsu Official Site, 'Chihô Rinkatsu Giren Sôkai' ['A General Meeting of the Regional Forestry Activisation Assembly Members' League'], in *Katsudô Hôkoku* [*Activity Report*]. Available from http://matsuokatoshikatsu.org/index1.html
113	Matsuoka Toshikatsu Official Site, 'Tokubetsu Kokkai no Kaikaishiki' ['Opening Ceremony of the Special Diet'], in *Katsudô Hôkoku* [*Activity Report*]. Available from http://matsuokatoshikatsu.org/index1.html
114	ibid.

115 Matsuoka Toshikatsu Official Site, 'Nihongata Chokusetsu Shiharai o dô subeki ka' ['How Should Japan-Style Direct Payments Be?'], in *Katsudô Hôkoku* [*Activity Report*]. Available from http://matsuokatoshikatsu.org/index1.html
116 Matsuoka Toshikatsu Official Site, 'Jyunêbu kara Kaette Kimashita' ['I Returned from Geneva'], in *Katsudô Hôkoku* [*Activity Report*]. Available from http://matsuokatoshikatsu.org/index1.html
117 Matsuoka Toshikatsu Official Site, 'Nihonhan Chokusetsu Shiharai no Saishûan Happyô', ['Announcement of the Final Draft of the Japanese Edition Direct Payments'], in *Katsudô Hôkoku* [*Activity Report*]. Available from http://matsuokatoshikatsu.org/index1.html
118 *Nikkei Weekly*, 14 November 2005.
119 'Taifû to Jishin de Nôgyô ni Daihigai' ['Great Damage to Agriculture in the Typhoons and Earthquake'], *Nôsei Undô Jyânaru*, No. 58, December 2004, p. 20.
120 Matsuoka Toshikatsu Official Site, 'Shûchû Gôu no Higaichi o Chôsa Shimashita' ['I Investigated Areas Damaged by Concentrated Heavy Rain'], in *Katsudô Hôkoku* [*Activity Report*]. Available from http://matsuokatoshikatsu.org/index1.html
121 Matsuoka Toshikatsu Official Site, 'Oguni-machi Gikai' ['Oguni Town Assembly'], in *Katsudô Hôkoku* [*Activity Report*]. Available from http://matsuokatoshikatsu.org/index1.html
122 Matsuoka Toshikatsu Official Site, 'Kuni no Moto, Chisan Chisui' ['The Foundation of the Country, Food Control and Riparian Works'], in *Katsudô Hôkoku* [*Activity Report*]. Available from http://matsuokatoshikatsu.org/site002//public/041.html
123 Matsuoka Toshikatsu Official Site, 'Kyûshû no Shûchû Gôu Higai Taisaku on Kinkyû Giron ['Urgent Deliberation on Localised Torrential Downpour Damage Countermeasures in Kyushu'], in *Katsudô Hôkoku* [*Activity Report*]. Available from http://matsuokatoshikatsu.org/site002//public/041.html
124 Tatebayashi, *Giin Kôdô*, p. 132.
125 *ibid.*
126 Personal communication, Professor Ellis Krauss, June 2005.
127 See also Chapter 6 on 'The Identical Twins of Nagata-chô'.
128 Curtis, *The Logic of Japanese Politics*, pp. 119–20.
129 Elsewhere this committee is listed as the Special Committee on Aviation Countermeasures (Kôkû Taisaku Tokubetsu Iinkai). See *Seikan Yôran*, 2004, Spring Edition, p. 564.
130 Matsuoka Toshikatsu Official Site, 'Kôkû Gyôkai no Anteika Shien o Yôsei' ['Requesting Support for the Stabilisation of the Aviation Industry'], in *Katsudô Hôkoku* [*Activity Report*]. Available from http://matsuokatoshikatsu.org/site002//public/041.html
131 'Matsuoka Toshikatsu Shi ni Kiku (Aso Gun Ishi Renmei Shiryoo no Peiji)' ['Ask Mr Toshikatsu Matsuoka (The Data Page of Aso County Doctors Federation)']. Available from http://www.geocities.jp/e_osan/ishirenmei_aso03_T_Matsuoka.html
132 Matsuoka Toshikatsu Official Site, '"Shoni Kyukyu Denwa Sodan Jigyo" Jitsugen e!!' ['Realising "The Paediatric Emergency Telephone Consultation Project"!!'], in *Katsudô Hôkoku* [*Activity Report*]. Available from http://matsuokatoshikatsu.org/sit002//public/053.html
133 '"Shoni Kyukyu Denwa Sodan Jigyo" Jitsugen e!!'. Visit: http://matsuokatoshikatsu.org/sit002//public/053.html
134 *Manichi Daily News*, 31 March 2005.
135 Matsuoka Toshikatsu Official Site, 'Shûgiin Yosan Iinkai "Riji" ni Shûnin' ['Taking up the Post of "Director" of the Budget Committee in the House of Representatives'], in *Katsudô Hôkoku* [*Activity Report*]. Available from http://matsuokatoshikatsu.org/site002//public/048.html
136 Under Prime Minister Koizumi, the members of the Executive Council are officially elected in the following way: 14 members are publicly elected from the LDP Lower House membership; six members are publicly elected from the LDP Upper House membership; and 11 members are designated by the LDP president. *Jiyû Minshutô Sômukai* [*The LDP Executive Council*]. Visit: http:ja.wikipedia.org/
137 *Yomiuri Shinbun*, 24 December 1992.
138 'Matsuoka Toshikatsu: Purofuiru', *Seikan Yôran*, 1995, First Half Year Edition, p. 269.

139 Matsuoka Toshikatsu Official Site, 'Jizokuteki na Rakunô Shien o' ['Continuing Support for Dairy Farming'], in *Katsudô Hôkoku* [*Activity Report*]. Available from http://matsuokatoshikatsu.org/index1.html
140 Activities in other forestry leagues are examined in Chapter 7 on 'Electoral Vicissitudes', as are Matsuoka's promotion of environmental causes through league activity.
141 Matsuoka Toshikatsu Official Site, 'Tokubetsu Kokkai ga Owarimashita' ['The Special Diet Ended'], in *Katsudô Hôkoku* [*Activity Report*]. Available from http://matsuokatoshikatsu.org/index1.html
142 See Chapter 6 on 'The Identical Twins of Nagata-chô'.
143 Matsuoka Toshikatsu Official Site, 'Giin Renmei Katsudô' ['Diet League Activities']. Available from http://www.matsuokatoshikatsu.org/wite002//public/033.html
144 Matsuoka Toshikatsu Official Site, 'Anzen, Anshin de Sugureta Nihon no Nôsanbutsu o Sekai ni' ['Safe, Secure and Prominent Japanese Farm Products to the World'], in *Katsudô Hôkoku* [*Activity Report*]. Available from http://www.matsuokatoshikatsu.org/site002//public/051.html. See also Chapter 5 on 'Exercising Power as a *Nôrin Zoku*'.
145 Matsuoka Toshikatsu Official Site, 'Chiiki no "Sakaya San" Sonzoku no Tame ni' ['For the Existence of Regional "Sake Shops"'], in *Katsudô Hôkoku* [*Activity Report*]. Available from http://matsuokatoshikatsu.org/site002//public/053.html
146 'Matsuoka Toshikatsu no Rirekisho'. Available from http://www.matsuokatoshikatsu.org/site002//public/008.html.

5

EXERCISING POWER AS A *NÔRIN ZOKU*

Matsuoka followed the classical career pattern for a *zoku*. His long-standing membership of PARC and Diet committees on agricultural and forestry, his attainment of the top executive positions in the key committees as well as his subcabinet positions on agriculture, forestry and fisheries earned him membership of the LDP's agriculture and forestry 'tribe'. Acquisition of formal policy positions over a period of time indicated an accumulated level of expertise and influence in a particular policy domain as well as the possession of close relations with the ministry responsible for administering that sector. As Matsuoka aspired to senior executive positions in the party and leadership positions in the government, he was aiming to use his status as a 'tribe Diet member' (*zoku giin*) as a means of furthering his ambitions to even higher office.

BECOMING A ZOKU

It is difficult to pinpoint when Matsuoka actually became a *nôrin zoku*. According to some commentators, he had the right to be called a *nôrin zoku* right from the start of his political career 'because he had received support from the late Tamaki Kazuo and because he represented the traditional locality of Kumamoto'.[1] Certainly, by the mid 1990s, Matsuoka's power to plunder the pork barrel had become widely known as a result of the projects built in his own district funded by the UR countermeasures package.[2] By 2000, 'while being a middle-ranking Diet member elected four times, his career history as a *nôrin zoku* stood out'.[3] Matsuoka allegedly monopolised agricultural, forestry and fisheries public works subsidies as chairman of the Lower House AFF

Committee.⁴ In 2000, Nokyo's National Council recognised Matsuoka as a key member of the next generation of agricultural leaders.⁵

Others, however, have not been quite as willing to accord Matsuoka the status of a *nôrin zoku*. He was described as 'not a traditional *nôrin zoku*, because he was not supported by the MAFF. He did not get the backing of the MAFF to stand in politics'.⁶ Furthermore, Matsuoka never had an easy relationship with MAFF officials, which was the norm for *zoku*. As Nakanishi comments, the 'best *nôrin zoku* are those who speak for the MAFF'.⁷ In practice Matsuoka only defended the MAFF's interests when they aligned with his own. A typical example was his support for the MAFF in forcing through Nagasaki Prefecture's Isahaya Bay Drainage Project.⁸ A DPJ executive observed that 'Matsuoka was joked about as "the caretaker of MAFF interests" (*Nôsuishô no shôeki no bannin*) because he resisted all opposition to the Isahaya Bay project, which reclaimed part of the Ariake Sea in Isahaya Bay'.⁹ The Ariake Sea was a nearly land-locked body of water bordered by the prefectures of Nagasaki, Kumamoto, Saga and Fukuoka. The ¥250 billion public works project filled in Isahaya Bay by reclaiming the land to create farmland and a large reservoir. The project area was surrounded by a 7 km-long main dyke whose gates were closed in 1997 to keep out seawater in order to facilitate the fill-in work.¹⁰ Matsuoka was reported as saying that 'as long as the LDP exists, we will not open the dyke (drainage) gate(s)'.¹¹ He also commented, 'there are only a few people who oppose it [the Isahaya Bay project] locally. They are only doing it for their own benefit. They live in the hills and worry about damage to the water, and making a living by taking photos of mudskippers'.¹² Locals, however, complained that the initial justification for the project (reclamation in order to create farmland) changed to 'water damage countermeasures policy'.¹³ In fact 'the only real purpose seemed to be to complete a large-scale public works project and so the official objective of the project did not really matter'.¹⁴

'GODFATHER' OF THE *NÔRIN ZOKU*

Despite Matsuoka's disputed *zoku* status, it is clear that Matsuoka was no ordinary *nôrin zoku*. In fact, by 2000, Matsuoka was regarded not just as a *zoku*, but as a *zoku* boss. He had become known as the 'new godfather'¹⁵ and 'Don' (as in Don Corleone) of the *nôrin zoku*.¹⁶ Being a 'Don' was equivalent to being a *nôrin zoku* boss, the most influential of veteran lawmakers. It bestowed extensive powers in a range of different domains. As 'a dominant *nôrin zoku* figure, he had influence over related groups, in budget acquisition, and in personnel affairs relating to politicians and bureaucrats'.[17]

Many of the references to Matsuoka's being a 'godfather' were to his leading position in the field of forest policy. He was called the 'Don' of forestry administration (*rinya no don*)[18] and 'an influential tribe Diet member' (*yûrokyu zoku*) from the Forestry Agency'.[19] One journalist in charge of MAFF issues said, 'because he is the only Diet member from the Forestry Agency, it is fair to say that in regard to forestry issues, he is a godfather-like figure'.[20] Nakanishi *et al.* wrote in 2002, 'Diet member Matsuoka has rapidly expanded his influence over the last few years as an influential tribe Diet member from the Forestry Agency and has come to be called "the boss of forestry". His influence extends from the budget and personnel affairs to the distribution of projects by the Forestry Agency'.[21] Itô agreed that on forestry administration, Matsuoka 'had no equal'.[22]

INFLUENCE OVER MAFF BUREAUCRATS

As a *zoku*, Matsuoka was expected to be both a protector and a beneficiary of the MAFF. As one of the gatekeepers of the political process, Matsuoka's job was to shepherd MAFF-drafted policy measures and bills through the party and the Diet as well as to lobby for the ministry's budget as a member of the ministry's supporters' group (*ôendan*).[23] In exchange, ministry officials, through the exercise of their discretionary powers, could arrange favours and benefits that would become the patronage Matsuoka dispensed to clients and supporters, and which were important ingredients in Matsuoka's electoral and political survival. Ideally, the relationship between Matsuoka and the MAFF should have been one of equality and mutual dependence.

However, the Matsuoka–MAFF connection did not quite fit this pattern. Because Matsuoka had strong ideas, MAFF officials said they had difficulties in dealing with him, and that he was not easy to talk to.[24] It was far from unusual for Matsuoka to shout and put pressure on MAFF executives in divisional meetings, which were held almost every day when the Diet was in session.[25] He was regarded 'as someone who did what he liked, and while, from the MAFF's perspective, he could be a reliable person, he could also be really annoying'.[26]

As far as the ministry was concerned, Matsuoka had only two positive attributes. First, as 'a bureaucratic OB, he studied policy extensively'.[27] Second, he 'was a convenient person for the MAFF to organise the Etô-Kamei faction',[28] which was dominated by LDP 'Old Guard' politicians and which 'had many loud-mouthed agriculture and forestry "tribe" members'.[29] These included Yatsu Yoshio, the faction's secretary-general and one-time MAFF minister, and Furuya Keiji who acquired his *jiban* from his *nôrin zoku* father Furuya Keiyû.[30] The Etô-Kamei faction inherited the mantle of the leading *nôrin zoku* faction from former Prime Minister Suzuki, whose faction (Kôchikai)[31] had been a *nôrin zoku* stronghold. After Suzuki retired in 1990, Etô Takami took over as leader of the *nôrin zoku*.[32] This group was known as the 'fighting faction' (*butôha*) amongst the *nôrin zoku*.[33]

When it came to links with bureaucrats, Matsuoka had extremely close connections to some officials in the MAFF. The term 'Matsuoka children' was even used to describe the ministry.[34] Matsuoka's personal connections in the MAFF spanned both career and non-career officials.[35] His *senpai* (seniors) were former Administrative Vice-Minister, Tanaka Hironao (who entered the MAFF in 1956), former Director-General of the Food Agency, Ishihara Mamoru (who entered the MAFF in 1970, close to when Matsuoka entered it), former Livestock Department Director, Nagamura Takemi (who entered the MAFF in 1972, but who resigned over the BSE problem) and others. Matsuoka also had a close relationship with the former Director-General of the Hokkaido Forestry Management Bureau, Ogawa Yasuo (who entered MAFF in 1968), and who was called 'the Boss of Hokkaido Forestry'.[36] Matsuoka reportedly made the best use of these 'Matsuoka children'.[37]

Matsuoka also had extremely intimate relations with non-career officials such as a former assistant divisional director of the Agricultural Structure Improvement Bureau, Satô Masato. Satô had a cosy relationship with the company building 'Refresh Villages' using UR countermeasures money provided under pressure from Matsuoka, *zoku giin* and agricultural groups.[38] Satô was reportedly at Matsuoka's beck and call in relation to the expenditure of the UR countermeasures funds.[39] Matsuoka was also close to a party official in LDP headquarters (a Mr Y), forming what was known as the 'Matsuoka—the LDP's Mr Y—the MAFF's Satô' line.[40] As a former MAFF executive explained

> [i]t is true that Satô was close to Matsuoka. Satô was a *jimukan* from Hokkaido and a dazzling and dynamic type of person. He was quite proficient at his work and obtained and handled budgets skilfully. His boss evaluated him highly. Since he just worked on structural improvement projects, he was rather puffed up with pride. He became a sort of 'structural improvement

zoku'. He did not listen to what his division chief said, and he conducted everything by himself. According to rumour, he selected projects in a self-willed manner, and was entertained by a large number of companies and prefectural government officials. However, three to four years ago, corruption in structural improvement projects came to light. A MAFF investigative committee was launched, and punishments were imposed. In consequence, MAFF officials conducting structural improvement were all replaced, and Satô was transferred to a local office.[41]

After Satô was punished and transferred to the Tokai Agricultural Administration Bureau (Tôkai Nôsei Kyoku),[42] Matsuoka tried to get him returned to headquarters (the MAFF main ministry in Tokyo). He repeatedly told the Director-General of the Structural Improvement Bureau, Yamamoto Tôru, in the presence of others, to return Satô to the bureau in the MAFF.[43] However, another ex-MAFF Diet member opposed Satô's return, and the plan failed.[44] Instead of being returned back to the MAFF in Tokyo, Satô was transferred to the Kanto Agricultural Administration Bureau so Matsuoka could save face to a certain extent.[45] However, it was through his relationship to Satô that Matsuoka was able to wield so much influence over the allocation of the UR countermeasures package.

MAFF officials also had long memories about the way Matsuoka behaved when he was deputy minister in 2001. Matsuoka saw the deputy minister's position as an opportunity to throw his weight around his old ministry and to subject the ministry to his power. He wanted to create a more hierarchical relationship, in which officials in the ministry were subordinated to politicians in the LDP. Matsuoka's behaviour naturally created a lot of resentment amongst officials in the MAFF. Surprisingly perhaps, it also created resentment amongst other *nôrin zoku* because it overturned customary decision-making norms and the traditional working relations between the party and the bureaucracy.[46]

Matsuoka's treatment of MAFF officials while he was deputy minister was commonly attributed to various grievances that he had held during his time in the ministry. One official reasoned that because Matsuoka was a *gikan* while in the MAFF, he gave the *jimukan* a hard time when he became deputy minister.[47] As a Forestry Agency OB explains

> Matsuoka was a *gikan* who graduated from Tottori University. Even though he was a high-ranking *gikan* (*jôkyûshoku*), he was often dismissed and treated coldly by career bureaucrats (*jimukanryô*) in the main ministry. So behind his yelling at the bureau chiefs who once looked down on him, there is a bitterness from that time (when he was in the Forestry Agency).[48]

According to some MAFF officials, Matsuoka had very strong views on agricultural policy and was infamous for calling up MAFF bureaucrats and yelling down the phone at them.[49] Even the WTO section of MoFA received many phone calls from him. It 'was so easy for Matsuoka to threaten officials, he could do it before breakfast'.[50] Few MAFF officials could have anticipated his eventual appointment as minister in 2006.

POLICY INTERVENTION

As an agricultural 'tribe' Diet member, Matsuoka exercised considerable influence over agricultural and forestry policy. At one time he held all the main PARC agricultural and forestry committee executive posts, which put him in a position to exercise power at critical stages of the policymaking process. His two most active and influential posts as a *nôrin zoku* were as chairman of the Agricultural Basic Policy Subcommittee and as secretary-general of the Agriculture, Forestry and Fishery Products Trade Investigation Committee.[51] In the subcommittee, Matsuoka played a pivotal role in the making of all aspects of rice policy and rice policy reform. In the trade investigation committee, Matsuoka was a key figure in formulating Japan's position in agricultural trade negotiations.

Through his committee executive posts, Matsuoka also earned membership of the *nôrin kanbu*, which gave him broad powers over all important agricultural and forestry policies. This made him a target for petitioning groups of all kinds across a range of policy areas. He regularly hosted groups of petitioners in his parliamentary office.

For example, in July 2005, Matsuoka received a delegation from Kumamoto Central Union of Agricultural Cooperatives, which made a number of policy requests. Following the visit, Matsuoka publicly committed himself to 'protecting Japanese agriculture for safe and anxiety-free food'.[52] In August of the same year he received a delegation from Kyushu forestry-related groups. They spoke to him about a budget proposal for the 2006 supplementary budget, which would provide compensation for the damage caused by heavy rain. Matsuoka agreed that he would tackle disaster restoration as an important issue.[53] In October, he received a number of representatives from agricultural groups. They wanted to present a number of requests relating to countermeasures for wheat and soybeans produced in 2006.[54]

Matsuoka's *giin gaikô*

As secretary-general of the trade investigation committee Matsuoka gained international notoriety as one of Japan's leading *nôrin zoku* through his conduct of *giin gaikô*. Because the LDP had to develop and hold its own position on agricultural trade policy matters independently of the government (meaning the MAFF) and the Koizumi administration, Matsuoka saw overseas delegations as part of his executive role in the committee. He claimed even to have spent his own money in exchanging opinions with a large number of countries concerned.[55]

Wherever he went, Matsuoka conducted a type of parallel diplomacy, designed not only to provide backing for the official Japanese government negotiating position but also to press WTO officials and representatives of foreign governments to his and the LDP's position on agricultural trade. This position was rabidly anti-free trade and pro-protection. Matsuoka was unrelenting in pushing his opposition to agricultural trade liberalisation at all points. According to one representative of a foreign trading power, 'he was not backward in saying how he wanted to run the world. He was absolutely committed to his [constituents'] cause. Discussions with him were conversations that went nowhere. He was like a travelling salesman who offered the same message all the time.'[56] His message ran along the following lines

> [a]gricultural production and food self-sufficiency are very important for Japan. This is especially true for cereals (meaning rice) because of the global shortfall. Cereal production can be expected to decline for every degree of global warming. The environment and food have become big problems. This makes the issue of 'bearers' of production in Japan a significant theme. It is imperative that we foster bearers in the current situation of the global retreat of Japanese agriculture. Urban dwellers have strong expectations of agriculture, and I've always backed the idea of urban dwellers' cultivating farm plots and their becoming 'quasi-farmers' (*jun nôgyôsha*) and for city people to have connections with agriculture and rural society. This is one way of resolving the problem of the shortage of cultivatable land. The cities and rural areas will become as one through agriculture. The multifunctionality of agriculture and rural areas and agricultural, mountainous and fishing villages are connected to the totality of people's lives and get the support and understanding of all people. We're being gradually pushed by foreign production, which must not be allowed to harm domestic production. If we consider the problem of food safety, because our country's agricultural products are superior in terms of safety, quality and taste, we should aim for exports. We are world leaders in product improvement and agricultural technology.[57]

In early February 2003 Matsuoka exchanged opinions with the Australian ambassador to Japan, John McCarthy, about the WTO negotiations. As Matsuoka put it

[t]he stance towards the WTO negotiations between Japan and Australia is very different. For Australia and the United States, food is just another traded commodity. However, for Japan and the great majority of European countries, consideration is given to the aspect of food as playing a role in land and environment conservation. Their opinions are different on the point whether a country should make farm products free trade products or not. I made sure that the Australian ambassador understood the policy of the Japanese side.[58]

In February 2003 Matsuoka also participated in an International Assembly of the Parliamentarians' Association for Agriculture and Fisheries (PAAF) in Seoul. The league brought together parliamentary members from 46 countries around the world. Matsuoka was serving as both chairman and deputy chairman of the association. The meeting was held because the WTO was entering the critical phase of establishing 'modalities' (agricultural trade liberalisation criteria)[59] in March. The second proposal for the 'modalities' was due to be announced in March 2003. Members attending the conference 'reconfirmed their intention to join together to appeal to the WTO to realise fair and justified trade rules enabling different agricultures to co-exist, and not to destroy agriculture in each country given that farming was based on differences in natural conditions and historical backgrounds in each country.'[60]

Earlier, in Bangkok, Matsuoka had attended the Asia Population Development Conference, where he had exchanged opinions with others about the problem of pressure on food security from advancing globalisation and future increases in global population, including in Asia. The conference ended with a resolution to consider the perspectives of population and environmental problems in WTO negotiations.[61]

Matsuoka's most assiduous courting was of WTO officials. He wore a path between Tokyo and Geneva in his endeavour to convince WTO officials of the need to protect Japanese agriculture. In the 2003 new round of agricultural trade negotiations, 'the Japanese with the best-known names in Geneva were Sakurai Shin [the chairman of the trade investigation committee] and Matsuoka Toshikatsu.'[62] Sakurai and Matsuoka became well known in the WTO as 'arrogant *nôrin zoku*'.[63] They were called 'Sakura to Matsu' (Cherry and Pine).[64]

In February 2003, Matsuoka conferred with WTO Director-General, Panitchpakdi Supachai, and Chairman of the Agriculture Special Sessions, Stuart Harbinson, when they visited Japan to participate in a WTO mini-ministerial-level conference in Tokyo. Matsuoka advanced claims for inserting considerations relating to the environment and population problems into the trade-negotiating framework. However, Supachai and others would only acknowledge that 'we are fully aware that each country's claim is different'.[66]

At one point during Supachai's visit to Japan, Sakurai, who was chairman of the trade investigation committee,[67] and Matsuoka, who was secretary-general, went for direct talks with Supachai. They gatecrashed the Imperial Hotel where he was staying, despite the fact that Supachai had refused an interview on the grounds that he was busy.[68]

In early March 2003, in the lead-up to the final WTO negotiations on the modalities, Matsuoka visited Geneva and Paris together with Sakurai as representatives of the 'Diet Members' Group to Support WTO Negotiations' (WTO Kôshô Shien Giindan) under the direct control of the LDP PARC chairman. Matsuoka claimed that he was appointed head of the delegation because of his 'connections with top-level executives of the WTO and the countries of the European Union and [his] achievements in Diet members' diplomacy over a long period especially in this field'.[69]

In Geneva Matsuoka and Sakurai once again held talks with WTO Director-General Supachai and exchanged opinions with the ambassadors of the United States, Australia and European Union. In the discussions with the ambassadors, whom Matsuoka described as representing 'only a minority of WTO member nations',[70] he argued that a negotiating framework 'that failed to consider the earth's environmental problem and the food problem of poor nations harmed the benefits of the countries in the world in the long run'.[71]

In the discussion with Supachai, Matsuoka and Sakurai issued a strong demand that the director-general take the interests of Japan and other countries that imported farm products into account in the negotiations.[72] Matsuoka pitched the debate in terms of a dispute about how to deal with conserving the earth's environment as the point at issue in agricultural trade negotiations between the European Union and Japan on the one side, and commercial food-exporting countries such as the United States and Australia on the other.[73] He asserted that trade in agricultural products 'cannot be liberalised in the same way as liberalising trade in industrial products'.[74] Farming had more than an economic function since it helped to protect the environment and prevent natural disasters.

In fact Sakurai and Matsuoka attracted a lot of negative press as a result of their visit and their tactics. They were described as 'storming' Geneva and arguing furiously with Supachai,[75] 'saying "we are putting ourselves out on a limb [by directly approaching you] on this one" and spouting statements such as "it is a crisis of the existence or death of the Japanese race"'.[76] Sakurai said, 'Japan should not compromise in the international task of liberalising trade in farm products because the survival of (the Japanese) people is at stake'.[77] Matsuoka made a similar comment, arguing that 'trade in agricultural products cannot be liberalised in the same way as liberalising trade in industrial products because of the multifaceted functions of agriculture in a nation's economy. In Japan, rice paddies play the same roles as dams in that both can prevent natural disasters'.[78] Supachai was shocked, saying 'I thought Japan was an advanced industrial country, but what comes here are just loud-mouthed agricultural and forestry tribe members'.[79] A Japanese popular weekly magazine commented

> [a]t the end of the day, the activities of the over-the-top agricultural and forestry 'tribe' members are, after all, no more than a political performance for the benefit of domestic farmers and agriculture-related groups. Sakura and Matsu have no power to influence foreign policy....They don't have the energy or force of the previous agricultural and forestry tribe members who could boast 300 rice Diet members.[80]

In fact, the director-general of a bureau in METI was quoted as saying, 'they think they are in charge behind the scenes, but in reality, they've become the laughing stock of the international negotiations.'[81]

On his return from the Geneva trip, Matsuoka attended a meeting of the trade investigation committee study team and reported back the results of the conference with the WTO director-general, the ambassadors of the European Union and others. With respect to the agricultural trade negotiation framework, Matsuoka and Sakurai affirmed the importance of even stronger cooperation with the countries of Europe in the agricultural trade negotiation process.[82]

In May 2003, Yatsu Yoshio (who was head of the trade investigation committee study team) and Matsuoka once again trekked to Europe, this time for WTO non-farm products (forest and fishery products) market negotiations. Their purpose was to convey their perspective on Japan's standpoint on market access negotiations for non-farm products. Matsuoka exchanged opinions with the Chairman of the WTO Non-Agricultural Market Access Negotiating Group, Ambassador Girard, who said he planned to promote debate on the issue for the benefit of all member nations including Japan.[83] Matsuoka made the usual points about the need for co-existence and co-prosperity, and a realistic settlement of non-agricultural market access issues.[84] After returning home, he and Yatsu reported back to the trade investigation committee.

When the WTO Doha Round reached its anti-climatic ministerial-level conference in Cancun, Mexico, in September 2003, Matsuoka was on hand, pushing the agricultural protectionist line. He gave a speech to the Inter-Parliamentary Union (IPU) meeting, which was being held at the same time. His speech was in English and, in it, he appealed for understanding of Japan's agricultural trade negotiating position as head of the Japanese joint Lower and Upper House Representatives' delegation. He reported that he received 'big praise from the participants of each country'.[85] When he returned to Japan he, along with MAFF Minister Kamei Yoshiyuki and chairman of the trade investigation committee (Sakurai) gave an account of the WTO meeting in Cancun to a meeting of the *sansha kaigi*.[86]

The intrepid trippers, Sakurai and Matsuoka, went back to Geneva in December 2003 to conduct further 'Diet members' diplomacy'. Once again they conferred with WTO Director-General Supachai and Chairman of the General Board of Directors, Carlos Pérez del Castillo, and others. Matsuoka reiterated the Japanese point of view, which, he claimed, was the same as some countries in Asia and the European Union.[87] Essentially, this was the standpoint that there should be a 'coexistence of diverse agricultures', which acknowledged the importance of the conservation of the earth's environment. Matsuoka demanded that 'future discussions at the WTO should not lean towards only the ideology of food exporting countries' side'.[88] On their return home, Matsuoka and Sakurai once again reported back to a meeting of the trade investigation committee.

In March 2004, under the theme 'Internationalisation and Japanese Agriculture', Sakurai and Matsuoka reported on the 'Diet members' diplomacy' they had been conducting to the Agricultural Basic Policy Subcommittee.[89] Matsuoka also gave a lecture on the results of the Diet members' diplomacy that they had been undertaking in relation to the WTO agricultural negotiations to the 'Young Diet Members Agricultural Policy Study Association' (Wakate Giin Nôsei Benkyôkai).[90]

In the same month, as leader of a Lower House delegation, Matsuoka attended the Steering Committee of the 'Parliamentary Conference on the WTO' held at the headquarters of the Inter-Parliamentary Union in Geneva. The purpose of the conference was to supervise the activities of the WTO. The gathering discussed how to strengthen its influence over international commercial issues starting with the WTO.

On the same trip, Matsuoka conferred with the chairman of the WTO Agriculture Negotiation Group, as well as the chairman of the Non-Farm Products Market Access Negotiation Group, and other WTO leaders. He made representations to the effect that export promotion measures such as export subsidies by industrialised nations should be immediately removed, and that Japan was unable to comply with any further liberalisation without consideration being given to forestry and fishery products.[91]

In June 2004, Matsuoka made three trips in a crescendo of agricultural trade diplomacy, two to the United States and one to South America on the WTO agricultural trade negotiations. Each time, Matsuoka was dispatched by the Trade Investigation Committee along with Chairman Sakura, and Acting Chairman Yatsu. In the United States they conferred with the chairman of the Senate Committee on Agriculture and the chairman of the House Agriculture Committee in order to emphasise the importance of rice paddy agriculture in Japan, particularly with respect to conserving land and the environment. In Matsuoka's view, the representatives of America's farm sector that they met gained a considerable understanding of the importance of rice paddies in Japan.[92] The meetings were held on the understanding that Matsuoka would report to the US Ambassador to Japan, Howard Baker, on his return to Tokyo. Matsuoka, Sakurai and Yatsu later held discussions with Ambassador Baker on the outline agreement of the WTO agriculture negotiations scheduled for July 2004.[93]

Matsuoka's visit to the countries of South America started in Brazil, which was one of the pivotal players in the G20 group (the major developing nations) within the WTO. His aim, once again, was to achieve wide recognition of Japan's negotiating position at the G5 ministerial-level conference of the WTO (consisting of five major nations and regions—including the United States, the European Union and Brazil). This was due to be held immediately before the presentation of the draft of the outline agreement by the WTO Committee on Agriculture Chairman, Tim Groser, scheduled for early July 2004.[94]

In Matsuoka's discussion with Brazil's foreign and agriculture ministers, he immediately identified Japan and Brazil's common interests at the WTO: the complete abolition of agricultural export subsidies, the drastic retrenchment of the domestic agricultural support policies of the United States and others, and the problems for developing nations such as Brazil in securing export quotas to developed nations. Matsuoka also underlined the importance of rice paddies in Japan. Both ministers agreed that the claims made by Japan were completely consistent with the claims made by Brazil. The Japanese and Brazilian sides confirmed that they would negotiate in cooperation. Both the Brazilian ministers said that they would act as goodwill ambassadors for Japan at the G5 in relation both to rice tariff problem and the environmental preservation functions of rice paddies.[95]

After Brazil, Matsuoka went on to Chile and Argentina where he met their foreign and agriculture ministers. Both countries supported Japan's claims at the WTO agriculture negotiations, and all promised to engage in mutual cooperation in order to resolve the WTO agricultural negotiation problem.[96]

In July 2004, Matsuoka was a member of a larger group of LDP Diet politicians who made the trip to Geneva, the LDP WTO Agriculture Negotiations Diet Members' Group (Jimintô WTO Nôgyô Kôshô Giindan). The group was dispatched in the lead-up to the announcement of the WTO General Council's Doha Agenda work program (the 'July package') containing, amongst other things, a 'Framework for Establishing Modalities in Agriculture'. The group included all the top guns of LDP agricultural policymaking. Its leader was Sakurai as chairman of the trade investigation committee. Others included Norota Hôsei (CAPIC chairman), the head of the Study Team (Futada Kôji), the chairman of the Agriculture and Forestry Division (Nakagawa Yoshio), and the vice-chairman of the PARC.[97]

In January 2005, Tim Groser visited Japan. He exchanged opinions on the WTO agricultural negotiations with Matsuoka and other executives of the Trade Investigation Committee at LDP headquarters. Groser had come to Japan with the establishment of the modalities for the next WTO Ministerial Conference in mind. Matsuoka and his colleagues strongly pressed Groser for modalities that took into account the co-existence of diverse agricultures and a balance of interests between agricultural importing and exporting countries. He was reminded by Matsuoka that the Japanese government and LDP treated important items such as rice in a separate framework. The LDP agricultural leaders appealed for sufficient guarantees for a number of sensitive items, and for the establishment of rules to lower tariffs that applied to sensitive items less than those that applied to general items.[98]

In April 2005, it was decided that Matsuoka would chair the group leading delegations to South America (Brazil and Argentina) on FTA issues. The following month he reported back to the committee on his trip to South America.

In June 2005, Matsuoka met again with the Australian ambassador, this time Murray McLean, who came to Matsuoka's office in the Diet building. As

Matsuoka recalls, he got the ambassador to understand the reality of Japanese agriculture. He emphasised that the mission of the LDP was to protect Japan's farm sector in world trade rules, and that he worked like a beaver as a representative of the LDP's mission in this respect.[99]

In July 2005, the executives of the Trade Investigation Committee, including Matsuoka (along with Sakurai and Yatsu) visited Geneva again. They conferred with Groser with a view to getting special treatment for sensitive products in the light of the scheduled issuing of the committee chairman's draft for negotiations. Matsuoka and his colleagues explained Japan's standpoint. They said that Japan could not expand the tariff quota for rice while consumption was declining and requested Groser's understanding of Japan's position. They reported that they had received sufficient acknowledgement from other countries as to the sensitive nature of rice for Japan, adding that there was no necessity for any improvement in market access for rice in order to conclude negotiations.[100]

Back home, Matsuoka (representing the LDP) attended a meeting with representatives of agricultural groups from foreign countries hosted by Zenchû. The representatives agreed that trade rules should be formulated that allowed the co-existence of diverse agricultures.

In September 2005, Matsuoka, as the representative of the Japanese Lower House, travelled to Geneva again, this time to attend a management committee meeting of the IPU's council relating to the WTO. Matsuoka delivered another speech in English in which he put Japan's position on agricultural trade, including the need for trade rules that prioritised environmental preservation under the title 'how agricultural trade rules should be'.[101] After Matsuoka's return to Japan, he attended the LDP's Trade Investigation Committee to give his report on the negotiations that he had conducted in Geneva, including with the head of the IPU's management committee, the chairman of the WTO's Committee on Agriculture and the assistant director-general of the WTO. In Matsuoka's report, he declared that he had told the WTO officials that market access, export subsidies and domestic support were the three areas for simultaneous resolution and that Japan had lowered its domestic support more than other countries (for example, Britain had increased its domestic support), and he reported that both officials showed understanding of what he had said.[102]

Following his successful re-election to the Lower House in September 2005, Matsuoka publicly committed himself to a position on the WTO agricultural trade negotiations. In his own words, this position would 'take the circumstances of our country's agriculture into consideration, and would assert what Japan needs to assert in international society and do our best for a harmonious settlement'.[103]

In October, Matsuoka once again flew to Geneva in order to attend the Five Interested Parties (FIPs—the United States, European Union, Australia, India and Brazil) meeting. In response to the proposal for uniform tariff reductions on agricultural products, he again reasserted Japan's position, which held that sensitive products such as rice should be excluded from tariff reductions. He pointed out that recent proposals from developing countries such as Africa were close to the Japanese position, but there remained a big gap between the United States and European Union proposals, and the question remained whether there would be agreement at the WTO Ministerial Conference held at the end of the year in Hong Kong.

Immediately prior to his departure for Geneva, Matsuoka attended the National Council's 'WTO Agricultural Negotiations Emergency Countermeasures, Basic Agricultural Policy Establishment National Representatives Gathering', and took on board the views expressed at that meeting. He attended the same gathering in the following month, and as secretary-general of the LDP's Trade Investigation Committee, he reported on his recent trade diplomacy in Geneva. Also, as chairman of the Agricultural Basic Policy Subcommittee, he explained the contents of the direct payments system. He noted that it was a problem in which all farmers were intensely interested. He made a full report on the WTO agricultural negotiations that he had undertaken in the party's trade investigation committee later that morning.[104]

With all this frenetic travelling and advocacy, Matsuoka certainly made sure that the LDP's hardline opposition to agricultural trade liberalisation was widely heard around the world. How much this contributed to actual negotiation outcomes is hard to gauge, but at least it advantaged him personally. He was able to demonstrate his commitment to farmers and to agricultural organisations in his constituency, which supported his re-election.

Promoting Japan as an agricultural exporter

Matsuoka has become a central figure in Japan's agricultural export offensive, a cause that he pursued as a form of disguised protectionism. When Matsuoka talked to two Japanese housewives in a highly publicised meeting in April

2004, he explained to them the importance of increasing the food self-sufficiency rate in the same breath as the LDP's plan to export Japanese food overseas.[105]

Matsuoka later embarked on a personal crusade for agricultural exports.[106] He obtained the all-out approval and consent for his initiative from Prime Minister Koizumi. In fact, in late 2004, the prime minister cited the anecdote of Japanese apples being sold for ¥1000 a piece in China.[107] Japanese rice has also become popular as a 'brand food' in the countries of East Asia including China.[108] Koizumi later held a meeting with apple growers, and unveiled a vision to increase exports of farm products to the ¥1 trillion a year level. He invited pioneer farm operators to the Kantei to exchange views. Learning that one of the attendees was running a farm underground in an office building in Ōtemachi, the central business district in Tokyo, the prime minister visited it in February 2005, praising the unique enterprise saying: 'Agriculture is a new industry. Agriculture has limitless possibilities'.[109] Koizumi was reported to be pinning his hopes on agriculture as the 'trump card' to turn around the construction-based rural employment structure, and to revamp his own image as someone who had 'turned a cold shoulder to rural areas'.[110]

In February 2005, during question time in the Lower House Budget Committee, Matsuoka made a pitch for the LDP, government and concerned groups to unite in an agricultural farm export offensive instead of staying in the usual defensive mode. He commented, 'it is time for Japanese agricultural products to shift from defence to offence'.[111] His view was that

> 'Made in Japan' agriculture, forestry and fishery products should be actively treated as export items since these products are fully competitive internationally. Like industrial goods, 'made in Japan' agriculture, forestry and fishery products are outstanding 'products', which have passed through the 'baptism' of the severely selective eyes of Japanese consumers. Although there is the criticism that these goods are 'high cost', the exceptional quality of some agricultural products such as apples and pears makes them increasingly in demand in some niche markets in Asia and elsewhere, where affluent consumers seek good-tasting delicacies.[112]

Matsuoka's activities in this area culminated in two delegations of politicians which he led to Beijing in January and June 2005 in order to make a pitch for exports of high quality Japanese rice to China, including some of the top Japanese brands of rice from Niigata and Kumamoto prefectures. Japan had still to obtain approval for rice exports to China. Even though selling rice to China would be like sending coals to Newcastle, Matsuoka argued that the kind of rice Japan wanted to export to China was top quality. Rice exports would, therefore, constitute a kind of 'niche' marketing. Matsuoka was quoted as saying, 'Japanese agricultural products should be the equivalent of the Mercedes-Benz or Rolls Royce of the automobile sector'.[113] Obviously he was making a pitch to the wealthy consumers of Asia. Moreover, if Japanese agriculture became an export industry, it would give the impression that farming in Japan was an internationally competitive industry worth promoting. It would go someway towards countering those who constantly criticised Japanese agriculture for being inefficient and lagging in productivity compared with other Japanese industries and with farm sectors in other developed countries—especially those that exported into the Japanese market. Matsuoka was very happy to see that his policy activities somehow contributed to 'the expediting of agricultural, forestry and fishery exports' being included as one of the major elements in the government's 2005 New Basic Plan, which provided an agricultural policy blueprint for the next 10 years.

In April 2005, the Japanese government launched a government-private sector council called the National Council for Promoting Exports of Agricultural and Marine Products. Its objective was to push exports of domestic farm products, aiming to double the value of agricultural exports over five years. Ideas included developing types of crops that catered to overseas markets. The council brought together representatives of the MAFF, METI, MoFA, local governments, agricultural cooperatives and food manufacturers. At the plenary session to launch the council, the prime minister noted, 'Japanese agricultural products, which are expensive but tasty, are fully exportable. Agriculture is a promising industry'.[114] He urged the farm sector to switch from its defensive approach to an aggressive one.[115]

In June 2005, Matsuoka received a delegation of agricultural groups from Kumamoto in his Diet office. They sought the implementation of fruit tree countermeasures. Matsuoka acknowledged that in Kumamoto, fruit tree agriculture was a key industry. He agreed that it was important to adopt a policy that would reward farmers' efforts. His way of approaching the issue, however, was to 'expand "agricultural policy on the offensive" (*seme no nôsei*) through farm product exports as well as expanding the demand for domestic fruit.'[116]

Following his successful re-election in 2005, Matsuoka once again took up the 'agriculture as an export industry' theme. He declared that Japan must 'put into effect an "agricultural policy on the offensive" in order to establish Japan's high-quality agricultural products as a large-scale export industry.'[117]

POLICY INTERFERENCE

Whereas intervening in policymaking via the PARC's agricultural and forestry committees largely facilitated Matsuoka's representation of sectional interests, his interference in the administrative decisions of the bureaucracy enabled him to guide benefits to specific localities and to obtain favours for particular clients. It was in the area of public works and associated construction contracts that the relationship between Matsuoka and the MAFF was observed to be the closest and even one of 'adhesion' (*yuchaku*).[118] The direction of mediation flowed from companies to politicians and then to government officials,[119] prompting some commentators to ask whether Matsuoka was a Diet representative or a political broker.[120]

In order for Matsuoka to undertake activities as a political broker, it was 'indispensable to have a "fat pipe" that controlled government offices'.[121] One political reporter described how Matsuoka interfered in the MAFF's administrative affairs

> Even in relation to small matters, his secretary always rings up the MAFF division in charge. Matsuoka then invites the divisional director or bureau director to an expensive Japanese-style restaurant. If they don't do what he says, he browbeats them. But if they are obedient, he suddenly changes and tames them with food and drink. In the fiscal 1999 supplementary budget, for example, Matsuoka made them allocate almost 20 per cent of the structural improvement works budget for the whole country to Kumamoto Prefecture, saying 'allocate ¥10 billion from the agricultural, forestry and fisheries budget to Kumamoto Prefecture. It's serious because hothouses were destroyed by a typhoon.'[122]

As a *zoku*, Matsuoka's powers to undertake policy interference were considerably enhanced because the *nôrin zoku* were the most influential actors in steering MAFF-drafted policies and bills through the PARC's committee process. They were the kingpins on whom MAFF officials were most dependent and to whom they were the most indebted. *Zoku* status bestowed much greater access to bureaucrats and thus recognition by those seeking favours that Matsuoka was a key person to approach. As a *zoku*, Matsuoka wielded unparalleled influence in mediation activities, and hence acquired the ability to collect the most money in exchange for favours.

Generally speaking, policy intervention and policy interference led to very different political and policy behaviours. In pursuit of sectional interests, Matsuoka conducted public lobbying activities, in party committees, in Diet members' leagues, *vis-à-vis* bureaucrats, and *vis-à-vis* party executives and the government leadership. In contrast, in pursuit of concessions and favours for himself and his clients, Matsuoka conducted private lobbying or petitioning activities mainly vis-à-vis bureaucrats. Policy intervention was overt and even propagandised (Matsuoka acted as a policymaker in formal policymaking contexts and claimed public credit for what he did), while policy interference was generally covert and lacking in transparency (Matsuoka acted as a broker or mediator in the pursuit of benefits for certain localities or favours for individuals, be they company executives, group leaders, local politicians, friends, or relations or whatever). When locals, including businessmen, came to Tokyo to petition for favours, they needed an agent or 'broker' who could act for them, someone who could intercede with bureaucrats. Matsuoka was the person who got things done for them. This meant interceding with the MAFF in areas of the ministry's allocatory or regulatory discretion; it was a part of Matsuoka's activity that did not usually see the light of day, and only recently became subject to attempted government regulation.[123]

The more public policy interference conducted by Matsuoka largely involved his leading delegations from his regional area to administrative offices in Tokyo, and his claiming of credit for the delivery of public works to his electorate. In December 2003, Matsuoka accompanied a delegation from the Central Kyushu Regional High Standard Road Promotion Association (Naka Kyûshû- Chiiki Kôkikaku Dôrô Sokushin Kiseikai)—chaired by the mayor of Aso Town—on a visit to MLIT. Their purpose was to request that the ministry construct the highway cutting across the centre of Kyushu. The association consisted of the municipalities in Kumamoto and Oita prefectures along the road. The delegation met the MLIT administrative vice-minister, chief engineer, technical officer, director-general of the Road Bureau and others individually. While expressing understanding of the fact that conditions for public works were severe, as the 'voice' of the people living in the area, the delegation told the MLIT officials that they looked forward to the early construction of the road and requested the cooperation of the ministry.[124]

In a similar episode in the following year, Matsuoka received a delegation from Nishihara Village assembly. Nishihara Village was in Aso County, and its assembly members continued to think of Matsuoka as their political representative, even though he had lost the seat of Kumamoto (3) in the 2003 election and switched to representing the Kyushu regional bloc.[125] It was

Matsuoka's reputation for bringing public works projects back to his local constituency that drew the delegation to Matsuoka's office. They demanded the provision of subsidies to construct a gymnasium in Nishihara Municipal Junior High School as well as prefectural road 206. Matsuoka commented on his website that

> [t]he provision of regional social capital is the hope of residents in that town. Although realisation of the demand is doubtful under the Koizumi administration's 'uniform budgetary cutback for all ministries and agencies', the construction of the gymnasium and prefectural road 206 are essential projects, which all residents of Nishihara Village are hoping for and want to realise at any cost. I promised to do all I could to realise this demand as soon as possible.[126]

Matsuoka's record of public works achievements reveals many egregious examples of the fruits of his policy interference. In Matsuoka's own constituency, in addition to projects funded by the UR countermeasures package, there were other equally infamous cases.

> In Soyô Town located in the foothills of Mount Aso in Aso County, the town administration spent ¥13 billion (of which ¥5.7 billion came from central government subsidies) on rebuilding a primary school that only had 67 students. This was nothing but a project that aimed to generate profit for particular construction companies, and had nothing to do with raising the standard of education in the town at all. It was a representative example of the construction politics (*doken seiji*) that thrived not just in Kumamoto (3), but in the whole of Japan.[127]

In the same town, Matsuoka supported successive mayoral elections of Gotô Keiki, one of his affiliated local politicians, by helping to secure subsidies for the construction of the new Soyo Town Hall in 1999 and the rest and recreation facility called Soyô Kaze (Light Breeze) Park, managed by the Soyô Kaze Yûgaku (Study Away Association), a limited liability company (*yûgen gaisha*) that was financed totally by Soyo Town.[128] Both Matsuoka and his faction boss, Kamei, were present at the lavish opening ceremony of the new Soyo Town Hall 'displaying the closeness of the incumbent mayor to two prominent members of the Lower House'.[129] Until the 2003 Lower House election when Matsuoka's support rate in Soyo Town fell to 51 per cent, he obtained consistently high support rates in this town (77 per cent in the 2000 elections, as shown in Table 3.2).

In addition, Matsuoka has also been an avid promoter of a major dam project, the Kawabe dam in Sagara Village in Kumamoto Prefecture, and he proudly claims to be quite important as a leader of the movement to promote the Isahaya Bay project.[130] His involvement with the construction industry has been described as so close that 'he appears to hold two posts: as a *nôrin zoku* and as a member of the construction "tribe" (*kensetsu zoku*).'[131] His reputation for promoting public works is such that '[e]ven many MAFF people and other *nôrin zoku* frown on the comments and conduct of Matsuoka, with the comment, "although Matsuoka talks in grandiose terms, in fact he spends money recklessly (*baramaki*) on farmers and on companies purely for his own election"'.[132]

There are, however, two sides to Matsuoka's policy interference. On the positive side, he has been an active sponsor of particular projects in particular districts. On the negative side, he has used his influence to prevent particular projects in particular districts from going ahead. The need for a plurality has put a premium on his exerting both kinds of influence. As Arai Satoshi, MAFF OB and former DPJ member of the Lower House for Hokkaido (3) in 1993-2005 explains

> [t]he pressure from the *zoku giin* began to intensify when I was an assistant divisional director in the MAFF in the early 1980s. When my seniors were in the same position in the MAFF, politicians did not stick their noses into the details of subsidised public works projects, but they began interfering even at the town and village level. Looking at this local scene, I began to think, 'if we leave this as it is, the Japanese political administration will start to destroy Japan itself'. Therefore, I try as much as possible now not to involve myself in MAFF-related areas even after I became a Diet member. The pressure from the *zoku* became even more intense under the SMD system. Under the MMD system, Diet members restrained each other, and it did not become very unfair. Now in order to get one's opponents to lose, candidates have begun to say things like 'do not conduct a project in that town' and 'don't use companies that support his opponent'.[133]

Matsuoka became expert at blocking budget allocations to those who opposed him by ordering MAFF officials to stop subsidies to his opponents. His 'arrogance was such that he pressured relevant places so that the leaders of opposing factions—his rivals for election in Kumamoto (3)—didn't get any subsidies.'[134] In this way, he could also reward or punish counties and towns according to the votes he received from them. For example, Matsuoka requested the MAFF to stop public works and providing budgetary funds to municipalities where he obtained lower votes or where people had expressed opposition to him. A blacklist was circulated in the MAFF's Structural Improvement Bureau of those municipalities where Matsuoka instructed the ministry officials to terminate public works projects. The existence of such a list was revealed by a MAFF official working in the bureau. He disclosed that 'shortly after he became an assistant divisional director at the Structural Improvement Bureau in 1998, his immediate superior casually instructed him along the following lines, "[t]here are the names of towns and villages Matsuoka *sensei* told us. Do you know the names?'"[135] Apparently, the relevant MAFF officials had earlier visited Matsuoka's office in the Diet members building in Nagata-chô. They had taken along the list of places where various subsidised projects were going to be undertaken in the municipalities of Kumamoto (3). The list of projects was part of the 1998 MAFF budget and they explained the list to Matsuoka. The list showed which municipality was going to receive what subsidies (i.e. the geographic distribution of subsidies).[136] Matsuoka then demanded that projects in numerous municipalities on the list not be executed. The officials made up a list of the blacklisted municipalities and distributed it to the relevant posts in the MAFF.[137] The reason why Matsuoka blacklisted particular localities was not clear to the MAFF officials. They assumed that one reason could be that Matsuoka's voting rate in those particular municipalities was low compared with other municipalities and that the head of the municipality had antagonised Matsuoka.[138]

One project in Ichinomiya Town, Aso County, was on the blacklist. The MAFF official in question called up the Kyushu Agricultural Administration Bureau and the Kumamoto Prefecture Agricultural Administration Department telling them the details of Matsuoka's demand and asking them to give up the project.[139] When asked to confirm or deny the existence of the blacklist, Matsuoka stated, '[e]ven if the sun rises from the west, there is no such case'.[140] A similar denial was issued by the administrative vice-minister of the MAFF at the time, Takagi Yûki.

The circumstances of the abandoned public works project in Ichinomiya Town, however, were revealed by other sources. According to Nôminren

> [a] local newspaper wrote that a direct sales facility for farm products was about to be created as a town project in Ichinomiya Town in Aso County. Because the town mayor did not pay his compliments to Mr Matsuoka for the project, Mr Matsuoka pressured the prefecture and Ministry of Agriculture, Forestry and Fisheries to trash the budget for the project.[141]

The project would have been funded from the MAFF's structural improvement budget as a mountain village promotion project. Prefectural officials reportedly pressured town assembly members who supported mayor Ichihara Norita (who was anti-Matsuoka), to persuade the mayor to pay his respects to Matsuoka.[142] As Ichihara himself explains

> [i]n 1992, Matsuoka asked me to join his *kôenkai*. Since I was under the good offices of another Diet representative, I declined the request. In the mayoral election that year, I was defeated by an opponent supported by Matsuoka. After this opponent became the mayor of the town, he

was diligent in constructing large-scale facilities in the town. However, in 1998, the mayor was arrested for bribery in relation to a meal-providing centre, and I took up my old position. Probably early in April, a government official of Kumamoto Prefecture Aso Office began saying 'would you please call on Mr Matsuoka to pay your respects?' I asked 'for what reason?', then the official said 'because otherwise we cannot obtain approval for mountain village development works'....I thought 'nonsense' and did not visit Matsuoka. Then, the budget was actually stopped. During the time of the Matsuoka faction mayor, the budget went ahead normally.¹⁴³

Matsuoka apparently demanded that the Structural Improvement Bureau 'reserve its resources' and gave the Ichinomiya Town project as an example. In response, the Kyushu Agricultural Administration Bureau reserved the budget for the Ichinomiya Town project in a private notification to the prefectural budget.¹⁴⁴

The whole episode 'provides a glimpse into Matsuoka's attitude as a Diet member: he treats warmly those who ingratiate themselves with him, but he takes away the livelihood of those people who disobey him completely.'¹⁴⁵ Matsuoka applied pressure using the subsidised project as bait and government officials as mere pawns, although in a series of interviews, Matsuoka claimed that 'he had not done such a thing at all.'¹⁴⁶

The fact that MAFF officials acted as Matsuoka's agents in not allocating the budget for particular projects in particular municipalities reveals his powers of policy interference. Matsuoka's personal intervention became the guideline that MAFF administrators followed.¹⁴⁷ Such interference presented 'a clear case of "privatisation" (*shibutsuka*) and "monopolisation" (*rôdan*) of bureaucratic administration by *zoku giin*'.¹⁴⁸ Even though he was a *zoku*, Matsuoka was still a backbencher, and it is clear that he had free and direct access to bureaucrats in the ministry and could make demands on them. This constituted an unusually direct line of contact between individual backbenchers and individual bureaucrats, which is normally outlawed in parliamentary cabinet systems where politicians deal with the bureaucracy only through the relevant ministers.

NOTES

1 Nakanishi Akihiko and Special Reporting Group, 'Suzuki Muneo, Matsuoka Toshikatsu', p. 105.
2 See Chapter 4 on 'Exercising Power as a *Nôrin Giin*'.
3 Itô Hirotoshi, 'Matsuoka Toshikatsu Daigishi no "Maboroshi no Hon" to Nôsuishô Baiomasu Jigyô to no Fushigi na Kankei' ['Matsuoka Toshikatsu Diet Member's "Phantom Book" and Its Strange Connection to the MAFF's Biomass Business'], *Zaikai Tenbô*, January 2003, p. 53.
4 Itô, 'Heisei Jiken Fuairu: Nôrin Jigyô Hojokin o Dokusen Suru Matsuoka Toshikatsu', p. 65.
5 '"Nôrin Giin" mo Kôkeisha Fusoku?' ['"Agriculture and Forestry Diet Members" Also Lack

Successors?'], *Nôsei Undô Jyânaru*, No. 30, April 2000, p. 1.
6   Personal interview, MOF official, January 2003.
7   Nakanishi, 'Matsuoka Toshikatsu', p. 28.
8   Hasegawa, 'Kanjûdanomi no Hazama de Shundô', p. 24. See also below and Chapter 7 on 'Electoral Vicissitudes'.
9   Nakanishi, 'Matsuoka Toshikatsu', p. 28.
10 *The Japan Times*, 24 May 2005.
11 Nakanishi, 'Matsuoka Toshikatsu', p. 28.
12 *Mainichi Shinbun*, 17 May 1997.
13 Visit: http://piza.2ch.net/giin/kako/987/987905181.html
14 *ibid.*
15 Itô, 'Matsuoka Toshikatsu Daigishi', p. 53.
16 Nakanishi and Special Reporting Group, 'Suzuki Muneo, Matsuoka Toshikatsu', p. 94.
17 'Za Sankuchuari', p. 59.
18 Nakanishi and Special Reporting Group, 'Suzuki Muneo, Matsuoka Toshikatsu', p. 100; Nakanishi and Journal Reporter Group, 'Matsuoka Toshikatsu to Iu Giwaku Nin', p. 179.
19 Nakanishi and Journal Reporter Group, 'Matsuoka Toshikatsu to Iu Giwaku Nin', p. 179.
20 '"Nishi no Muneo"', p. 38.
21 Nakanishi and Journal Reporter Group, *op.cit.*, p. 179.
22 'Matsuoka Toshikatsu Daigishi', p. 53.
23 Itô, 'Heisei Jiken Fuairu: Nôrin Jigyô Hojokin o Dokusen Suru Matsuoka Toshikatsu', p. 66.
24 '"Muneo no Bôrei"', p. 28.
25 Ayukawa, 'Jimintô de mo Shinkô suru "Matsuoka Hazushi"', p. 20. See also Chapter 6 on 'The Identical Twins of Nagata-chô'.
26 Itô, 'Heisei Jiken Fuairu: Nôrin Jigyô Hojokin o Dokusen Suru Matsuoka Toshikatsu', p. 64.
27 Nakanishi, 'Matsuoka Toshikatsu', p. 28.
28 This became the Kamei faction after Etô retired from politics at the time of the 2003 Lower House election. Kamei himself left the LDP in August 2005, when he failed to secure LDP endorsement in the Lower House election. He became a member of the Kokumin Shintô (People's New Party) and now serves as its acting head. See also Chapter 7 on 'Electoral Vicissitudes'.
29 Nakanishi, 'Matsuoka Toshikatsu', p. 28.
30 *ibid.*
31 The Kôchikai was an LDP faction originally founded by Ikeda Hayato, and was subsequently led by Ôhira Masayoshi, Suzuki Zenkô and Miyazawa Kiichi, all four of whom served as prime minister. Its leadership was then passed on to Katô Kôichi, followed by Ozato Sadatoshi and then Horiuchi Mitsuo.
32 The Etô-Kamei faction's position as the *nôrin zoku*-dominant faction was attributed to the electoral demise of some prominent agriculture and forestry 'tribe' Diet members from other factions in the 2000 elections as well as the previous departure from the LDP of leading agricultural policy experts, who had experience of being MAFF Minister such as Hata Tsutomu, Kanô Michihiko and Tanabu Masami. These developments reportedly gave rise to a dearth of human resources amongst the LDP's *nôrin zoku*, with the result that the Etô-Kamei faction came to the fore. 'Za Sankuchuari', p. 60.
33 'Kaibunsho ga Tobikau Inshitsusa Nûsui "Jinjii Kûsû" no Uchimaku' ['The Insidiousness of Mysterious Documents Flying About: Inside Information on MAFF "Human Resource Battles"'] *Shûkan Daiyamondo*, 20 April 2002, p. 56.
34 Nakanishi and Journal Reporter Group, 'Matsuoka Toshikatsu to Iu Giwaku Nin', p. 183.
35 The career class is made up of those officials, like Matsuoka, who have passed the Level 1 entrance exam for the public service, while the non-career class are those who have passed the Level II and III exams. They are known as middle-ranking (*chûkyûshoku*) officials.
36 Nakanishi and Journal Reporter Group, 'Matsuoka Toshikatsu to Iu Giwaku Nin', p. 183.
37 *ibid.*, pp. 183-184.
38 Hasegawa, 'Nôsuishô o Haishi seyo', p. 37.

EXERCISING POWER AS A *NÔRIN ZOKU*   145

39 Nakanishi and Journal Reporter Group, *op.cit.*, p. 183.
40 *ibid.*, pp. 183-184.
41 *ibid.*, p. 183.
42 It is also reported that he was sent to the Kyushu Agricultural Administration Bureau. Hasegawa, 'Nôsuishô o Haishi seyo', p. 37.
43 Hasegawa, 'Nôsuishô o Haishi seyo', p. 37.
44 Nakanishi and Journal Reporter Group, 'Matsuoka Toshikatsu to Iu Giwaku Nin', p. 183.
45 Hasegawa, 'Nôsuishô o Haishi seyo', p. 37.
46 Ayukawa, 'Jimintô', p. 20.
47 Personal comment, Ministry of Foreign Affairs official, March 2005.
48 'Kinkyû Nyûin shita', p. 28.
49 *ibid.*
50 '"Muneo no Bôrei"', p. 28.
51 See Chapter 4 on 'Exercising Power as a *Nôrin Giin*'.
52 Matsuoka Toshikatsu Official Site, 'Hisaichi no Genchi Chôsa ni Hairimasu' ['Participating in On-The-Spot Investigation of a Disaster Area'], in *Katsudô Hôkoku* [*Activity Report*]. Available from http://matsuokatoshikatsu.org/index1.html
53 Matsuoka Toshikatsu Official Site, 'Gomeifuku o O'inorimasu' ['I Pray for his Happiness in the Next World'], in *Katsudô Hôkoku* [*Activity Report*]. Available from http://matsuokatoshikatsu.org/index1.html
54 Matsuoka Toshikatsu Official Site, 'Yûsei Mineika Tokubetsu Iinkai no Shidô' ['The Special Committee on Postal Privatisation Starts'], in *Katsudô Hôkoku* [*Activity Report*]. Available from http://matsuokatoshikatsu.org/index1.html
55 Matsuoka Toshikatsu Official Site, 'Tô Nôrinsuisanbutsu Bôeki Chôsakai de WTO Kôshô no Jôkyô Kaiseki' ['Situational Analysis of WTO Negotiations at the Party's Agriculture, Forestry and Fishery Products Trade Investigation Committee'], in *Katsudô Hôkoku* [*Activity Report*]. Available from http://matsuokatoshikatsu.org/index1.html
56 Personal communication, Department of Agriculture, Fisheries and Forestry (Australia) official, June 2005.
57 This is a summary of Matsuoka's views revealed in an interview and reported on *Rensai Kikaku* [*Serial Project*]. Available from http://www.nca.or.jp/shinbun/20040213/nouiin040213_2_rensai.html
58 Matsuoka Toshikatsu Official Site, 'WTO Kôshô ni tsuite Chûnichi Ôsutoraria Taishi to Iken Kôkan' ['Exchanging Opinions with the Australian Ambassador to Japan About the WTO Negotiations'], in *Katsudô Hôkoku* [*Activity Report*]. Available from http://matsuokatoshikatsu.org/site002//public/003.html
59 The modalities are a rough negotiating framework showing in what areas agriculture should be liberalised, but largely without specific numbers. In Matsuoka's own words, 'the modalities are applied to all WTO member nations and regions, and decide the numerical value of the reduction in domestic protection such as lowering tariffs and domestic subsidies by what extent over what years.' Matsuoka Toshikatsu Official Site, 'Gurôsâ WTO Nôgyô Iinkai Gichô to Kaidan' ['A Talk with Groser WTO Committee on Agriculture Chairman'], in *Katsudô Hôkoku* [*Activity Report*]. Available from http://www.matsuokatoshikatsu.org/site002//public/059.html
60 Matsuoka Toshikatsu Official Site, 'Nikkanchû Kokusai Nôgyô Kaigi o Kaisai' ['Holding a Japan-Korea-China International Agricultural Conference'], in *Katsudô Hôkoku* [*Activity Report*]. Available from http://matsuokatoshikatsu.org/site002//public/003.html
61 Matsuoka Toshikatsu Official Site, 'Ajia Jinkô Kaikatsu Kaigi ni Sanka (Bangkok)' ['Participating in the Asia Population Development Conference (Bangkok)'], in *Katsudô Hôkoku* [*Activity Report*]. Available from http://matsuokatoshikatsu.org/site002//public/003.html
62 'Za Sankuchuari', p. 59.
63 *ibid.*
64 *ibid.*, p. 58.
65 *ibid.*, p. 59.

66 Matsuoka Toshikatsu Official Site, 'Supachai WTO Jimukyokuchô to Mendan' ['Talking Personally with WTO, Director-General Supachai'], in *Katsudô Hôkoku* [*Activity Report*]. Available from http://matsuokatoshikatsu.org/site002//public/003.html
67 Sakurai was also from the Etô-Kamei faction. He lost his Lower House seat in the 2000 elections because of a scandal, but cleared himself of disgrace by getting back into the Diet in the House of Councillors in 2001 as a member for the PR (national) constituency.
68 'Za Sankuchuari', p. 59.
69 Matsuoka Toshikatsu Official Site, 'Ôzume no WTO Kôshô e Shûgiin yori Daihyôdan Haken' ['Dispatch of the Delegation from the House of Representatives for the Final Phase of the WTO Negotiations'], in *Katsudô Hôkoku* [*Activity Report*]. Available from http://matsuokatoshikatsu.org/site002//public/041.html
70 Matsuoka Toshikatsu Official Site, 'Jyunêbu Hômon no Seika o Hôkoku' ['Reporting the Results of the Geneva Visit'], in *Katsudô Hôkoku* [*Activity Report*]. Available from http://matsuokatoshikatsu.org/site002//public/041.html
71 'Jyunêbu Hômon'. Available from http://matsuokatoshikatsu.org/site002//public/041.html
72 Matsuoka Toshikatsu Official Site, 'WTO Kôshô Nikkan Daihyô Giin Jyunêbu e' ['Japan and South Korea Assembly Delegation Group for WTO Negotiations Go to Geneva'] in *Katsudô Hôkoku* [*Activity Report*]. Available from http://matsuokatoshikatsu.org/site002//public/003.html
73 'WTO Kôshô Nikkan Daihyô Giin Jyunêbu'. Available from http://matsuokatoshikatsu.org/site002//public/003.html
74 *The Japan Times*, 5 March 2003.
75 'Za Sankuchuari', p. 59.
76 *ibid*.
77 'LDP Claims Survival of Japanese at Stake in WTO Farm Talks', *Kyodo News*. Available from http://www.japantoday.com/e/?content=news&cat=1&id=251497
78 'LDP Claims Survival of Japanese at Stake'. Available from http://www.japantoday.com/e/?content=news&cat=1&id=251497
79 'Za Sankuchuari', p. 59.
80 *ibid*.
81 *ibid*.
82 'Jyunêbu Hômon no Seika o Hôkoku'. Available from http://matsuokatoshikatsu.org/site002//public/041.html
83 Matsuoka Toshikatsu Official Site, 'WTO Hi Nôsanhin Shijô no tame Saido Hôô' ['Revisiting Europe for WTO Non-Farm Products Market Negotiations'], in *Katsudô Hôkoku* [*Activity Report*]. Available from http://matsuokatoshikatsu.org/site002//public/041.html
84 'Yatsu Yoshio Shûgiin Giin, Matsuoka Toshikatsu Shûgiin Giin to Jirâru Gichô to no Kaidan no Kekka Gaiyô' ['The Summary of Conference Result among House of Representative Member Yatsu Yoshio, House of Representative Member Matsuoka Toshikatsu and Chair Girard']. Available from http://www.rinya.maff.go.jp/kouhousitu/wto/files/0305ym.htm
85 Matsuoka Toshikatsu Official Site, 'IPU (Rekkoku Gikai Dômei) de Shûsan Daihyô toshite Supîchi' ['Speech as the Representative of the House of Representatives and House of Councillors at the IPU (Inter-Parliamentary Union)'], in *Katsudô Hôkoku* [*Activity Report*]. Available from http://matsuokatoshikatsu.org/site002//public/041.html
86 'Nôgyô Kankei Seisaku Kettei no Ashidori', *Nôsei Undô Jyânaru*, No. 52, December 2003, p. 31.
87 Matsuoka Toshikatsu Official Site, 'WTO Jimukyokuchô ra to Kaidan' ['Talks with the WTO Director-General and Others'], in *Katsudô Hôkoku* [*Activity Report*]. Available from http://matsuokatoshikatsu.org/site002//public/048.html
88 'WTO Jimukyokuchô ra to Kaidan'. Available from http://matsuokatoshikatsu.org/site002//public/048.html
89 'Nôgyô Kankei Seisaku Kettei no Ashidori', *Nôsei Undô Jyânaru*, No. 54, April 2004, p. 29.
90 *ibid*.

EXERCISING POWER AS A *NÔRIN ZOKU*   147

91 Matsuoka Toshikatsu Official Site, '"WTO ni kansuru Giin Kaigi" Unei Iinkai e Nihon kara Shûgiin Daihyôdan o Haken' ['The Dispatch of the House of Representatives' Delegation from Japan to the Steering Committee of the "Parliamentary Conference on the WTO"'], in *Katsudô Hôkoku* [*Activity Report*]. Available from http://matsuokatoshikatsu.org/site002//public/053.html
92 Matsuoka Toshikatsu Official Site, 'WTO Nôgyô Kôshô ni mukete Tô Hakken ni yoru Giin Gaikô o Tenkai' ['Developing Diet Members' Diplomacy Through Dispatch by the Party for the WTO Agriculture Negotiations'], in *Katsudô Hôkoku* [*Activity Report*]. Available from http://matsuokatoshikatsu.org/site002//public/054.html
93 Matsuoka Toshikatsu Official Site, 'Bêkâ Bei Chûnichi Taishi to Kaidan' ['Conversation with the U.S. Ambassador to Japan, Ambassador Baker'], in *Katsudô Hôkoku* [*Activity Report*]. Available from http://matsuokatoshikatsu.org/site002//public/054.html
94 Matsuoka Toshikatsu Official Site, 'WTO Kôshô e muke Nanbei Shokoku e Giin Gaikô Tenkai' ['Development of Diet Members' Diplomacy to the Countries of South America for the WTO Negotiations'], in *Katsudô Hôkoku* [*Activity Report*]. Available from http://matsuokatoshikatsu.org/site002//public/054.html
95 Matsuoka Toshikatsu Official Site, 'WTO Kôshô, Nanbei Shokoku Giin Gaikô Gaiyô' ['Summary of Diet Members' Diplomacy to the Countries of South America for the WTO Negotiations'], in *Katsudô Hôkoku* [*Activity Report*]. Available from http://matsuokatoshikatsu.org/site002//public/056.html
96 ibid.
97 'Nôgyô Kankei Seisaku Kettei no Ashidori', *Nôsei Undô Jyânaru*, No. 57, October 2004, p. 31.
98 'Gurôsâ WTO Nôgyô Iinkai Gichô to Kaidan'. Available from http://www.matsuokatoshikatsu.org/site002//public/059.html
99 Matsuoka Toshikatsu Official Site, 'Chûnichi Ôsutoraria Taishi Raisho' ['The Australian Ambassador to Japan Comes to the Office'], in *Katsudô Hôkoku* [*Activity Report*]. Available from http://matsuokatoshikatsu.org/index1.html
100 'Nôgyô Kankei Seisaku Kettei no Ashidori', *Nôsei Undô Jyânaru*, No. 63, October 2005, p. 31.
101 Matsuoka Toshikatsu Official Site, 'Tokubetsu Kokkai Kaikai kara Jyunêbu e' ['From the Opening Session of a Special Diet to Geneva'], in *Katsudô Hôkoku* [*Activity Report*]. Available from http://matsuokatoshikatsu.org/index1.html
102 Matsuoka Toshikatsu Official Site, 'Jyunêbu de WTO Giin Gaikô no Hôkoku' ['Report of WTO Diet Member's Diplomacy in Geneva'], in *Katsudô Hôkoku* [*Activity Report*]. Available from http://matsuokatoshikatsu.org/index1.html
103 Matsuoka Toshikatsu Official Site, 'Matsuoka Toshikatsu Daigishi kara Minasama e' ['To Everyone from Matsuoka Toshikatsu Diet Member']. Available from http://www.matsuokatoshikatsu.org/site003//public/077.html
104 Matsuoka Toshikatsu Official Site, 'WTO mo Chokusetsu Shiharai mo Ôzume' ['The Final Wrap Up of Both the WTO and the Direct Payments'], in *Katsudô Hôkoku* [*Activity Report*]. Available from http://matsuokatoshikatsu.org/index1.html
105 *Deirii Jimin* [*Daily LDP*], available from http://www.jimin.jp/jimin/daily/04_04/21/160421b.shtml
106 Matsuoka Toshikatsu Official Site, 'Oishikute Anshin na Nippon no Nôrinsuisanbutsu o Kaigai e!' ['Delicious and Quality Assured Japanese Agriculture, Forestry and Fisheries Products for the Overseas Market!'], in *Katsudô Hôkoku* [*Activity Report*]. Available from http://www.matsuokatoshikatsu.org/site002//public/053.html
107 *Tokyo Shinbun*, 9 March 2005.
108 'Oishikute Anshin na Nippon no Nôrinsuisanbutsu o Kaigai e!'. Available from http://www.matsuokatoshikatsu.org/site002//public/053.html
109 *Tokyo Shinbun*, 9 March 2005.
110 ibid.
111 'Oishikute Anshin na Nippon no Nôrinsuisanbutsu o Kaigai e!'. Available from http://www.matsuokatoshikatsu.org/site002//public/053.html

112 *Nikkei Weekly*, 29 August 2005.
113 *The Japan Times*, 21 January 2005.
114 *Nihon Keizai Shinbun*, 28 April 2005.
115 *ibid*.
116 Matsuoka Toshikatsu Official Site, 'Yushutsu Sokushin de Juyô Kakudai' ['Expanding Demand by Promoting Exports'], in *Katsudô Hôkoku* [*Activity Report*]. Available from http://matsuokatoshikatsu.org/index1.html
117 'Matsuoka Toshikatsu Daigishi kara Minasama e'. Available from http://www.matsuokatoshikatsu.org/site003//public/077.html
118 Hasegawa, 'Nôsuishô o Haishi seyo', p. 35.
119 Nakanishi and Journal Reporter Group, 'Matsuoka Toshikatsu to Iu Giwaku Nin', p. 178.
120 *ibid*.
121 *ibid*., p. 183.
122 Nakanishi and Special Reporting Group, 'Suzuki Muneo, Matsuoka Toshikatsu', p. 103. The same source revealed information gained from the Japan Communist Party newspaper (*Akahata*), published on 4 January 2000 to the effect that Kumamoto Prefecture was allocated ¥9.3 billion or 17 per cent of the fiscal 1999 supplementary budget for structural improvement and mountain village development, while Hokkaido, in second place, obtained 10.6 per cent, or ¥5.8 billion. While this distribution was influenced by the terrible damage done by a typhoon,' the public agreed that the "power" of Matsuoka contributed to this distribution'.
123 See Chapter 6 on 'The Identical Twins of Nagata-chô'.
124 Matsuoka Toshikatsu Official Site, 'Naka Kyûshû Ôdan Dôrô' ['For the Early Realisation of the Central Kyushu Crossing Road'], in *Katsudô Hôkoku* [*Activity Report*]. Available from http://matsuokatoshikatsu.org/site002//public/048.html
125 See Chapter 7 on 'Electoral Vicissitudes'.
126 Matsuoka Toshikatsu Official Site, 'Chiiki no Shakai Shihon Seibi wa Jûmin no Negai' ['The Provision of Regional Social Capital is the Hope of Local Residents'], in *Katsudô Hôkoku* [*Activity Report*]. Available from http://matsuokatoshikatsu.org/site002//public/053.html
127 Hasegawa, 'Kanjûdanomi no Hazama de Shundô', p. 25.
128 Mayor Gotô invited a former assistant police inspector who was in charge of investigating election violations in Soyo Town to become director and vice-president of the company. When the mayor stepped down as president, the former assistant police inspector replaced him. Hasegawa, 'Jimin "Gajô" no Chikaku Hendô', p. 27.
129 Hasegawa, 'Jimin "Gajô" no Chikaku Hendô', p. 27.
130 'Matsuoka Toshikatsu: Purofuiru', *Seikan Yôran*, 1998, Latter Half Year Edition, p. 188.
131 'Hini Kaku "Matsuoka Toshikatsu Daigishi" no Patoron no "Yappari"' ['"The Expected" from the Dignity-Lacking Patrons of "Matsuoka Toshikatsu Diet Member"'], *Shûkan Shinchô*, 13 December 2001:58.
132 Nakanishi, 'Matsuoka Toshikatsu', p. 29.
133 Nakanishi and Special Reporting Group, 'Suzuki Muneo, Matsuoka Toshikatsu', p. 105.
134 Itô, 'Heisei Jiken Fuairu: Nôrin Jigyô Hojokin o Dokusen Suru Matsuoka Toshikatsu', p. 64.
135 Hasegawa, 'Nôsuishô o Haishi seyo', p. 36.
136 *ibid*.
137 *ibid*.
138 *ibid*.
139 *ibid*.
140 *ibid*.

141 Visit: http://www.nouminren.ne.jp/dat/200208/2002081202.htm. The 30th December 1999 issue of *Akahata* also reported this affair.
142 Itô, 'Heisei Jiken Fuairu: Nôrin Jigyô Hojokin o Dokusen Suru Matsuoka Toshikatsu', p. 65.
143 Nakanishi and Special Reporting Group, 'Suzuki Muneo, Matsuoka Toshikatsu', pp. 103-104.
144 Itô, 'Heisei Jiken Fuairu: Nôrin Jigyô Hojokin o Dokusen Suru Matsuoka Toshikatsu', p. 65.
145 Nakanishi and Special Reporting Group, *op.cit.*, p. 104.
146 Itô, 'Heisei Jiken Fuairu: Nôrin Jigyô Hojokin o Dokusen Suru Matsuoka Toshikatsu', p. 66.
147 Hasegawa, 'Nôsuishô o Haishi seyo', p. 36.
148 *ibid.*, p. 35.

6

THE IDENTICAL TWINS OF NAGATA-CHÔ

Matsuoka's motto is 'straight truth and take great care of those who take care of you'.[1] An example of 'taking care of those who take care of you' is Murakami Kôsuke, who served for two years as head of the Kumamoto Prefecture Agricultural Policy Department until 2000, and who became 'policy advisor in charge' (*seisaku tantô komon*) of the Kumamoto Prefecture Central Union of Agricultural Cooperatives and related prefectural Nokyo federations. He said, 'Matsuoka looked after me a lot, and I am grateful to him.'[2]

As for Matsuoka's commitment to 'straight truth', when, in 2001, journalists from the *Bungei Shunjû* started investigating a rumour of favouritism in NHK where Matsuoka's son worked, Matsuoka called up one of the journalists. Besides issuing a detailed denial of the allegations, he said

> [h]ow old are you, where were you born, do you have parents, where do they work?....You are searching for private information. I have the right to ask you the same questions!....Is this your life's work? If any funny articles get out, I'm going to run you down! You had better be prepared! Understand?....You are all worse than cockroaches! You cockroach rascal! I hope you realise this.[3]

After spitting out these words with such force that it seemed that saliva would come flying out of the receiver, he slammed the phone down. In a subsequent interview a year later, Matsuoka said, '*Bungei Shunjû* are always publishing lies'.[4]

An *Asahi* journalist also reported having received threats of legal action from Matsuoka after he interviewed MAFF Production Bureau Chief Sugata Kikuhito and Agriculture and Livestock Industries Corporation (ALIC) Chairman of the Board of Directors Yamamoto Tôru (who had previously been the director-general of the MAFF Structural Improvement Bureau). Yamamoto was the director-general of the Forestry Agency at the time of the Yamarin affair, in which Matsuoka allegedly received a ¥2 million bribe.[5] ALIC was also the corporation that paid the BSE subsidy to meat traders, in which further bribery allegations were made against Matsuoka, stemming from the fact that both Suzuki Muneo and Matsuoka were directly involved in implementing this subsidy and both were closely connected to the meat industry.[6] Matsuoka accused the journalist of being "'a parasitic worm who preys on the MAFF", to which the journalist retorted "*You* are the parasitic worm who preys on the MAFF"'.[7]

THE TERRIBLE TWOSOME

Many reports have surfaced in the Japanese media about Matsuoka's relationship with Suzuki Muneo, and the way in which he and Muneo supported each other in politics and policymaking. Matsuoka was widely known in Nagata-chô as the Diet member who was closest to Muneo.[8] He was described as the 'No. 1 follower' (*hitotsu no kobun*)[9] of the politician who, at one time, 'was considered a future candidate for prime minister'.[10]

Matsuoka and Suzuki's first meeting goes back to 1975, not long after Matsuoka joined the Forestry Agency. At the time, Matsuoka was executive manager of a MAFF divisional assistant directors' study group. MAFF Minister Nakagawa was invited to lecture to that group and Suzuki came along as his political affairs secretary. Matsuoka recounted that 'we sat at a desk, side by side. We were sworn brothers'.[11] Matsuoka frequently talked about the fact that he and Suzuki drank with Nakagawa. He had even boasted earlier that although he was working in the Forestry Agency, he would become the minister's secretary when Nakagawa became MAFF minister.[12]

A slightly different story was recounted by a person connected to the Forestry Agency. He suggested that it was during Matsuoka's time as chief of the Forestry Management Station in Teshio that Matsuoka realised the opportunities that could be had through his association with Nakagawa and Muneo. As the official recalls

> [d]uring his time as forestry office chief, he became close to Muneo, who was just out of the Fisheries Agency and working as Nakagawa Ichirô's secretary. At the time, Nakagawa was a rising politician in Hokkaido, and furthermore, had a firm footing amongst the godfathers of the Fisheries Agency *zoku*, and so for a young chief-of-office from the central agency, it was a

good opportunity for promotion. I have heard that it was during this period as forestry office chief that Matsuoka decided to aim to become a politician.[13]

Matsuoka's sworn friendship with Muneo began during this period.[14] Ever since that time, according to some commentators, 'Matsuoka and Muneo worked together as a duo, running around Kasumigaseki and Nagata-chô and dashing along the highroad to success'.[15]

Various terms have been used to describe the Muneo-Matsuoka relationship. Journalists have coined the terms 'identical twins of Nagata-chô' (*ichiransei sôseiji*)[16] as well as 'fraternal twins' (*niransei sôseiji*) and 'sworn brothers' (*gikyôdai*)[17] in an attempt to capture the closeness and similarities between Matsuoka and Muneo and their *modus operandi* as politicians. An agricultural and fisheries-connected Diet member observed, 'they are just like twin brothers. What they say and the way they talk are identical.'[18] Matsuoka was also called a 'brother-in-law' of Suzuki, and Matsuoka himself called Suzuki 'more than a sworn friend relation' (*meiyû ijô no aidagara*)[19] and a 'sworn friend of 30 years standing' (*sanjûnenrai no meiyû*).[20] Nôminren questioned whether it could trust Matsuoka—Muneo's follower—on rice policy matters even though he was chairman of the Agricultural Basic Policy Subcommittee at the time.[21] The farmers' group implied that given Matsuoka's close association with Muneo, he would turn a deaf ear to farmers in spite of being a 'boss of the *nôrin zoku*'.[22]

Summing up the relationship, one veteran political journalist said: 'Muneo and Matsuoka are really very similar. They are like copies of each other. From their political methods and fundraising to their threatening tone, who draws influence from whom, they are exactly alike in everything.'[23] Matsuoka's political methods 'go beyond the unreasonable because they are the same as Muneo's, and they have been made fun of as the "threatening duo"'.[24] Matsuoka has been called the 'wild boy' of the agriculture and forestry tribe (*nôrin zoku no abarenbô*),[25] while his mate has been labelled the 'department store of suspicion' (*giwaku no depâto*)[26] and even worse, the 'general trading company of suspicion' (*giwaku no sôgô shôsha*).[27] One political reporter commented

> [a]ccording to a MAFF official, Matsuoka became bad (*waruku naru*) around 1995 after he became chairman of the Agricultural Basic Policy Subcommittee. When he was first elected, Matsuoka was nice and honest, but in his second term, he began to go along with Muneo after he was elected for a second time in 1993 with LDP endorsement. It was then that his behaviour degenerated.[28]

People asked how someone such as Matsuoka, who had only been elected four times (in 2001) and who had held only two government positions (parliamentary vice-minister and deputy minister) could manage to wield so much political power. Why had he, as someone whose national name value was inferior, attracted attention?[29] Much of it was attributed to Matsuoka's association with Suzuki and how he modelled his political behaviour on Suzuki's. What Matsuoka learned from Suzuki was that a sure-fire way to secure money and votes and to realise his ambitions in politics was to become an influential politician (*yûryoku seijika*) who could deliver public works projects to his local district. It 'was Matsuoka's guiding of the budget to Kumamoto, his use of influence over public works in the prefecture and his provision of patronage to agriculture and forestry-related groups that gave Matsuoka a presence in the agricultural sector that was more than expected.'[30] One of Suzuki's most famous sayings was '*chihô e no reiki yûdô de wa nai, "kôsei haibun" de aru!*', meaning 'it is not guiding benefits to the regions, it is "fair distribution"!'[31] In fact, Suzuki had in common with Kamei and Tanaka Kakuei an infamous reputation as a *rieki yûdô seijika*.[32] Kamei is reported to have said unashamedly: 'What's wrong with guiding benefits [to local regions]? We're doing politics for the people.'[33]

THE 'SPECIAL ACTION SQUAD'

In the Diet, Muneo and Matsuoka were known as the 'special action squad'.[34] The two of them exercised their power in various divisions and committees, whilst mutually complementing each other.[35] For over ten years, Matsuoka allegedly dominated Japanese agricultural policymaking along with Suzuki. He was able to 'control agricultural policy using the forceful political power of "the MM (Matsuoka-Muneo) duo" as a weapon'.[36] Matsuoka and Muneo were known to have 'joined hands as *nôrin zoku* for some time'.[37]

Matsuoka and Muneo's *modus operandi* was to 'make deals with producers and companies (about what they wanted) beforehand. They then implemented the deal by forcing it through the relevant division. They cleared the party procedures by force, took the credit for policy, and then obtained the division's consent to leave the matter entirely to their own discretion'.[38] As one LDP executive elaborated

> Muneo and Matsuoka completely controlled the Agriculture, Forestry and Fisheries Division. When a meeting was held, they would bring Diet member followers called 'the special action

squad' and make the squad speak in a way that was convenient for them. Especially when the budget and rice price were decided, the squad not only spoke out but also blocked the remarks of those Diet members who were opposed to the opinions of Muneo and Matsuoka. The squad threatened the Diet members by saying 'if you say that sort of thing, we will make you lose the next election. We will go to your electoral district and expose today's statement'. The squad even said to government officials that they 'would get them fired'.[39]

In trying to explain how Muneo and Matsuoka were able to wield so much power as 'the worst tribe Diet member duo',[40] one veteran political reporter went back to the splintering of the LDP in June 1993. Muneo and Matsuoka's generation was advantaged by the split in the party because members of the LDP's 'comprehensive agricultural policy faction' (*sôgô nôseiha*), such as Hata Tsutomu and Ishiba Shigeru, left the party. Hata was one of the two leading lights amongst *nôrin zoku* at the time. By 'stepping into the vacuum, Muneo and Matsuoka gained power as the mainstay *nôrin giin*'.[41] As Matsuoka was not a member of the breakaway group from the Takeshita faction, leaving the LDP was not an option for him. He was a member of the Mitsuzuka faction at the time, which came through the Fukuda Takeo-Abe Shintarô line. As the mass media commented 'Muneo *giin* and Matsuoka *giin* inherited the rights and interests of the "tribe Diet members."'[42]

An influential MAFF executive provided a similar explanation

> [i]n the past, Nakagawa Ichirô and Watanabe Michio were *nôrin zoku*. After that, Katô Kôichi and Hata Tsutomu took over. In the time of Katô and Hata, they listened properly to the opinions of both farmers and the MAFF and understood the need for compromise. However, those Diet members, who hold power now, play quite different roles. Compared with 10 years ago, these current Diet members attach greater importance to the opinions of the producer side and try to accommodate their demands just as they are. The names of these Diet members are Mr Matsuoka, Mr Muneo and Mr Etô Takami. Mr Matsuoka has the experience of holding the positions of Agriculture and Forestry Division Chairman and Agriculture Basic Policy Subcommittee Chairman (this committee decides rice production adjustment). His advantage is that he has connections with dominant figures such as Mr Nonaka Hiromu and Mr Muneo, and they are skilful in controlling people and parliamentary proceedings. In 1993, Hata and others left the LDP, and the LDP broke up. The LDP slid down to the opposition party, and the Matsuoka class gained strength in the vacuum. The declining power of Ministry of Finance greatly influenced this movement. Recently, *sensei* (Diet members) are showing their influence by saying 'we ignored the intention of Ministry of Finance. We will control things'. Since they will make Ministry of Finance (in the financial crisis) spend money, deficit bonds increase. Previously there was a very natural discussion along the lines that 'if we formed a budget for rice, budgets could not be spent on other things'. It would be better if they were a little bit more intelligent in their approach, but...[43]

By drawing on each other's influence, Muneo and Matsuoka allegedly manipulated policies in the way they wanted and strengthened their political power.[44] Matsuoka's political style of yelling at the bureaucrats in LDP divisional meetings was just like Muneo's and earned him the nickname 'Muneo of the West'.[45] Matsuoka would raise his voice, threaten people and shut them up.[46] As a MAFF OB explained

> [t]aking his cue from his sworn friend Muneo, saying 'I will listen to the voices of consumers and producers', Matsuoka intervened in leaf tobacco and rice price decisions, and yelled at bureaucrats. Since he's become powerful, the Agriculture and Forestry Division itself has taken on an abnormal atmosphere. At my place, a phone call came from the current bureau director-general in an exhausted voice. Today he also called saying that he got yelled at by Matsuoka, who said, 'you're an idiot'. He's just like Muneo.[47]

Muneo was also known to get angry and shout a lot.[48]

> [an] official in the Fisheries Agency committed suicide, it is said, because of repeated harassment from Muneo. The official was seen as a hardworking and good man, a law-abiding citizen who did not want to bend the rules for Muneo, but who had to endure threats such as 'I'm going to wreck your life' and 'I'll make sure you don't get promoted'.[49]

Suzuki was also known as 'the behind-the-scenes foreign minister' who threatened MoFA officials and forced his views through during the Okinawa summit in 2000.[50] When LDP Diet member, Hirasawa Katsuei, openly called Suzuki the 'behind-the-scenes' foreign minister on a TV show, Suzuki later accosted him in the corridor of the Diet members' offices, saying in *yakuza*-like tones, 'Oi… Hirasawa-kun, what do you mean by my being the behind-the-scenes foreign minister?'[51] Hirasawa shouted back, 'How am I wrong?'[52] Just as the two were about to launch into each other, someone came along and so the fistfight did not amount to anything.[53]

Political journalist, Yamamura Akiyoshi, commented on Muneo and Matsuoka's bullying tactics

> [s]ix or seven years back, the LDP's Agriculture and Forestry Division was divided into pro and anti-Muneo factions, and with the backing of Nonaka's power, the Suzuki-Matsuoka combination would yell at, belittle and get rid of Diet members who disagreed with them. They would bring the MAFF officials in charge over to their side and manipulate things to their liking. Already in the last three or four years, there is no one who goes against them.[54]

Nonaka, Suzuki and Kamei (Matsuoka's faction boss) all had in common the fact that they grew up in poverty, which reportedly made them into 'tough and shrewd political players'.[55]

Matsuoka and Muneo would regularly call on each other when they thought they might need backup. When Muneo got into strife for interfering in MoFA affairs (in this case, influencing the ministry not to permit NGOs to participate in an Afghanistan aid donors' conference in Tokyo), Matsuoka directly attacked Muneo's main political critic (Hirasawa) in the Executive Council of the LDP, alleging that Hirasawa's statements had 'slandered the party itself'.[56] When journalists approached Matsuoka directly for comment, he retorted: 'What? You're so rude!'[57]

One of the techniques of the terrible twins was to send the 'shock troops' under their command (about 10 other Diet member-followers) to back each other up. In October 2000

> Matsuoka turned up at the Foreign Affairs Division with more than 10 of his followers. This was at Suzuki's request in order to back him up on the issue of sending surplus rice to North Korea, an idea originating with Nonaka, a Suzuki backer, who was deputy LDP secretary-general at the time. Suzuki told Matsuoka to say, 'rice support for North Korea is important'. This would help solve the rice surplus problem at the time. After Muneo argued that rice support was necessary for the progress of Japan-North Korea relations, Matsuoka stated: 'We understand the sentiment of abducted families, but we want the decision of Minister of Foreign Affairs, Kôno Yôhei, as the majority.' Whereupon, his followers said, in a previously arranged chorus, 'that's right, that's right, that's right'. The voices of those who opposed rice support for North Korea were drowned out. The general position of the party was decided after the divisional meeting. Half a million tonnes of rice were subsequently sent to North Korea.[58]

On another occasion, Suzuki provided backup for Matsuoka. In October 2001, Matsuoka asked Muneo to join a dinner party one evening in Akasaka. Executives of the Kumamoto Prefecture construction industry association were holding a meeting to petition LDP Diet members representing Kumamoto Prefecture. Six Diet representatives were there.[59] Strictly speaking this was just a dinner party for members of the Diet from Kumamoto Prefecture.

> [when] one of the association executives, who was known to oppose Matsuoka, asked about the Diet politician (Kaneko Yasushi) who had stood as an Independent in the 2000 general election in Kumamoto (5) and who had defeated the LDP candidate, both Muneo and Matsuoka were outraged. Matsuoka shouted at the executive that he would destroy his company and make a show of his power. The executive retorted, 'go ahead and destroy it!', whereupon a shouting match ensued, which continued for 10 minutes. It was touch and go for a while, not forgetting that Matsuoka was a member of a karate club in his high school days.[60]

DIVISION OF LABOUR

As a *zoku* Diet member, Suzuki secured a strong foothold in both the MAFF and MoFA.[61] Nokyo's National Council admitted

> on the policy front, particularly in regards to price decisions on wheat, sugar beet and raw milk for processing, which are all closely connected to his local region in Hokkaido, he would go to all the LDP subcommittee meetings and by violently pressuring bureaucrats, he would guide policies. Suzuki was not a member of the agriculture and forestry executive, but his influence was such that crop and dairy prices could not be decided without his agreement.[62]

However, Suzuki gradually made room for Matsuoka as a *nôrin zoku*, with a division of labour gradually appearing between the two.

> His and Matsuoka's respective spheres of influence were the Ministry of Foreign Affairs and the Ministry of Agriculture, Forestry and Fisheries respectively. Matsuoka was to the MAFF what Suzuki was to MoFA. The two of them controlled the LDP's Foreign Affairs Division and the Agriculture and Forestry Division. Suzuki used to run the Agriculture and Forestry Division, but he stopped coming seven or eight years after he started to dabble in diplomatic affairs. He basically left the running of the division to Matsuoka and one other LDP *nôrin giin*, Futada Kôji from Akita, thinking that it would be in safe hands. Muneo probably thought 'I can leave the role to this man'.[63]

Muneo's former protégé, who was an official in MoFA, called Suzuki the 'Rasputin of the Foreign Ministry'[64] in his memoirs. As Reed writes

> Suzuki was remarkably powerful. He was more in control of the Ministry of Foreign Affairs than the Foreign Minister, at least with respect to Russia, Africa, expenditures on the northern islands, and bureaucratic personnel decisions. He directed MOFA expenditures to companies that contributed to his campaign and probably directed a significant amount directly into his own pocket.[65]

It is said, however, that Matsuoka's power over the MAFF surpassed that of Suzuki over MoFA.[66] Matsuoka's 'threatening attitude frightened the people in the MAFF and in construction companies.'[67] This did not always make Matsuoka very popular. Government officials within the MAFF, construction contractors and others frowned on his intimidating behaviour.[68] An influential construction company executive in Matsuoka's electoral district elaborated on Matsuoka's behaviour under Suzuki's influence.

> Matsuoka's face has gradually become evil. Just after he was elected, he still retained a young and pleasant impression. Doesn't he realise this himself? His face does not look like a Diet member. I have seen tens of House of Representatives members so far, but Matsuoka is

completely different from other politicians. He has a different quality altogether. He wants to do everything by himself. At any rate, he is not satisfied unless everything centres on him. He says: 'I don't approve of anything unless I permit it'. This is Matsuoka's style. Wherever there are public works, Matsuoka pokes his nose into the majority of them. Several years ago, Matsuoka controlled public construction works in Kumamoto Prefecture. Fundamentally, Diet members should not intrude into public construction works in a prefecture. I advised him, 'if you scatter money to only one company, only one out of ten companies can make a living. If you treat only one company nicely, you will have nine companies for enemies. If you continue that, you will be totally surrounded by enemies.' After I gave him this advice, Matsuoka said 'I understand', but...[69]

In 2000 a weekly magazine conducted a questionnaire asking the question 'which politician do government officials want to lose the election?' According to the results, Matsuoka's close political ally, Suzuki came in first, while Matsuoka himself came in eighth as someone who 'uses personnel matters and the budget for his own interests'.[70]

THE MATSUOKA-MUNEO FUND-GATHERING AXIS

Matsuoka had in common with Muneo a strong ability to collect political funds, especially an ability to obtain a great deal in company contributions.[71] Matsuoka's pulling power in terms of funds, along with this *oyabun* (patron) Suzuki, was the real origin of the saying '[i]n the east Suzuki, in the West Matsuoka'.[72] Matsuoka tapped into a very lucrative vein of political funding by deploying the same kind of methodology as Muneo, that is, blatant influence peddling and exercising influence wherever he could—from obtaining central and local government public works funding to the selection of companies to do the public works.[73]

The pattern of Matsuoka's fund gathering reveals several conspicuous aspects. First, he demonstrated extraordinary money-collecting power early on in his Diet career. In 1994, as a fourth-year Diet Member, he raised about ¥10 million from two fund-raising parties (*hagemasukai*, or *seiji shikin pâtî*). This was rare for such a relatively junior Diet member.[74]

Second, Matsuoka's fund-gathering ability improved along with his experience in the political world, that is, as he gained seniority and status as a Diet member and as he acquired important agricultural and forestry committee posts and sub-cabinet positions. There were quantum leaps in donations from 1996, especially between 1996 and 2000, when Matsuoka progressed to third and fourth-term Diet membership and gained a footing as a 'backbone' Diet member (*chûken giin*).⁷⁵ Highly significant in this regard was Matsuoka's successive holding of various important posts such as MAFF parliamentary vice-minister (in 1995) and his appointment as chairman of the UR-Related Countermeasures Subcommittee. As already pointed out, Matsuoka was deeply involved in the distribution of the ¥6.01 trillion in UR countermeasures expenditure. Then in December 2000, Matsuoka was made MAFF deputy minister in the second Mori Cabinet and firmed up his position as a *nôrin zoku*.⁷⁶

Third, Matsuoka's fund collecting was comparatively large relative to other Diet members, not only those of equivalent experience, but also those of considerable influence and standing. According to political funds for 1996 data for 1996 collected by the *Asahi Shinbun*, the total amount Matsuoka received from the LDP No. 3 Electoral District Branch, via his *kôenkai* and via his political funds management group, the Matsuoka Toshikatsu New Century Politics and Economics Discussion Association, amounted to ¥280 million, which was well in excess of the average for all politicians elected to the Lower House in that year (¥160 million).⁷⁷ Amongst 384 successfully elected SMD Diet members in 1996, Matsuoka was ranked forty-first.⁷⁸ For a politician elected only three times, this was extraordinary. Those higher on the list were all well-established, famous Japanese politicians, most of whom were faction leaders, party leaders, and current, former or future prime ministers. Up there, of course, was also Muneo, who was one of the LDP's top fund raisers 'and the only one of the top five not a faction leader.'⁷⁹

The same kind of data in 1997 listed the top 50 political fund management organisations of LDP Diet members. Matsuoka's political fund management organisation, the Matsuoka Toshikatsu New Century Politics and Economics Discussion Association, was ranked twenty-sixth with around ¥79 million. This was the income from fund-raising parties alone.⁸⁰ Once again, all those above Matsuoka on the list were senior LDP figures including past, present and future prime ministers, as well as factions and the New Party Harbinger (Shintô Sakigake). It was, therefore, an extraordinary amount for a third term Diet member who was not a faction leader, and who was nowhere near a ministerial or prime ministerial post, or the party leadership.

In 2000, Matsuoka collected ¥104.13 million in company donations and was ranked 13th on the list of Diet politicians in terms of amount collected.⁸¹ While this was short of first-placed Suzuki with ¥250 million, it exceeded that collected by the leader of a political faction, former Prime Minister Mori.⁸²

Fourth, Matsuoka's career connections with the Forestry Agency, his personal influence as the 'Don' of forestry policy, and his links with the forestry industry proved highly productive in terms of financial contributions. The New Century Politics and Economics Discussion Association received large quantities of funds from forestry-related political groups. In 1998 alone, looking at just the major contributions from this source, Matsuoka collected ¥4.4 million from the National Mountain and Forest Roads Political League (Zenkoku Chisan Rindô Seiji Renmei), ¥2 million from the National Forestry Civil Engineering and Construction Industry Political League (Zenkoku Shinrin Dobuku Kensetsugyô Seiji Renmei), ¥1.88 million from the Japan Timber Industry Political League (Nihon Ringyô Seiji Renmei), and ¥1.24 million from the Forestry Proprietors' Political Association (Ringyô Keieisha Rinseikai). These amounts also included payments for tickets to fund-raising parties. In addition, the National Mountain and Forest Roads Political League contributed ¥8 million to the LDP Kumamoto Prefecture No. 3 Electoral District Branch represented by Matsuoka. Although these groups made donations to many politicians, they were particularly generous to Matsuoka.[83]

Other investigations revealed similar funding links between Matsuoka and forestry-related organisations. The National Political Federation of Forest Civil Engineering and Construction Companies (Zenkoku Shinrin Dobuku Kensetsugyô Seiji Renmei) donated ¥14 million to Matsuoka, the National Political Federation of Afforestation and Forestry Roads (Zenkoku Chisan Rindô Seiji Renmei] ¥8.46 million and the National Timber Industry Federation (Zenkoku Mokuzai Sangyô Renmei) ¥4 million.[84] These groups made up more than half of the number of political groups contributing to Matsuoka's political funding organisation.[85] Matsuoka was assiduous in attending meetings of the Japan Association of Forestry Civil Engineering Leagues (Nihon Ringyô Doboku Rengô Kyôkai), which claimed that protecting forests also protected national land, especially if there were a lot of damage from typhoons. Protecting forests also raised issues of flood control and forestry roads (that is, public works). In 2000, Matsuoka received a total of ¥70 million in donations and party tickets from forestry-related industries.[86]

Matsuoka also had financial ties to extra-departmental groups (*gaikaku dantai*) of the Forestry Agency, which were dependent on the agency in terms of contracted business. Of particular interest was the sum of ¥360,000 donated to Matsuoka in 1996 by the Japan Forest Technology Association (Nihon

Ringyô Gijutsu Kyôkai), which was simultaneously receiving large amounts of subsidies from the MAFF (amounting to as much as ¥40 billion in 1996 and 1997).[87] Making such political donations was, strictly speaking, contrary to Article 22 of the Political Funds Regulation Law (*Seiji Shikin Kiseihô*), which bans bodies that receive grants from the state, such as subsidies, from making donations to political campaigns for a year.[88]

In another case, Matsuoka received a large political contribution from an organisation that was involved in bid-rigging (*dangô*), that is, illegal collusion amongst companies in order to share out government contracts. The organisation in question was a public interest corporation (*kôeki hojin*) under the jurisdiction of the Forestry Agency called the State-Owned Forests Survey Works Cooperation Association (Kokuyû Rinya Sokuryô Jigyô Kyôryokukai). Qualification for admission to the association was being a Forestry Agency OB, meaning that it was a 'descent from heaven' (*amakudari*) corporation. The association received an exclusion advice from the Fair Trade Commission because of undertaking repeated *dangô* in tenders for surveys and investigations of state-owned forests and making unfair profits. Matsuoka received a total of ¥9.42 million in contributions over five years in the period 1996–2000 from this association. JCP House of Councillors member Ôgata Yasuo took up the issue in the Economy, Trade and Industry Committee of the Upper House. He denounced 'a public-interest corporation involved in *dangô* and making unfair profits. Some of these unjust profits were "returned" to Matsuoka, who had strong influence over forestry administration.'[89]

In 2000, a leading article appeared in the *Asahi Shinbun* on 24 March stating that two incorporated foundations attached to the Forestry Agency were donating money to a Forestry Agency OB who was a Diet member. The organisations were the Forestry Benefit Association (Rinya Kôzaikai)[90] and the Forestry Civil Engineering Consultants (Ringyô Dôboku Konsarutantsu). Both were *gaikaku dantai* of the Forestry Agency, in this case incorporated foundations (*zaidan hôjin*) charged with forestry research, surveying and planning. Most of the executives and staff of the organisations were Forestry Agency OBs, making them *amakudari* corporations of the Forestry Agency. The benefit association donated ¥480,000 each year to Matsuoka's political funding group in the period 1996–98 (see also Table 6.1), and in 1996 bought ¥400,000 worth of party tickets.[91] The consultants' group contributed ¥480,000 each year in 1996 and 1997.[92] In total, these two groups donated

¥2.8 million to Matsuoka between 1996 and 1998. The *Asahi Shinbun* took up the ethics of a cycle whereby *kôeki hojin* that obtained jobs from the MAFF then gave political donations to OB Diet members.[93]

Moreover, while both were *kôeki hôjin* as incorporated foundations, 'they had investments in private companies that also made financial contributions to Matsuoka in direct conflict with the 1996 cabinet decision about "the standards of permission for establishing and guiding public interest corporations"'.[94] In addition, both incorporated foundations sub-contracted works worth ¥2.1 billion (in the case of Rinya Kôsaikai), and ¥140 million (in the case of Ringyô Dôboku Konsarutantsu) to these companies. As DPJ Lower House member, Ishii Kôki, observes, 'this kind of three-sided financial connection represents a typical politics-bureaucracy-industry triangle.'[95] In fact, the 'collusive structures' centring on the MAFF's agricultural civil engineering bureaucrats and the land improvement and rural development industries are replicated in the Forestry Agency.[96]

A far more serious case of adhesion reputedly centred on a forestry company called Kyôrin Consultants (Kyôrin Konsarutantsu), in which various forestry public interest corporations had invested, with 25 per cent financed by the Rinya Kôzaikai and 18 per cent by an incorporated association (*shadan hôjin*), the Japan Forestry Technology Association (Nihon Ringyô Gijutsu Kyôkai).[97] Although technically a private company, Kyôrin Consultants employed Forestry Agency OBs as executives and obtained jobs from the bureaucratic agencies. It was effectively a subsidiary company of Forestry Agency *amakudari* corporations. As a bureaucratic consulting firm and subsidiary company of Forestry Agency *gaikaku dantai* and *amakudari* corporations, it was part of a typical pyramid structure that spread down from the Forestry Agency to *amakudari* corporations, then to subsidiary *amakudari* companies and finally to private subcontractors of *amakudari* corporations. The web of networks spread across the regional areas of Japan.[98]

Matsuoka had this pyramid system at his beck and call, and was given political donations by the *amakudari* corporations, the subsidiary companies and even the private subcontractors.[99] A previous chairman of the subsidiary company Kyôrin Consultants, Nakamura Yasushi, later became a policy secretary of Matsuoka's.[100] Nakamura channelled donations to Matsuoka from the Japan Forest Technology Association, which held 18 per cent of the shares of Kyôrin Consultants, via Kyôrin, which was technically a private company and therefore, in this way slipped under the Political Funds Regulation Law (which banned public organisations from making political donations).[101]

Fifth, contributions from companies and organisations (*kigyô, dantai*) in the construction industry made up a large proportion of donations to Matsuoka's political funding organisation.[102] Matsuoka's political funds revenue and expenditure report (*seiji shikin shûshi hôkoku*) in 2000[103] revealed significant contributions from construction companies.[104] On the contributors list were major general contractors (*zenekon*), civil (agricultural) engineering companies (*doboku kaisha*) and construction companies (*kensetsu kaisha*) that received contracts for the projects in which Matsuoka was involved.[105] Companies and places of business on the list numbered just under 500.[106] Most were, in short, construction companies (*doken gyôsha*), or construction material suppliers such as glass companies.[107] The 'donations varied from around ¥100,000 to ¥5 million at the most, but it demonstrated how loyal to Matsuoka the small and medium-sized civil engineering construction industry was, and how big their expectations were of him.'[108]

One company executive said, '[p]ublic works projects make up 90 per cent of our business. With this recession, I'd say all of the civil engineering and construction companies in Kumamoto are in much the same position'.[109] Another said, '[s]tructural improvement projects from the MAFF come to us thanks to Mr. Matsuoka. He's doing his best to ensure that construction continues for both the Kyushu bullet train and the Kawabe River dam projects'.[110] For example, Matsuoka received ¥33.85 million from 42 companies that were contracted by MLIT to construct the Kawabe River Dam.[111] Another person connected to the political world commented that 'Mr Matsuoka advocates increases in the national debt and expansion of the supplementary budget, and there is no doubt that behind his opposition to decreases in the budget [under Prime Minister Koizumi], there is the "pressure" of donations'.[112]

Looking at the overall picture, over the six years between 1995 and 2001, a total of 483 companies and organisations contributed more than ¥50,000 a year to the Matsuoka Toshikatsu New Century Politics and Economics Discussion Association.[113] Including donations of under ¥50,000, a total of ¥211,860,160 was contributed to Matsuoka's political funding group from companies and organisations located across 22 prefectures.[114] The top 10 companies and organisations contributing to Matsuoka's political funding group in order of amount between 1996 and 1999 are listed in Table 6.1.[115] As the table shows, 60 per cent of the top-ten listed donors were construction companies in Kumamoto Prefecture.

Most of the companies on the list also contributed to the LDP Kumamoto Prefecture No. 3 Electoral District Branch, of which Matsuoka served as the branch representative. In 2000, Matsuoka collected about ¥920,000 from two political fund-raising parties, and the LDP's Kumamoto Prefecture No. 3 Electoral District Branch collected ¥104 million donated by about 500 companies connected to public works such as civil engineering and construction material companies.[117] According to rumour, a construction company also shouldered Matsuoka's secretary's salary.[118] A spokesman from one of the companies on the list of Matsuoka donors said 'a contribution amounting to hundreds of thousands of yen per year is insignificant for us. We believe Matsuoka *sensei* has contributed to the development of forestry. The fact is, we contract forestry-related work'.[119]

Table 6.1 **Top 10 contributors to Matsuoka Toshikatsu New Century Politics and Economics Discussion Association**

Rank	Name of company and organisation (location)	Amount contributed (¥)
1	Mitsuhashi Company (Shibetsu City, Hokkaido Prefecture)	2,000,000
2	Sugimoto Construction (Aso Town, Kumamoto Prefecture)	1,792,000
3	Jotoh Logistics Warehouse (Ozu Town, Kumamoto Prefecture)	1,720,000
4	Japan Conservation Engineers (Minato Ward, Tokyo)	1,680,000
5	Forestry Benefit Association (Bunkyo Ward, Tokyo)[116]	1,440,000
6	Dai Ichi Kiko (Kumamoto City)	1,260,000
7	Ishizaka Company (Kikuka Town, Kumamoto Prefecture)	1,200,000
8	Mori Industry (Aso Town, Kumamoto Prefecture)	1,152,000
9	Renda Gumi (Hondo City, Kumamoto Prefecture)	1,100,000
10	Otsubo Construction Industry (Hakusui Village, Kumamoto Prefecture)	1,080,000

Matsuoka's New Century Politics and Economics Discussion Association also benefited from contributions from 115 individuals, amounting to a total of ¥44,355,000 between 1996 and 1999. These contributions were approximately 25 per cent of the total donated by companies and organisations. However, the number of individuals was still large. The top 10 contributors by total over the period 1996-2001 are listed in Table 6.2.[120]

The main contrast between Table 6.1 and 6.2 is the fact that individual contributors came from a greater range of geographic locations (50 per cent were from Kumamoto, and 20 per cent were from Hokkaido) compared to the company and organisations that contributed to Matsuoka.

As head of the LDP branch in Kumamoto (3), Matsuoka received the bulk of contributions, which were collected in small amounts but from large numbers of construction-related companies in Kumamoto Prefecture.[121] Even Matsuoka admitted that 'although individual contributions are small, the company contribution is large overall. Contributions are the result of the fact that various people support me.'[122]

In the LDP's Kumamoto No. 3 electoral district revenue and expenditure report, companies from Hokkaido, where Suzuki's electoral district was located,

Table 6.2 Top 10 individual contributors to Matsuoka Toshikatsu New Century Politics and Economics Discussion Association

Rank	Name of individual (location)	Amount contributed (¥)
1	S. N (Kushiro City, Hokkaido Prefecture)	2,500,000
2	M. N (Kushiro City, Hokkaido Prefecture)	1,000,000
3	T. U (Enzan City, Yamanashi Prefecture)	500,000
4	K. S (Akita City, Akita Prefecture)	480,000
5	T. S (Ito City, Shizuoka Prefecture)	480,000
6	M. N (Miyagawa Village, Mie Prefecture)	480,000
6	M. A (Kikuka Town, Kumamoto Prefecture)	480,000
6	K. M (Menda Town, Kumamoto Prefecture)	480,000
9	K. I (Aomori City, Aomori Prefecture)	400,000
10	H. K (Yatsushiro City, Kumamoto Prefecture)	360,000
10	S. M (Kikuchi City, Kumamoto Prefecture)	360,000
10	T. H (Iwaki City, Fukushima Prefecture)	360,000
10	H. M (Niigata City, Niigata Prefecture)	360,000
10	T. N (Tsunagi Town, Kumamoto Prefecture)	360,000

were also prominent.[123] In fact, a large number of companies contributed to both Matsuoka and Muneo. Table 6.3[124] lists companies contributing more than ¥50,000 per year and the amounts received by Muneo between 1995 and 1999 and by Matsuoka between 1996 and 1999. The amounts were donated to Muneo's 21st Century Policy Research Association and Matsuoka's New Century Politics and Economics Discussion Association. A total of 24 companies had the same company name and place. Most were forestry or construction-related companies located in Hokkaido Prefecture.

First place on the list was Mitsuhashi Company, which was a leading timber company in Hokkaido Prefecture. It also made annual contributions of ¥500,000 to the LDP's Kumamoto No. 3 Electoral District branch every year even after 2000.[125]

Matsuoka's revenue and expenditure report was similar to Muneo's report in that there was an overwhelmingly large number of contributions around ¥100,000. Moreover, many of the same names contributing ¥100,000 were listed in Matsuoka's revenue and expenditure report almost every year.[126]

In 2005, details of Matsuoka's political funding income report for 2004 were revealed. The Matsuoka Toshikatsu New Century Politics and Economics Discussion Association was ninth on the list of the top 10 recipients of donations from fund-raising parties. Those ahead of Matsuoka's political funding group were LDP factions (old Hashimoto faction (first), Mori faction (fifth), Horiuchi faction (seventh) and Kamei faction (eighth)), political parties (the DPJ (third)), an industry league (the Pharmaceutical Industry Political League (sixth)), and two leading LDP politicians (Hiranuma Takeshi (second) and Nakagawa Hidenao (fourth)).[127]

STEALING FROM THE PUBLIC PURSE?

In 2000, the Kumamoto Prefecture No. 3 Electoral District Branch collected ¥12,100,000 in contributions from individuals.[128] Of this, ¥5,000,000 was donated by a single individual, Yoshii Junichi, who had been with Matsuoka since his first election in 1990. Yoshii came to be employed in the publicly funded position of No. 1 secretary after Matsuoka was first elected in 1990. He stayed in that position until 2000.

The exact amount that Yoshii donated was ¥5,127,427. His annual salary in 2000 was estimated to be ¥8,850,000, and so ¥5,000,000 represented a sizeable slice of his annual income. Investigative journalists dug around and came up

Table 6.3   Companies contributing to Matsuoka and Muneo

Company name	Location of head office	Amount contributed to Matsuoka (¥)	Amount contributed to Muneo (¥)
Mitsuhashi Company	Shibetsu City, Hokkaido Prefecture	2,000,000	2,500,000
Yamarin	Obihiro City, Hokkaido Prefecture	720,000	600,000
Taiyoo Development	Yatsushiro City, Kumamoto Prefecture	720,000	360,000
Ikoma Gumi Corporation	Asahikawa City, Hokkaido Prefecture	480,000	1,800,000
Kuramoto Sangyo Co.	Ikutahara Town, Hokkaido Prefecture	360,000	720,000
Matsumoto Gumi	Hakodate City, Hokkaido Prefecture	240,000	1,800,000
Daido Industry Development	Shirataki Village, Hokkaido Prefecture	240,000	360,000
Hokusei Co.	Tobetsu Town, Hokkaido Prefecture	140,000	500,000
Takaya . Doken Co	Iwamizawa City, Hokkaido Prefecture	130,000	120,000
Saito Gumi	Ashoro Town, Hokkaido Prefecture	130,000	1,940,000
Kikuchi Construction	Teshio Town, Hokkaido Prefecture	120,000	720,000
Kikuchi Gumi	Takinoue Town, Hokkaido Prefecture	120,000	480,000
Kyo Construction Industry	Ebetsu City, Hokkaido Prefecture	120,000	600,000
Sanjo Construction	Kitami City, Hokkaido Prefecture	120,000	1,760,000
Yamamoto Gumi	Teshio Town, Hokkaido Prefecture	120,000	360,000
Kagoshima Construction	Teshio Town, Hokkaido Prefecture	120,000	120,000
Morinaga Gumi	Asahikawa City, Hokkaido Prefecture	120,000	240,000
Ishiyama Gumi	Teshio Town, Hokkaido Prefecture	120,000	360,000
Daiho Corporation	Chuo Ward, Tokyo Hokkaido Prefecture	120,000	720,000
Tanimura Gumi	Iwamizawa City, Hokkaido Prefecture	120,000	630,000
Tsujihiro Gumi	Fukui City, Fukui Prefecture	120,000	660,000
Tanaka Industry	Shibetsu City, Hokkaido Prefecture	120,000	1,800,000
Shimada Construction	Abashiri City, Hokkaido Prefecture	120,000	2.500,000
Marufuku Construction	Kagoshima City, Kagoshima Prefecture	100,000	120,000

with some interesting facts. Apparently Yoshii had temporarily retired after the June 2000 general election, but was almost immediately re-employed in July 2000. When he resigned he received a retirement allowance of approximately ¥6 million. When asked whether he had contributed his retirement allowance to Matsuoka, he acknowledged, 'I suppose that's the case…I quit once because there was an election on'.[129]

When Matsuoka was approached directly at the Diet members' dormitory in Tokyo to confirm or deny this, he answered, 'I don't know. This is the first I've heard of it. I don't know.'[130] However, when he was told that Yoshii had admitted it, he said, '[t]hat's a procedural matter. That's what I've heard. I don't know anything about matters concerning Yoshii.…I don't know anything about the retirement fund issue.'[131] Following this encounter with journalists, he got straight into his car, still carrying the bag of garbage that he was intending to throw into the rubbish.[132]

The same night journalists accosted Matsuoka in the Diet members' dormitory, they received a phone call from Yoshii. He elaborated on the details of receiving the retirement allowance, and, regarding the donation to Matsuoka, he said, 'it's something that I did of my own accord as a means of settling things and taking responsibility now that I was re-employed.'[133]

A similar suspicion arose that Matsuoka generally raked off the allowances paid to his state-supplied secretaries. According to a person connected to the local political world, Matsuoka would keep the allowances of his state-funded first and second secretaries, and then redistribute the money to five or six secretaries including his privately funded secretaries.[134] When asked about this point also, he totally denied it, saying 'that's not the case at all. We don't do that.'[135] It would seem, however, that raking off secretaries' allowances is a semi-norm in Nagata-chô.[136] More than 80 per cent of Diet member's offices do it, according to Arima Harumi, a political commentator with experience as a Diet member's secretary.[137]

FACTION-BUILDING

Muneo and Matsuoka thought they were so politically successful that they could attract a tribe of followers who could always be bribed by political funding into following politically powerful Diet members. They had an eye on generational succession and the reorganisation of factions, and they worked together to gather young Diet members and hold a study group.[138]

In August 2000, rumours circulated that the traitor Muneo was aiming to create a separate faction of his own. Information came to light about a meeting held at Ishingô, a Chinese restaurant in Akasaka, Tokyo. A total of 25 people turned up, 14 from the Etô-Kamei faction, to which Matsuoka belonged, and 11 people from the Hashimoto faction to which Suzuki belonged. The prospective name given to the group was the Suzumatsukai, taking the initial syllables of both Muneo's and Matsuoka's names. It was the beginning of a new faction under Suzuki's leadership but people did not find the name very inspiring. Suzumatsu was very close to another Japanese word 'suzumushi' meaning a 'cricket', and reference was made to the 'two crickets' singing in unison.[139]

The new grouping was supposed to take over from the group called the 'Society to Create Tomorrow's LDP' (Jimintô no Asu no Tsukurukai), which was an embryonic grouping attracting younger Diet members after the 2000 general election when the Mori-Nonaka executive structure in the LDP loosened.[140] One of the participants commented

> I do not care for this society. Its members are only flattered by the mass media. We are conducting activities in the Suzumatsukai with the intention of creating the mainstream of the next regime. For the name of the association, there was a discussion to call it the Suzumatsukai, taking the names of both Suzuki and Matsuoka. However, some thought that "the name was uninspiring and a bit too conspicuous" and so the name was not decided. It was designed to be an association in preparation for the formation of the Suzuki faction, and aimed to draw in other members by adding politicians from other factions.[141]

The new grouping was due for official inauguration by September 2000, adding Diet members from other factions.[142]

After the meeting 'a reporter asked Matsuoka, "Will you create a faction?", to which he replied, "We are considering that option. It may end up being that way."'[143] Matsuoka was known to be a follower (*kobun*) of PARC Chairman Kamei. Asked about this, he gave the following explanation.

> To start with, we need to back up the Nonaka-Kamei executive regime. Next, we will create the Kamei faction in opposition to YKK (Yamasaki, Katô and Koizumi). Then sooner or later, we aim to create the Suzuki faction. My plan is to stabilise the LDP through cooperation between the Shisuikai faction and the Hashimoto faction.[144]

When asked whether he wanted Suzuki to become prime minister, Matsuoka said, 'Mr Kamei is my boss. First, I want Mr Kamei to do his best to become prime minister…Since Mr Suzuki is also a competent politician, he will naturally become a candidate for prime minister later'.¹⁴⁵ Suzuki was more reluctant publicly to acknowledge the existence of the embryonic new faction. He affirmed that he was from the Hashimoto faction and a close advisor of former LDP Secretary General Nonaka, while his friend Matsuoka was a close advisor and follower of PARC Chairman Kamei.¹⁴⁶

In spite of Matsuoka and Muneo being labelled 'the identical twins in Nagata-cho', one point of difference between them really stood out. Muneo was well known for distributing political funds to Diet members and prefectural assembly members in order to build-up a tribe of followers. His style was aggressively to distribute money to other politicians. Politicians who received money from Muneo formed the core of his 'Munemune Kai' (Munemune Association). In 1996 (an election year) he reportedly gave ¥7.1 million to this group, ¥1.2 million in 1997, ¥2.7 million in 1998, ¥2.5 million in 1997, ¥2.7 million in 1998, ¥2.5 million in 1999 and ¥2.3 million in 2000.¹⁴⁷ Matsuoka was a member and reportedly received ¥7.5 million in funding from the association over a period of two years.¹⁴⁸ The same kind of money flows were not matched by Matsuoka. He was less keen on distributing his own funds.¹⁴⁹ Despite this, a report surfaced in late 2001 that outside the top factions, only Matsuoka and Suzuki had distributed *mochidai* (rice cake money) to young Diet members, which suggests that Matsuoka was being reckless beyond his means.¹⁵⁰

Documents presented to MIAC by Matsuoka's political funding group confirmed that he received donations from Suzuki's political funding groups.¹⁵¹ The Special Investigation Department of the Tokyo District Public Prosecutor's Office also uncovered the fact that Suzuki did not record all the necessary financial details in his revenue and expenditure reports, including outlaying funds for activities that were not political activities such as contributing funds for the cost of construction of Matsuoka's own house.¹⁵²

According to one source, amounts flowing from Muneo to Matsuoka between 1990 and 2000 amounted to ¥44 million.¹⁵³ The ¥12.5 that Matsuoka received from Muneo over the period 1996–2000 was more than six times the amount that Matsuoka received from Kamei in 1998.¹⁵⁴ Then, in 2001–2002, Matsuoka received funding worth ¥7,050,000 from Muneo.¹⁵⁵

Discrepancies were observed in Muneo's recorded outgoings compared to Matsuoka's recorded incomings. In one case, Matsuoka did not record Muneo's contribution at all.¹⁵⁶ Overall, the amounts were so large, they confirmed the existence of the Munemune Kai, in short, the Suzuki Muneo faction.¹⁵⁷

Moreover, they also underlined the very special relationship between the Matsuoka and Muneo. No Diet politician contributed so generously to another without regard to some purpose or other. Muneo reputedly 'used his money and his power ruthlessly...using the familiar tactics of money and intimidation.'[158] He had 'twice been voted the most corrupt politician in all Japan.'[159] As a construction company executive observed

> [t]here are two paths. One is a shorter way, but there is a cliff in the way. If the cliff collapses, the path is dangerous. The other path takes time, but it is safe. If someone is an ordinary politician they don't take the dangerous path. However, I got the impression that Matsuoka and Muneo are advancing along the most dangerous, shortest distance path.[160]

BORDERLINE BRIBERY: MONEY FOR FAVOURS

Various political contribution scandals suggest that what Matsuoka engages in goes well beyond the boundaries of political brokerage. It skirts the fine line between legal and illegal activity and, on some occasions, arguably crosses over the line: 'Matsuoka is asked to mediate and accept a bribe in the guise of a political funding contribution in return'.[161] Several examples of this kind of activity have come to light during Matsuoka's political career. Such activities can only be described as 'borderline bribery', or 'money for favors'. They have been a staple of Matsuoka's Diet career.

Matsuoka's clientelism has extended well beyond activities that could be called 'policy interference'. The deals he has engineered on behalf of clients have sometimes been matters of public policy over which bureaucrats exercised discretion, but he has also mediated in a much wider circle that just bureaucrats. The common feature of such activities is that they are conducted in order to solicit money from individuals. Matsuoka has sold his influence as a mediator to various clients who solicited favours (engineered by Matsuoka) for money. He has provided his clients with inside information, with powerful contacts, with loans, with government contracts, with exemption from the application of specific administrative regulations, or simply his continuing services as a broker.

In 1993, when Matsuoka was a second-term Diet member, he called the King of Real Estate, Sasaki, from a telephone booth at Haneda Airport, saying, 'I'm about to go to Kumamoto. Can you lend me ¥30 million?'[162] Sasaki's answer was, '"*Sensei*, since I learned a lesson from Company T, I cannot lend you money any more." Matsuoka said, "you can't lend me the money?" and I declined by saying "not this time". Then he said, "there is nothing I can do about it" and hung up the phone.'[163] On this occasion, Matsuoka was clearly seeking a loan for himself, just as he earlier had mediated a loan from Sasaki to Company T.

The dividing line between a loan and a financial contribution is, in Matsuoka's case, often quite blurred. In 1996, Sasaki received a witness summons to the '*Jûsen* (Housing Loan Company) Diet'. From about five days before the summons day, Matsuoka starting calling Sasaki. As Sasaki recalls

> Matsuoka called 4 or 5 times a day and said 'I borrowed, not received the money from you, right? Please do not talk about anything suspicious (in the Diet).' Indeed, there is no such case that Matsuoka received the money from me, and I thought he misunderstood. I recalled that I calmed him down by saying 'I know that, *sensei*'. Since this continued for three to four days, I thought Matsuoka was saying a peculiar thing.[164]

Matsuoka was subsequently named in a statement from a major financing company president's to the *jûsen* inquiry as having been 'asked a favour'.[165]

In 2002, Matsuoka had another encounter with Sasaki at the Capital Tokyu Hotel in Nagata-chô. As Sasaki recalls

> [w]hen I went to the restroom, Matsuoka called out 'long time no see', and we shook hands by chance. Then, Matsuoka was all smiles and said 'is there any good chance to make a bit of money?' I replied with a joke that 'if there were a good chance to make a bit of money, I would call at his office soon'.[166]

Other information suggests that Matsuoka's mediation for bribes was structural, meaning institutionalised in his political organisation. A Mr 'K', who was the former chief of Matsuoka's local office (and whose later position was described as office counsellor) reputedly controlled Matsuoka's electoral district. Numerous people with local political connections in Kumamoto testified

> [w]hat 'K' is doing is the same as what Satô Saburô[167] is doing in Katô Kôichi's office. It is a mystery why the mass media takes no notice of this matter. He ['K'] is a purely a broker secretary, and what he is doing is the same as the *sensei*. The year before last, on the occasion of the prefectural gubernatorial election, the LDP's federation of branches asked for cooperation from general contractors without knowing that they were already contributing money to Matsuoka. 'K' was said to have been enraged that they were interfering with their source of finance.[168]

THE FUJI BANK SCANDAL

In 1991, not long after Matsuoka was elected, his name surfaced in the Fuji Bank Fraudulent Loans Scandal. This 'drew attention very early on to Matsuoka's "big shot" ways.'[169] The Fuji Bank Akasaka Branch public relations section chief undertook ¥700 billion in illegal loans, with ¥260 billion vanishing into the night. 'Zennippan', a resort development company, was one of the companies financed by Fuji Bank.[170] Investigations into the affair revealed that Matsuoka received approximately ¥9 million from the resort development company (which he later returned).[171] The former president of Zennippan, Hanada Toshikatsu, looks back on Matsuoka's connection with the Urausu Resort Development in Urausu Town, Hokkaido Prefecture

> I met Matsuoka for the first time just after he became a Diet member. My first impression of Matsuoka was just as I saw him. People call Matsuoka a mini-Muneo and say he shouts in (LDP) divisional meetings. But, he was cringing to me, and I got the impression that he was just a typical government official. His personality was straightforward. I was introduced to Matsuoka by Sonoda Hiroyuki (House of Representatives member, representing Kumamoto) who brought him to a high-class Japanese-style restaurant in the Ginza in the daytime saying, 'please take care of Matsuoka'. Mr Sonoda introduced Matsuoka while we were eating and drinking. He said Matsuoka might be useful for something in relation to resort development. My association with Matsuoka went as far as that. Even in relation to the report that I shouldered the salary of Matsuoka's secretary, because Sonoda asked me to do it, I just paid. I paid not because Matsuoka asked me for the support. Well, it is true that I paid a total of ¥19.2 million (in political contributions, party tickets, secretary's salary), but in what order? Since the *Asahi Shinbun* investigated the matter and wrote about it in an article, it is true. Generally, since Matsuoka is timid, he cannot tell a lie. Although I told Matsuoka 'you do not have to pay the money back', he made a point of paying the money back in full by money transfer. For that reason, Matsuoka is timid.[172]

The scandal involved the removal of protected forests for the Urausu resort development. As Hanada continues

> I did not make any requests to Matsuoka in relation to the removal of protected forests for the resort development. Matsuoka himself explained, 'when I inquired [with the Forestry Agency], (the application was already submitted and) the arrangement of the content was already completed. I did not do anything more than that'. This is also true. However, since I ran into trouble by failing to meet the deadline for the bank financing contract, I asked Matsuoka to 'cooperate in a businesslike manner.' Matsuoka said, 'since the application has already been submitted, if the schedule is delayed, please let me know.' I did not say anything. He was just smiling. At such a time, a politician does not say anything. After that, there were no reports from Matsuoka regarding his inquiries to the Forestry Agency.[173]

The *Asahi Shinbun* (dated the 7 of November 1991) reported that 'Matsuoka made inquiries to the Forestry Agency regarding the removal of protected forests for Urausu Resort, asking questions such as "how are you handling it?" around June 1990.'[174]

THE BSE SCANDAL

In September 2001, the first outbreak of BSE occurred in Japan, and the whole of Japan fell into a BSE panic.[175] The origin of the BSE problem was traced to the 'dysfunctional' MAFF.

> It was insensitive to the BSE warning from the EU and triggered the disease because of cosy relations with and fear of *zoku giin*, and hatred between executives….Talented staff were neglected and forced to retire at an early age by the MAFF administrative vice-minister and his group. Although their precise number was unknown, problems with personnel even extended to the suicide of some staff.[176]

One MAFF OB argued that

> [c]orruption within the MAFF was the main reason why the ministry was unresponsive to the warning from the EU regarding the outbreak of mad cow disease in the UK and the information about human infection with Creutzfeldt Jacob disease (CJD). Agricultural administration was paralysed by mutual animosity between officials, the continuous occurrence of various scandals, the fear of *zoku* Diet members as well as the adhesion with them. The MAFF's state of absent-mindedness caused the outbreak of mad cow disease.[177]

The most problematic aspect of this whole issue was the purchase by the government and disposal by incineration of domestically produced beef, which allowed for the fraudulent substitution of foreign for domestic beef by meat trading companies. The beef buy-up policy was put in place even before the blanket inspection of cows for Mad Cow Disease was implemented, which began on 18 October 2001.[178] At first, MAFF Minister Takebe Tsutomu commented 'the beef before the inspection is also safe',[179] and the MAFF was negative about the government's purchasing the beef. However, the LDP's *nôrin zoku* refused to go along with this.[180] As a result, the MAFF did a complete policy switch. Its original position was that beef should be safe as long as the internal organs such as the brain and intestines were removed. This was behind 'Minister Takebe's foolish performance in eating fried beef (*yakiniku*) during the BSE debacle'.[181]

Muneo and Matsuoka were two of the principal dealmakers in a critical phase of compiling measures to cope with the BSE outbreak in late 2001.[182] When the outbreak first occurred, Muneo and Matsuoka pressured the MAFF to purchase beef from the meat companies that held a large quantity of stock. The LDP's BSE Countermeasures Headquarters (BSE Taisaku Honbu), which had been set up to deal with the problem, held a meeting on 17 October 2001. At the time, one cow had been found to have BSE, all beef deliveries had been stopped and meat companies had 13,000 tonnes of processed beef in stock. Beef traders, who were holding the slaughtered beef that had been excluded from the market, complained, 'if the MAFF's investigation of Japanese beef cattle proceeds, the price of the 13,000 tonnes of beef excluded from the market will fall to zero, and beef traders will be forced into an awkward situation.'[183]

The meeting of the countermeasures headquarters was attended by various *nôrin zoku*, the MAFF deputy minister, the MAFF Production Bureau director, the MAFF Livestock Department director[184] as well as other MAFF officials, together with officials from the Ministry of Health, Labour and Welfare, and the Ministry of Education, Science and Technology. The MAFF was loath to incinerate the meat because it looked like an admission that the beef was not safe.[185] Muneo, Matsuoka and others shouted that 'all the beef should be purchased before the inspection'.[186] According to a journalist attached to the MAFF, Matsuoka and Muneo said, 'the state has to bear the burden of the whole amount of the incineration fee and the purchase cost of the 13,000 tonnes of beef.'[187] Muneo argued (in words that were to become infamous),[188] '[t]he state just has to say they will take the 13,000 tonnes. That is all that needs to be done. All right? It's a simple solution. It's just a matter of ¥26 billion if it's ¥2000 a kg, or ¥13 billion if it's ¥1000 a kg. You can get Etô (Takami) *sensei* or somebody to make the budgetary measures. Got that?'[189] Muneo reportedly intimidated the MAFF bureaucrats present at the meeting,[190] including his shouting at Production Bureau Director-General Kobayashi Yoshio.[191]

Matsuoka also made his point strongly from the very beginning of the meeting, saying 'the state should purchase all the beef even if it costs more than ¥10 billion. In fact, even if it costs ¥100 billion as in the EU, we should do it. We should do as much as if not more than the EU.'[192] The essence of his subsequent remarks went as follows.

> Who will purchase the stock, the MAFF or the Ministry of Health, Labor and Welfare? I'll get everyone to decide here and now before you leave the room. Even the EU conducted intervention purchasing. Even though we said we would do the equivalent of the EU, why are you leaning in the direction of the Ministry of Finance? Doing roughly equal to the EU is the consensus of the division. The point is to produce an outcome. Just talking and listening is no good.[193]

According to one report, 'since there were deputy ministers from two ministries there, Matsuoka wanted them to make the state take care of it politically. Another thing Matsuoka wanted was for the prime minister to make a safety declaration in two days time.'[194] CAPIC advisor, Etô Takami, said to the assembled MAFF officials, 'I do not ask you to take responsibility. We should purchase under the ruling party's responsibility'.[195] When Muneo turned up at the meeting the following day, he backed Matsuoka up by saying the same things as Matsuoka had already said, '[a] safety declaration is exceedingly essential. It should be performed by prime minister.'[196] The 'MAFF's policy was overturned as if it were nothing, without a chance given to officials to put a counterargument.'[197]

The scene at the LDP BSE Countermeasures Headquarters in mid-October 2001 where Matsuoka, Muneo and Etô sent for MAFF bureaucrats and demanded the buying up of beef by the state was shown over and over again in the media.[198] Etô, Suzuki and Matsuoka became well known as the three key people at the centre of the emergency countermeasures project. Having received their instructions, the MAFF immediately instituted the all-head inspection of beef cattle, and the processed meat in the warehouses was purchased by the government and incinerated.[199]

However, it did not take long for the smell of corruption to hover over the beef deal. As one media source commented

> [a]s tribe Diet members, this [countermeasures project] may have been an 'honest and decent' thing to do, but if there were any begging from special industry members involved, it becomes a crime. Furthermore, there were many businesses bumping up their stock with imported and meat cut-offs and rushing to cut up beef before the buy-up began on 25 October.[200]

As it turned out, prior to the countermeasures meeting, the amount and manufacturers' asking price for the beef in stock had been passed on by a big-shot beef executive directly to MAFF officials in a meeting in the MAFF building. This big shot and representatives of three related meat groups had pre-decided the allotted amounts for each group in a room in the ministry, and passed this on secretly to the bureau in charge. The MAFF (Meat and Egg Division) said this was not in fact so, and also denied any pressure from tribe Diet members and beef traders.[201] MAFF Minister Takebe also declared openly, 'the MAFF is responding firmly to unreasonable opinions from politicians.'[202] Nevertheless, the MAFF official who drew up the beef buy-up scheme was reportedly opposed to it, suggesting that he gave in to pressure from the *zoku giin* led by Muneo and Matsuoka, who had taken on board the views of those in the meat industry.[203] The scheme provided for government purchase of 13,000 tonnes of frozen meat for ¥29.3 billion in a total budget of ¥200 billion for BSE countermeasures.[204]

Matsuoka was a member of the three-man LDP Europe Research Group (Ôshû Chôsadan) headed up by former MAFF Minister Yatsu. The group travelled to Europe to see how the Europeans had dealt with the BSE problem. When the group returned to Japan it reported back to a meeting of the Agriculture and Forestry Division and the BSE Countermeasures Headquarters.[205]

There was a strong suspicion of adhesion between Muneo and the meat industry concerning the system by which the MAFF enforced the purchase of domestically produced beef.[206] Everyone recognised that Muneo and Matsuoka accorded priority to the meat distribution industry in the BSE countermeasures. Suzuki's electorate was the Hokkaido PR bloc, while Matsuoka's was in Kumamoto Prefecture, both prominent beef-producing regions. Livestock producers were important sources of support for both politicians.[207] On top of this, major meat companies donated money to one of Muneo's political funding groups.[208] One popular magazine in Japan (*Shûkan Bunshun*) disclosed that 'Suzuki Muneo and "Muneo of the West" (Nishi no Muneo)—Muneo of the East and West—had orchestrated the scandal of the BSE beef buy-up'.[209]

The May 2002 issue of the magazine *Sentaku* commented, 'both are known for their intimate relations with major meat wholesale companies such as Hannan in Osaka City and Fujichiku in Nagoya City.'[210] The links went back to the time of Nakagawa Ichirô.[211] Former chairman of the Osaka meat company Hannan Corporation, Asada Mitsuru,[212] known as the 'Don' of the meatpacking industry, was reportedly a supporter of Nakagawa and when Nakagawa died, he became a supporter of Suzuki.[213]

Another source in the LDP elaborated

> [t]he Hannan and Fujichiku big meat groups originally had good relations with one-time *nôrin zoku* godfather, the late Nakagawa. After the death of Nakagawa, while his first son Shôichi kept his distance from the meat industry, it was Muneo Suzuki, who had been Nakagawa's secretary, who began to deepen friendly relations. Muneo invited Asada to his eldest daughter's wedding as the guest of honour. It is said that the two companies and Muneo were intimately bound up in each other's dealings.[214]

The close relationships between Asada and Muneo centred mainly on financial support, such as making donations, providing luxury cars, paying consultants' fees (¥300,000 for half a year)²¹⁵ and providing Suzuki with a *kôenkai* office, the Osaka Food Distribution Research Institute (Ôsaka Shokuhin Ryûtsû Kenkyûjo). Through Suzuki and his sworn friend Matsuoka, Asada had a direct pipeline to the political world.²¹⁶ According to one report

> 'Muneo of the West', Matsuoka, joined up with Muneo, Hannan, and Fujichiku. In the 1990 Lower House elections, Muneo introduced Matsuoka to Asada as his influential backer. Ever since, a division of labour apparently had been promised: Hannan has been Muneo, and Fujichiku has been Matsuoka.²¹⁷

Prior to the October 1996 election, Matsuoka went to pay his respects to Asada, and to ask for his help when running in the election.²¹⁸

Hannan Corporation was not only involved in the distribution of beef, but also in beef production, rearing 5,800 head of beef cattle, and producing more than 470,000 tonnes of meat annually. Total sales of the group exceeded ¥300 billion.²¹⁹ The extent of Hannan's involvement in beef production and distribution accords it considerable market power. It even 'has the power to influence beef prices.'²²⁰ Because of this power, an official of the Livestock Department of the former Livestock Bureau of the MAFF commented that the MAFF could not neglect the 'Asada Pilgrimage'.²²¹

Fujichiku leads the beef industry in the Nagoya area, and Asada was also an executive of the company. A person in the meat industry explained, '[t]here isn't a business that can stand up to the Hannan-Fujichiku alliance, and their presence is such that even the government administration acknowledges their superiority.'²²²

Asada's portion of the beef buyback scheme was 1700 tonnes—or more than 10 per cent.²²³ All up, a total of 40 groups nationwide were involved in the government-funded beef buy-up scheme, but of these, the three meat-trading groups in the Kansai region (Osaka, Aichi and Hyogo) offered suspiciously large amounts of beef for purchase when compared with other groups.²²⁴

The beef inspection regime became much stricter in early 2002, moving from sampling all lots to inspecting all boxes in order to check whether imported beef was definitely not included in the beef for incineration. Not surprisingly, six beef groups petitioned MAFF Minister Takebe for a relaxation in the inspection regulations and a return to less severe sample inspections. The groups at the centre of the request were the National Federation of Meat Industry Cooperative Associations (Zenkoku Shokuniku Jigyô Kyôdô Kumiai Rengôkai, or Zennikuren), the National Federation of Agricultural Cooperative Unions (Zenkoku, Nôgyô Kyôdô Kumiai Rengôkai, or Zennô) and others. Their representatives went to the MAFF to discuss the issue. Asada, who disliked appearing in official and public capacities, was involved in the request behind the scenes.[225]

A highly significant meeting took place in Minister Takebe's office in April 2002. A report that made mention of the meeting was compiled by the chairman of Zennikuren for the chairman of the Prefectural Federation of Meat Industry Cooperative Associations (Todofuken Shokuniku Jigyô Kyôdô Kumiai Rengôkai), its prefectural organisation. At the meeting in the minister's office, the meat industry lobbied the government for necessary countermeasures in response to the distressing situation in the meat marketing industry. It was opposed to the MAFF's announcement at the end of March that it would subject all beef boxes for sale for examination, switching from sample box to all-box testing. From the MAFF side, the minister, and others such as the Production Bureau director-general, the Livestock Bureau director-general and the Meat and Egg Division director were present. From the industry side, executives from Zennikuren and 19 other executives from companies such as the Japan Ham and Sausage Industry Cooperative Association participated. Of these, 14 were members of Zennikuren. According to one report, the group included big wigs from the meat industry such as Asada and Nagoya's Fujichiku's President Fujimura Yoshiharu.[226] A MAFF official recalled, 'they pressed Minister Takebe, saying "explain yourself about the all-box examination!!!" and "do you mean to treat the industry like criminals?!" Muneo reportedly made the arrangements for this group negotiation'.[227]

Because Asada and Fujimura were big wigs of the meat industry, who sponsored powerful MAFF Diet members, Minister Takebe could not easily reject the 'request activities' associated with these two.[228] However, he 'lost his temper at the meeting with the meat industry executives, saying: "Why do I have to be spoken to in such a way?!" He declared he would "do the box examinations no matter what!" Thus the meat industry's plan to stop the box examinations collapsed'.[229]

As it turned out, the major meat-wholesaling company, the Osaka-based Hannan Corporation with which Matsuoka and Muneo were deeply connected, was later found to have received the subsides for BSE illegally.[230] In August 2004, Asada pleaded guilty to swindling the government out of ¥5.03 billion through the beef buy-back scheme.[231] He pleaded guilty to conspiring with others to label imported and other types of ineligible beef falsely as domestic meat in order to qualify for government subsidies.[232] The ¥5.03 billion amounted to about a quarter of the ¥21 billion in BSE-related subsidies that the government paid out.[233] Asada reportedly gathered advance information on the government's program from politicians 'with whom he was friendly'[234] as well as from MAFF officials.[235] He purchased 94.5 tons of imported beef from Heisei Foods in Hiroshima Prefecture. This meat was then sold under the buyback plan as domestic beef. It originally came from a meat processor in Kumamoto City.[236] In total, Asada falsely labelled about 434 tonnes of unsold imported beef, which was ineligible for the buy-back, and ordered his group companies to procure more of it.[237] This was not the first occasion in which Asada had broken the law. He was arrested in 1987 for bribing a Livestock Industry Promotion Corporation (LIPC) official, Aoyama Yutaka. Veteran journalist, Mizoguchi Atsushi in his book *Emperor of Meat: A Man Who Made A Fortune, Asada Mitsuru* (*Shokuniku no Teiô: Kyofu o Tsukanda Otoko, Asada Mitsuru*), 'depicts Asada as a powerful political fixer. He is portrayed as a man who works secret deals with politicians such as former Liberal Democratic Party heavyweights Muneo Suzuki and Hiromu Nonaka, and is a close friend of Yamaguchi-gumi boss Yoshinori Watanabe.'[238]

The Tokyo Metropolitan Police Department conducted a secret investigation of Hannan and its links to politicians. While the investigation began with Suzuki as their prime target, their focus gradually shifted to Matsuoka. An executive of the Metropolitan Police Department stated, '[w]e have received instructions from the Tokyo District Public Prosecutor's Office in code saying, "give us M, whatever it takes", but there are some who are saying, "we are getting confused as to whether M is "Muneo" or "Matsuoka"'.[239]

The Special Investigation Department of the Tokyo District Public Prosecutor's Office maintained top-secret documents called 'The Politicians' File'. It contained records allegedly pertaining to Nonaka, Kamei, Suzuki and Matsuoka, including the personal connections and flow of money relating to these four. An OB of the Special Investigation Department commented, 'we exposed Kanemaru (Shin, former LDP deputy-president) for tax evasion, and then Takeshita (Noboru, former prime minister) died. There is a possibility that we might get two of the four (Matsuoka and Muneo) left in the file'.[240]

One MAFF OB was scathing about Matsuoka's role in engineering the domestic beef buy-up scheme, thereby arranging concessions for people in the meat industry who were a source of financial backing. In this respect, Matsuoka's skill reputedly far exceeded that of Muneo.[241] This was despite an assertion from

Matsuoka's Diet office that '[t]he BSE Countermeasures Headquarters created the system of the state buy-up of beef, not Matsuoka. Therefore, to say that he is receiving a portion of that money as political donations is groundless.'[242]

After his behaviour in the LDP's countermeasures committee, the MAFF began working towards a parting of the ways with Matsuoka, and even the LDP tried to get rid of him as a 'noise-maker'.[243] At the beginning of March 2002, the first meeting of Special Committee Concerning Securing Food Safety (Shoku no Anzen Kakuho ni kansuru Tokumei Iinkai), under the chairmanship of former Defence Agency Director-General, Norota Hôsei, was held. The committee was launched by the LDP to debate the review of food safety administration. In reality, the special committee represented the first attempt by the party 'to remove the noise-makers such as Muneo and Matsuoka'.[244] The special committee was an organisation under the direct control of PARC Chairman Asô. The special committee comprised senior LDP figures who set about appointing PARC chairs and deputy chairs not only for agriculture, forestry and fisheries divisions, but also for health, labour and welfare, environment, cabinet and other related divisions. An influential Diet member, who was a core member of the committee, explained that 'this is Matsuoka's removal.'[245]

THE SYRIAN EMBASSY AFFAIR

Matsuoka's name also surfaced in relation to a dubious affair involving the Syrian Embassy in Tokyo. A building in Azabu-Nagasaka-chô, which the embassy was leasing, was presented for auction in July 2001. The building's owner had gone bankrupt, and a real estate company made a successful bid for it at the auction. The company subsequently asked the Syrian Embassy to vacate the building. When the embassy refused, the company requested an eviction notice from the Tokyo District Court. The court decided that the eviction was not possible under the Vienna Convention. However, the High Court,[246] to which the real estate company appealed, argued that the compulsory execution (of the eviction) was possible, based on the fact that the person renting the building in question was an individual, a Mr Kabul, the temporary acting Syrian ambassador, and therefore, diplomatic extra-territoriality did not apply in this case.[247]

The High Court decision made the eviction possible and in December 2001, the compulsory execution of the eviction began. The public safety authorities inspected the rooms and peeled off every single piece of wallpaper, saying that Syria was an anti-American, Islamic country.[248] However, a Japanese person claiming to be connected to the Syrian side appeared at the real estate company and requested suspension of the execution. An hour later, another person called Izumi Hideki, who claimed to be the Diet secretary of Tanikawa Kazuo,[249] also appeared and demanded, '[s]top the compulsory execution! A political settlement has been made over this property.'[250] However, a court official 'indicated that the court had a handle on the fact, saying, "[t]he enforcement officers turned a deaf ear to that individual, and the building was vacated, as ordered."'[251]

It was later learned that Izumi had been fired from Tanikawa's office for embezzling tens of millions of yen 10 years earlier. There had been no contact between Tanikawa's office and Izumi since, and he had been requested to stop using name cards claiming to be Tanikawa's secretary.[252] Izumi later appeared suddenly at MoFA with Matsuoka and the temporary acting Syrian ambassador. At the time, Matsuoka was allegedly working for the temporary acting ambassador in relation to the Syrian Embassy's building problem. Matsuoka belonged to the Japanese-Syrian Friendship Diet Members' League (Nihon Shiria Yûkô Giin Renmei), which was practically defunct, but the Syrian side calculated that the services of Matsuoka as a broker were for sale and that he would be able to squeeze MoFA.[253] Matsuoka and Izumi pressed the administrative vice-minister, saying, 'we ask you to please do your best in regard to the Syrian case'.[254] However, their request was to no avail. MoFA had decided against the Syrian Embassy in relation to the leasing issue,[255] and as a result, the embassy felt betrayed by MoFA.[256] It decided to resort to power politics, seeing MoFA as an imaginary enemy.[257]

Following the court order for eviction issued by the Tokyo High Court, the Syrian Embassy also filed a special appeal. The Supreme Court rejected this appeal on 23 January 2002.[258] On the evening of 24 January, the Director-General of the MoFA Minister's Secretariat, Komachi Kyôji, and the Director-General of the Middle East and Africa Bureau, Shigeie Toshinori, were summoned to a dinner hosted by Matsuoka at an Akasaka restaurant in Tokyo, the Tsuruhachi. Waiting at the restaurant[259] were the Syrian Chargé D'Affaires, Mr Haida, and two Japanese people who claimed to be connected to the Syrian Embassy, one of whom was Izumi. Others at the meeting reportedly included *yakuza* and the embassy's Egyptian interpreter, who used a false name and who had acted as a go-between for Middle Eastern ambassadors and Japanese traders and brokers, and who had possible connections with *yakuza*. He had been involved in many embassy-related troubles.[260] As one journal reported

[a]fter an hour, Matsuoka arrived at the restaurant and the group decided to move to a larger private room. However, suddenly there was a commotion. Chargé D'Affaires Haida had asked Matsuoka to put in a good word to MoFA on the issue of the Syrian Embassy move, to which Matsuoka said, 'I was in my electorate until yesterday, and I just came back. I promise I'll do it' and apologised. Haida reported that he had been warned: 'You shouldn't rely on dodgy connections' and that his application was refused at the gate when he went to MoFA to register the number plate of his official car. After the translator had finished relating the story, Matsuoka exclaimed: 'What!?' He then made several angry phone calls on his mobile phone, leaving his guests waiting in the large room. Then, in just under an hour at about 9pm the Director-General of the MoFA Ministers Secretariat Komachi and the Director-General of the Middle East and Africa Bureau Shigeie appeared.[261]

Matsuoka reportedly shouted at the MoFA officials at the restaurant. He said to Shigeie

'[y]ou always tell lies, don't you?....You said you had an appointment with Nogami, but he said he didn't recall having made such arrangements. Were you not going to come if I didn't find out?' Shigeie could only reply ashamedly. At this point, Matsuoka broached the issue of the Syrian Embassy. Komachi replied: 'There is not much we can do about the judgement of the law...' But as if to cut him off, Matsuoka said harshly, 'There is something wrong here. This is a diplomatic issue. Don't you know how important Syria is to Japan?'[262]

After this, Muneo arrived with some of his close associates (MoFA division directors), including the Director of the MoFA Policy Planning Division, Uemura Tsukasa, former administrative secretary to the Minister of Foreign Affairs, Tanaka Makiko.[263] Matsuoka 'who had wanted to give the MoFA bureaucrats a scare, had summoned Suzuki, who was holding a thank-you party for the International Conference on the Reconstruction of Afghanistan at a nearby steak restaurant. Uemura reportedly turned pale upon seeing Komachi and Shigeie.'[264] Matsuoka told Suzuki of MoFA's clumsiness in dealing with the issue and the meeting ended at 11pm. With nothing resolved, however, the only purpose of the dinner party appeared to be Matsuoka's intention to demonstrate his influence. Komachi and Shigeie were later reshuffled from their posts.[265] When later questioned about why he had called Komachi and other MoFA officials to the restaurant, Matsuoka replied, '[a]t the meeting, the Syrian Embassy produced documents on the building in English. So we decided to call in specialists from the Ministry of Foreign Affairs.'[266] He also explained the situation (and his part in it) in the following way

[t]he previous chargé d'affaires at the Syrian Embassy warned me last year that this might escalate into a bilateral problem. So I talked several times to Foreign Ministry officials, such as Mr Shigeie, to look for ways to reach an amicable settlement. The Syrian side has blamed the Foreign Ministry for the consequence (forcible eviction). (The Syrian side) protested that its

flag had been taken away. They also said: 'Our president is angry because Syria's dignity has been undermined'....The intensity of the discussion on the Syrian side was serious. They showed their discontent and anger even over drinks. They were quite prickly.[267]

A later report revealed further contact between MoFA officials and Matsuoka and Izumi. The Parliamentary Secretary for Foreign Affairs, Matsunami Kenshirô, division chiefs and bureaux directors-general had a meeting with Matsuoka and Izumi in February after the restaurant incident.[268]

The episode raised questions about Matsuoka's true purpose in acting as a mediator for Syria.[269] Asked why he had become involved in the issue, Matsuoka replied that he belonged to the Japanese-Syrian Friendship Diet Members' League and was asked for guidance as Kabul was an acquaintance. He claimed that the first time he met Izumi was on the 24 January at the dinner. He denied that he had taken Izumi to MoFA before that, although he admitted that Izumi might have been there at the time.[270] He said that he had never heard of the 'political settlement' that Izumi demanded should prevent the execution of the eviction and also flatly rejected any suggestion that there was any 'giving and receiving of money'.[271] As a result of the affair, some commentators asked whether Matsuoka was taking over from Muneo in having MoFA under his thumb because Muneo was on the verge of sinking (into political oblivion) as a result of corruption scandals.[272]

THE YAMARIN SCANDAL

In 2002, reports surfaced that a Hokkaido logging company by the name of Yamarin,[273] which was at the centre of a political bribery probe, had paid Matsuoka ¥2 million in 1998. At the time, Matsuoka was chairman of the State-Owned Forests Problems Subcommittee. The payment was not recorded as income by his political funds management group.[274] Earlier (in 1996 and again in 1997), Yamarin had officially contributed ¥360,000 to Matsuoka.[275] Such funds were reported under the Political Funds Regulation Law.

In June 1998, information came to light that Yamarin had carried out illegal logging from around 1992 and the Forestry Agency was considering a severe administrative punishment for the company.[276] According to one report

> Yamarin went into a mountainous area in the middle of the night and cut trees down with chainsaw. As far as we know, Yamarin stole 7,062 trees. In fact, this had been a daily occurrence by Yamarin over a long period. The illegal logging had been exposed twice, and a total of 10 people had been arrested. At the same time, a total of nine executives from a local forestry office had been appointed to high positions in Yamarin. Someone formerly connected to the forestry

industry commented, 'Yamarin and the local forestry office are completely companions in crime. The public prosecutor ignored the issue. This collusion went as far as the local forestry office lending even its official seal to Yamarin, in order to send the logs to the market'.[277]

The Forestry Agency later imposed a seven-month administrative punishment on Yamarin for illegal logging in national forests. The sanction froze its qualification to participate in tenders for the purchase of logs from the Obihiro Forestry Management Branch Office, a local outpost of the Forestry Agency. Yamarin had no choice but to withdraw from bidding for the public sale of logs by the Obihiro office.[278] The Yamarin President Yamada Isao (85 years old) then petitioned Matsuoka[279] about the matter and gave him the ¥2 million donation. Two days after Matsuoka received the money he spoke with senior officials of the Forestry Agency and asked for leniency for Yamarin.[280] According to a person connected to the local political world

> Yamada served as the chairman of Suzuki Muneo's *kôenkai*. He has been deeply involved in politics up to now. If Yamarin were excluded from tenders for cutting down trees in state forests by the Forestry Agency, its business would not be viable. So Matsuoka desperately attempted to rally his strength by using his political power.[281]

The fact that Matsuoka had received a donation from Yamarin came to light in the process of the investigation of Muneo by the Tokyo District Public Prosecutor's Office.[282] Matsuoka's office denied receiving money from Yamarin, saying that the matter was still 'under investigation, but there were no incidents where he had asked for favours or had approached people'.[283] One of his secretaries

> [c]laimed that the money Matsuoka had received from Yamarin was a 'political donation' [which was, therefore, above board and not tied to any political favour]. However, the secretary could not confirm the date and amount of money involved, saying that all the relevant documents had been 'scrapped'.[284]

The secretary also said that Matsuoka 'had no recollection [of the donation] whatsoever'.[285] His 'office replied to a newspaper interviewer that Matsuoka had returned the money by the end of the year, but his office "did not remember" the date or the method of receiving the money.'[286] A sports newspaper, *Nikkan Sports*, commented that two days after receiving the donation on 4 August 1998, Matsuoka 'called the Forestry Agency director-general and appealed to him to take the "appropriate" steps for his punishment, but the director general refused.'[287] Another source disclosed that Matsuoka returned the money to Yamarin in early 1999 after illegal logging in government forests in Hokkaido became an issue in the Diet.[288]

Matuoka's receipt of political funds from Yamarin was allegedly part of a bigger deal involving Muneo, who was later prosecuted for accepting a ¥5 million bribe from Yamarin. Muneo accepted the bribe in exchange for seeking favourable treatment for the company from the Forestry Agency.[289] According to sources at the Tokyo District Public Prosecutor's Office, President Yamada met with Suzuki on 4 August and asked him to get the Forestry Agency off the firm's back. Four executives from Yamarin, including Yamada, visited Suzuki's office in the Kantei (Suzuki had just been appointed deputy chief cabinet secretary). They asked him to pressure Forestry Agency officials to sell Yamarin trees on national land after the end of the ban. The 'company wanted the agency to sell it an amount of trees equivalent to what it was unable to purchase under the ban and to do so outside the public bidding process.'[290] The ban on Yamarin's participating in tenders had been extended for a month and its main forestry businesses were greatly affected.[291]

At Suzuki's behest, President Yamada allegedly delivered ¥5 million in cash to one of Suzuki's secretaries in his Diet office on the same day. Having received the donation, Suzuki called a senior Forestry Agency official on the spot to ask for better treatment for Yamarin but the request was denied.[292] Later that same day, Yamada and other executives met with Matsuoka and gave him the ¥2 million. According to a statement by Yamada handed to the Lower House by Muneo in mid 2002, the Yamarin president recollected that on 4 and 5 August, he had also made a donation of ¥2 million to Matsuoka and ¥500,000 to Matsushita Tadahirô,[293] who was the MAFF parliamentary vice-minister at the time, and who represented the Kyushu PR constituency. Yamada's statement also said, '[the Public Prosecutor said to me] they were only after Suzuki. I understood this to mean that although I had given donations to Matsuoka and Matsushita on 4 and 5 August, their cases would not be subject to investigation'.[294] Even so, when the Muneo scandal broke, Matsuoka, clearly under strain, reportedly 'raised his voice to government officials and bureaucrats of special public corporations, and sometimes even to his fellow Diet members'.[295] Matsuoka and Matsushita were questioned by the Special Investigation Department and were reported to be 'trembling in fear'.[296] Officials in the Ministry of Justice were heard to say '[i]f anything, (Matsuoka's) aims are more obvious than Mr Suzuki's, and in terms of dirty money finding its way to him, he has great aptitude'.[297]

It was on 7 July 2002, three weeks after Muneo's arrest that *Sankei Shinbun* reported Matsuoka's threatening of Kitamura Naoto, an LDP Lower House Diet member representing Hokkaido (13), in a headline saying 'Matsuoka: Election Defeat Rather Than Crush Yamarin'.[298] Kitamura had reportedly whistle-blown to MAFF officials at that time that Matsuoka was pressuring the Forestry Agency regarding Yamarin. Matsuoka became angry because Kitamura had told some top officials in the MAFF that '[both Muneo and Matsuoka] had pressured the Forestry Agency on Yamarin's behalf'.[299] Matsuoka called up Kitamura and threatened him '[i]f you intend to smash Yamarin, I will make you lose in the next election.'[300] Matsuoka was already daggers drawn with Kitamura because Kitamura had defeated Muneo twice—in 1996, as a candidate from the New Frontier Party and again in 2000, as a member of the LDP, forcing Muneo to retain his Diet seat only by virtue of the LDP party list in the Hokkaido PR bloc. Kitamura had refused to move over for Suzuki as the LDP's endorsed candidate in Hokkaido (13).

In fact, the ties linking Yamada with Muneo and Matsuoka ran very deep. According to a person with connections to the Forestry Agency

> Yamarin was an influential company in the east of Hokkaido which developed all-out support for Nakagawa Ichirô from the time it was called Yamada Forestry [Yamada Ringyô]. President Yamada supported Muneo after Nakagawa's death, and even served as chairman of Muneo's supporters' organisation, but there was a head clerk in his company called Akahori. He was the person involved and present at the bribe at the deputy chief cabinet secretary's office, and is still the president of a company related to Yamarin. This person is actually a classmate of Matsuoka from Tottori University.[301]

Both Akahori and Matsuoka were in the Department of Forestry[302] at Tottori University, and both Matsuoka and Muneo were known to value friendships with old classmates.[303] According to another classmate

> [i]f I remember correctly, the two [Matsuoka and Akahori] were a year apart. Akahori was Matsuoka's junior, but he hardly came to university and had to repeat about four years. So they actually graduated about five years apart, but the Forestry Department of Tottori University has a very strong alumni network. The department established an alumni association called 'Sarenkai' in various places, its name taken from the crest that the Tottori sand dunes create. There are very few OBs in Hokkaido, so this probably made their sense of camaraderie all the stronger.[304]

According to a person related to Yamarin

> Yamarin was totally under the control of President Yamada Isao, his second son President Yamada Satoshi and the director at the time, Horiuchi. On the second floor of the headquarters,

their three desks were arranged facing the company employees, and everything was decided by these three, from important directions of the company projects to donations to politicians. Akahori was the only outsider out of all the Yamada family firms to rise through the ranks, and he was particularly trusted by President Isao. Akahori is a classmate of Matsuoka, so the relationship is pretty obvious. So in regards to the August 1998 lobby that has now become an issue, it was natural for Yamarin to go to Matsuoka.[305]

A journalist attached to the Tokyo District Public Prosecutor's Office also revealed

> [f]or Matsuoka who has served as the chief of a forestry office in Hokkaido, this region is like a 'second base of operations' (*konkyochi*), it also being Suzuki Muneo's sphere of influence. In reality, in Matsuoka's political funding reports, donations from Hokkaido forestry-related businesses are far greater than others.[306]

When Matsuoka was chairman of the Agricultural Basic Policy Subcommittee in 2002, he made a point of conducting an on-the-spot survey of Hokkaido himself as part of the LDP's input into the new Rice Policy Reform Outline (*Kome Seisaku Kaikaku Taikô*). Other agricultural heavyweights went to other prefectures. Furthermore, 'the [bribery] affair involved Matsuoka's closest business link where he also had a classmate, so of course, he was going to fret. I don't think it was a coincidence that he "took refuge" in hospital the day after the Yamarin reports.'[307]

Matsuoka's office admitted that Matsuoka was a former classmate of Akahori of Yamarin and that through Matsuoka's period of attachment in Hokkaido, the relationship was such that 'Matsuoka was supported by Akahori'.[308] However, the office elaborated

> [w]ith respect to the donation, it is said that (Yamarin) came to the office on 5 or 6 August 1998, but since it was a problem of people cutting down other people's trees, we would have not accepted it, even if there were an offer of a donation. We do not keep a record or list of visitors here at the office so (the visit and the donation) cannot be confirmed.[309]

In March 1999, Matsuoka reportedly travelled to Hokkaido to support the election of Hokkaido Prefecture assembly member, Yamada Rintarô, who was the eldest son of President Yamada, in his bid for a second term. One local political personage commented

> [t]he rally for Yamada Rintarô was splendidly conducted at Culture Hall in Obihiro City. First, the support video by Muneo was put on the screen, and subsequently, Matsuoka said: 'Well, I came to act on behalf of Suzuki Muneo. I am the first follower, Matsuoka'. At the rally, local forestry office executives lined up in a row in an anteroom, and all the executives kowtowed to Matsuoka. It seemed as if the rally was also a lobbying rally by forestry persons to Matsuoka. However, Matsuoka's support had the opposite effect to what was intended. The district was the electoral district of Nakagawa Shôichi (the first-born son of the late

Nakagawa Ichirô), who was an old enemy of Muneo. (In such a district), support from Matsuoka (Muneo's friend) had a rather negative impact on the election campaign. Because of Matsuoka's support, Yamada Rintarô lost the election.[310]

The scandal involving Yamarin was one of the scandals that ultimately felled Muneo. There was an order from the Supreme Public Prosecutor's Office to the Tokyo District Public Prosecutor's Office to 'get Muneo, no matter what'.[311] Suzuki had received an 'unlawful request' from Yamarin and obtained a substantial amount of money in return. Even 'though it was a formal political donation, like the Recruit scandal involving former Chief Cabinet Secretary, Fujinami Takao, a corruption case can be made if a donation can be linked to a specific request. The Special Investigation Department [of the Public Prosecutors Office] was clearly aiming for a case of mediation bribery.'[312]

Three of Suzuki's aides were also arrested 'on suspicion of failing to declare around ¥100 million in donations to Suzuki's political fund management group. All three were suspected of violating the Political Funds Control Law.'[313] They were all previously secretaries of Nakagawa Ichirô, and after Nakagawa committed suicide, they transferred to Suzuki when he successfully ran for Nakagawa's seat. Prosecutors were also poised to charge Suzuki himself with 'instructing his aides to conceal the donations.'[314]

Muneo left the LDP under a cloud in March 2002 but remained a Diet member in spite of a Diet resolution urging him to give up his seat.[315] He was arrested on 19 June 2002 on suspicion of the crime of accepting bribes for mediation (*assen shûwaizai*).[316] Asada, however, continued to act as his patron, providing him with a car to his office right in the middle of the scandal.[317] Moreover, Matsuoka was summoned as a witness in the investigation of Suzuki, and afterwards, he was given a 'thank you' party by Suzuki.[318]

The ramifications of Muneo's arrest went far and wide and also caught Matsuoka potentially in the net. There were reports that Matsuoka would be next to be taken to court for committing a mediation bribery crime in relation to both Yamarin and the BSE issue.[319]

The day after the scandal broke in the *Yomiuri*, Matsuoka disappeared. It was later disclosed that he had gone to hospital for haemorrhoid surgery.[320] The Tokyo District Public Prosecutor's Office investigated Matsuoka as a witness to the Yamarin affair and the circumstances of the Yamarin donation to Matsuoka but decided not to prosecute. It was a huge relief for Matsuoka. He telephoned one of his influential supporters in his electoral district and said, '[e]verything is over. We do not have to worry any more'.³²¹

At the same time, reports surfaced that an important private secretary of Matsuoka, 'K'—his 'Satô Saburô'—had escaped overseas.³²² Rumours also surfaced of a flood of politically defamatory literature about Matsuoka in Nagata-chô. Local news section reporters were rushing from place to place to get the information.³²³

Muneo's arrest and Matsuoka's patent difficulties had implications that went well beyond the Yamarin scandal itself. As a veteran political reporter from a national newspaper explained

> [i]n the city [newspaper] desk way of thinking, arresting a House of Representatives member over a ¥5 million bribery is a 'small incident', but it is 'a big incident' for Nagata-chô. This is because Diet members' daily political activities under certain circumstances are equivalent to 'accepting bribes for mediation'.³²⁴

The Suzuki case set a stricter benchmark for judging what did and did not constitute political bribery, which would be prosecuted under the Political Funds Regulation Law. As a political journalist explained

> [p]reviously, political pressure from *zoku giin* when they mediated for companies was not considered to amount to a crime of bribery if the money were legally processed in conformity with the Political Funds Regulation Law. It was a system for receiving money lawfully by which they solicited political donations from a large number of companies and groups in small amounts over a long term. However, by making Muneo's case a criminal case, even if the money provided by companies were reported to Ministry of Internal Affairs and Communications as political funds, the money was recognised as constituting bribery. This is an epoch-making decision by the Special Investigation Department. This is the essence of the Yamarin scandal…The major premise, on which the money that was received was skilfully and legally processed, collapsed. This great change had an impact on Nagata-cho.³²⁵

In 'arresting Suzuki, the Special Investigation Department of the Tokyo District Public Prosecutor's Office handed down a "no" to the old method of political funding'.³²⁶ Even though formal (and even reported) donations might be involved, a corruption case could be made if the donation could be linked to a specific request.³²⁷ Henceforth, the major premise of activities where the route of mediation went from companies to politicians to government offices collapsed.³²⁸

This was not the only repercussion from the Yamarin affair. People speculated about who would be taking the lead in the sorts of agricultural policy areas in which Suzuki specialised, such as price decisions, and about future power relations between the MAFF and *nôrin giin*.³²⁹ By leaving the LDP, a decline in Muneo's power in relation to bureaucrats and policy decisions was unavoidable, given his substantial influence over the MAFF and also over MoFA.³³⁰

While Muneo's arrest for accepting bribes from Yamarin caused huge ripples, in reality, however, it was Muneo's friend, his sworn brother, Matsuoka, who exercised enormous power over those connected to the Hokkaido forestry industry.³³¹ As the president of a reforestation company in Ohihiro, Hokkaido, commented

> Matsuoka, who is from the Forestry Agency, is an extremely influential presence for the Hokkaido forestry industry…He has experience as chief of the Teshio Forestry Office in Hokkaido when he was a young bureaucrat of the Forestry Agency. He was only in his mid–30s, but local businesses treated this chief-of-office from the central Forestry Agency as a precious guest. The executives in the forestry office and the central Forestry Agency were valued contacts for the forestry businesses, so of course, they were treated well.³³²

Nevertheless, after the Yamarin episode, Matsuoka hunkered down and went pretty quiet. His appearances on the TV program 'Sunday Project' (*Asahi* National Broadcasting), on which, at one time, he appeared regularly as a representative of the 'resistance forces' (*teikô seiryoku*) to Prime Minister Koizumi³³³ declined as criticism of Muneo increased.³³⁴ As the media commented, '[r]ecently, Matsuoka is quite silent in Nagata-chô. His and Muneo's high-handed methods have become unacceptable. The significance of Muneo's arrest was that their "methods" were possibly becoming illegal.'³³⁵ Somebody else said that

> [i]n a sense, Matsuoka was like a mudskipper in Isahaya Bay. The water level in the bay had sunk after the water gates on the dyke to the bay were built, and the mudskipper dried up after his environment was degraded. It is a really ironical consequence for Matsuoka, who supports the development of Isahaya Bay land reclamation by drainage (laughter).³³⁶

THE ACTIVITIES AND WHEREABOUTS OF MR 'A', OR IS IT MR 'K'? ³³⁷

When the Yamarin scandal broke and investigations were underway, the movements of one of Matsuoka's close advisors attracted attention in Kumamoto. This person (referred to only as a 'Mr A') apparently disappeared, and speculation was rife about whether he had truly disappeared or was in hospital. Mr A had reputedly supported Matsuoka since he first went into politics, and had, at one time, even served as chief of his Kumamoto electoral office.³³⁸ A staff member of Matuoka's office in Kumamoto City confirmed that the current (2002) job description of Secretary 'A' was 'LDP Kumamoto (3) Branch Office Chief'.[339]

According to a person knowledgeable about local politics in Kumamoto

> [Mr A] rides expensive cars, and is very influential. In particular, he is strong in the Kyushu Agricultural Administration Bureau, although he has a bossy tone and has many enemies. There were even accusations within Matsuoka's *kôenkai* that 'A' was mediating deals and meddling, and prefectural assembly members linked to Matsuoka protested loudly about it at a meeting.'[340]

Mr A's business card later showed that he was no longer Matsuoka's branch office chief, but described himself as a 'consultant' to Matsuoka's *kôenkai*.[341] An executive of Matsuoka's *kôenkai* said, 'I don't know what he is doing now. I don't like that guy. He reeks of concessions (*riken*). I always thought he was a dodgy type.'[342]

When journalists asked Matsuoka's Kumamoto office about Mr A, they said, '[h]e hasn't disappeared, but I don't know when he will next come to the office. I don't know whether he is the office chief. I don't even know whether he has the business cards of advisor [to Matsuoka].'[343] The Diet did not know of his whereabouts either.

In 2000, a report surfaced of a heated confrontation that took place in Kumamoto Castle Hotel between Araki Katsutoshi on the one hand, and Matsuoka and Secretary 'A' on the other.[344] Araki had formed the Matsuoka faction (i.e. those who followed Matsuoka) in the prefectural assembly. The faction was called the Matsushôkai, meaning the Matsuoka 'Winning' Association. Araki served as chairman of that association and was seen as the most influential supporter of Matsuoka in that region.[345] Araki was formerly co-president of Araki Group constructions, which was a joint-stock construction company. The Araki Group's main company headquarters and Araki's residence was in Shisui Town in Kikuchi County, which was located in Matsuoka's constituency of Kumamoto (3). Araki admitted that 'it might be possible that the Araki Group had raised money for Matsuoka.'[346]

At this meeting, however, Araki was upbraiding both Matsuoka and Mr A, his secretary. Araki was saying

> Matsuoka, are you making him do it, or is 'A'-kun doing it on his own accord? If we take money from companies, we will get a bad reputation. We can't get sufficient votes even though we're trying hard, and the reason is because you guys are taking money from business people. If this is the case, I can't support this.[347]

In short, Araki was berating both Matsuoka and Mr A for the way they were collecting political funds, although he did not touch at all on why he accused the two of such things.[348] The occasion was a breakfast meeting of the Matsuoka 'Winning' Association being held at the hotel. Of the 56 prefectural assembly members, 10 were present. Because Araki was angry, the whole place became deathly silent. Secretary 'A' made a statement denying that he had taken such action, but there was a short interval of silence. Matsuoka kept his mouth shut. Those present were impressed by Araki's courage, and word of the incident immediately spread to the construction company world in the prefecture. Araki became chairman of the prefectural assembly after this, and stopped being chairman of the Matsuoka 'Winning' Association. However, he remained the leading light of the Matsuoka faction.[349]

Later in 2002, 'details of a bribery case in Fukuoka District Court uncovered a memo written by the chairman of a construction firm in Fukuoka Prefecture. The memo revealed that ¥300 million in cash had been handed to a secretary of Matsuoka'.[350] The chairman was being charged with a different bribery case, and was later found guilty. The court judged, however, that the memo was highly reliable. Matsuoka's secretary later told the press, 'I don't remember whether I met him [the chairman]. That's absolutely groundless'.[351]

AFTERWORD ON SUZUKI

When Muneo split with the LDP in early 2002, the National Nokyo Council in a commentary summed up the particular attributes of Muneo, viz., 'putting pressure on bureaucrats, guiding benefits to local regions and collecting political funds in a way that invited suspicion.'[352] It added that the people had said a resounding 'No' to these kinds of political methods and that Suzuki had been virtually drummed out of the LDP because of what many saw as his objectionable behaviour.[353]

However, it takes more than political oblivion and a prison sentence to keep a politician like Suzuki down. After his release from prison, he was quoted as saying that a Diet member is the representative of his region and that there was nothing wrong in arranging favours and getting advantages for local districts.[354] His political career was resurrected in the 2005 Lower House election and he returned to the Diet as head of the New Party Mother Earth.[355] He succeeded only in the Hokkaido PR bloc, which means that he has no local district as such, but can work for industries and companies based in the prefecture.

Although it is difficult to gauge Suzuki's status and future at this point, given that it is customary for corrupt politicians to leave the LDP, get re-elected (as a kind of cleansing process) and come back as cleanskins, he may be able to revive his political fortunes. On the other hand, Suzuki's power will be limited by the fact that he is out of the LDP and is only a second-ranking PR politician from a minor party.

Moreover, MoFA has no wish to return to the bad old days. After Suzuki was re-elected, it issued a formal manual instructing the ministry's officials on how to handle Suzuki. The manual, entitled 'How to deal with Lower House member Muneo Suzuki', urged officials not to dine with Suzuki and to submit a report whenever they met him.[356] Suzuki and his aides reacted strongly to the ministry's moves, saying '[w]e will reveal the true state of the Foreign Ministry and ministry bureaucrats'.[357]

NOTES

1. 'Matsuoka Toshikatsu no Rirekisho'. Available from http://www.matsuokatoshikatsu.org/site002//public/008.html.
2. Hasegawa, 'Kanjûdanomi no Hazama de Shundô', p. 24.
3. 'Nishi no "Muneo" Matsuoka Toshikatsu wa Kisha "I" o "Gokiburi ika" to Kimetsuketa' ['"Muneo" of the West Matsuoka Toshikatsu Asserts Journalist "I" is "Lower than a Cockroach"], *Shûkan Bunshun*, 27 March 2003, p. 29.
4. 'Nishi no "Muneo" Matsuoka Toshikatsu', p. 29.
5. See below.
6. 'Kinkyû Nyûin shita', p. 28. See also below.
7. Hasegawa, 'Nôsuishô o Haishi seyo', pp. 35–6.
8. 'Nishi no Muneo: Matsuoka Toshikatsu o Torimaku amari ni Kuroi Jinmyaku' ['Muneo of the West: The Extremely Evil Personal Connections that Surround Matsuoka Toshikatsu'], *Shûkan Bunshun*, 21 March 2002, p. 167.
9. Visit: http://www.nouminren.ne.jp/dat/200208/2002081202.htm
10. Nakanishi and Journal Reporter Group, 'Matsuoka Toshikatsu to Iu Giwaku Nin', p. 184.
11. Nakanishi and Special Reporting Group, 'Suzuki Muneo, Matsuoka Toshikatsu', p. 99.
12. *ibid.*
13. 'Kinkyû Nyûin shita', p. 27.
14. *ibid.*
15. Nakanishi and Special Reporting Group, 'Suzuki Muneo, Matsuoka Toshikatsu', p. 99.
16. *ibid.*, p. 98.
17. 'Kinkyû Nyûin shita', p. 26.
18. Nakanishi and Special Reporting Group, 'Suzuki Muneo, Matsuoka Toshikatsu', p. 99.
19. *ibid.*, p. 94.
20. *ibid.*, p. 99.
21. Visit: http://www.nouminren.ne.jp/dat/200208/2002081202.htm
22. *ibid.*
23. Nakanishi and Special Reporting Group, 'Suzuki Muneo, Matsuoka Toshikatsu', p. 94.

24 Itô Hirotoshi, '"Muneo no Meiyû" no Arata na Taidô: Matsuoka Toshikatsu ga Shikakeru Kokka Purojekuto o Oe' ['The New Movements of "Muneo's Sworn Friend": Follow the State Project that Matsuoka Toshikatsu is Going to Do'], *Gendai*, June 2004, p. 287.
25 Itô comments that formerly amongst the *nôrin zoku*, there were many *abarenbô*, such as Watanabe Michio, but that they had become weaker with the demise of Japanese agriculture. Itô, 'Heisei Jiken Fuairu: Nôrin Jigyô Hojokin o Dokusen Suru Matsuoka Toshikatsu', p. 64.
26 Nakanishi and Special Reporting Group, 'Suzuki Muneo, Matsuoka Toshikatsu', p. 95.
27 'Kinkyû Nyûin shita', p. 26.
28 Nakanishi and Special Reporting Group, 'Suzuki Muneo, Matsuoka Toshikatsu', p. 103.
29 Itô, 'Heisei Jiken Fuairu: Nôrin Jigyô Hojokin o Dokusen Suru Matsuoka Toshikatsu', p. 64.
30 *ibid.*
31 Hashimoto Naoyuki, 'Letter from Yochomachi', Posted 11 September 2005, available from http://homepage.mac.com/naoyuki_hashimoto/iblog/C47j8131471...
32 In Kamei's electorate of Hiroshima (6), there are two stations where the Shinkansen stops. He is famous for building big roads that few cars use, as well as bridges and gigantic dams. These are criticised as being of no use to the local people, but constructed for the benefit of local general contractors with the country's money. Visit: http://www.ch-sakura.jp/bbs_thread.php?ID=224796&GENRE=sougou
33 Quoted in: http://picard.blog.bai.ne.jp/?eid=14991
34 Nakanishi and Special Reporting Group, 'Suzuki Muneo, Matsuoka Toshikatsu', p. 100.
35 'Kinkyû Nyûin shita', p. 26.
36 Nakanishi and Journal Reporter Group, 'Matsuoka Toshikatsu to Iu Giwaku Nin', pp. 183–84.
37 'Seijika o Katte ni Kenkyû Suru: Matsuoka Toshikatsu' ['Researching Politicians Arbitrarily: Toshikatsu Matsuoka'], *Yûkan Fuji News*, 9 August 2000, visit: http://ww.fujinews.com/today/2000-08/200000809/0809-08.htm
38 Nakanishi and Special Reporting Group, 'Suzuki Muneo, Matsuoka Toshikatsu', p. 100.
39 Nakanishi and Journal Reporter Group, 'Matsuoka Toshikatsu to Iu Giwaku Nin', p. 184.
40 Nakanishi and Special Reporting Group, 'Suzuki Muneo, Matsuoka Toshikatsu', p. 103.
41 *ibid.*
42 *ibid.*, p. 105.
43 Nakanishi and Journal Reporter Group, 'Matsuoka Toshikatsu to Iu Giwaku Nin', p. 184.
44 Nakanishi and Special Reporting Group, 'Suzuki Muneo, Matsuoka Toshikatsu', p. 94.
45 'Nishi no "Muneo"', p. 38. One local inn owner in Matsuoka's electorate also dubbed him 'Muneo Suzuki of the West' (*Nishi no Suzuki Muneo*). Hasegawa, 'Jimin "Gajô" no Chikaku Hendô', p. 27.
46 'Kinkyû Nyûin shita', p. 26.
47 *ibid.*
48 Kokita Kiyohito, 'Suzuki Muneo no Tsukurareta' ['How Suzuki Muneo was Made'], *Aera*, 18 February 2002, p. 19.
49 'Suzuki Muneo ni Dôkatsu sareta Jisatsu shita Nôsuishô Kyaria Kanryô' ['The MAFF Career Bureaucrat Threatened by Suzuki Muneo Who Committed Suicide'], *Shûkan Bunshun*, 21 February 2002, pp. 26–7.
50 Itô, 'Shinbun ga Zettai ni Hojinai', p. 82.
51 'Hirasawa Katsuei Vs Matsuoka Toshikatsu: "Makiko-Muneo" no Dairi Senso"' ['Hirasawa Katsuei Versus Matsuoka Toshikatsu: "The Makiko-Muneo" Proxy War'], *Shûkan Shinchô*, 21 February 2002, p. 45.
52 'Hirasawa Katsuei Vs Matsuoka Toshikatsu', p. 45.
53 'Hirasawa Katsuei Vs Matsuoka Toshikatsu', p. 45.
54 'Kinkyû Nyûin shita', p. 26.
55 *Tokyo Shinbun*, 3 August 2005.
56 'Hirasawa Katsuei Vs Matsuoka Toshikatsu', p. 44.

57 ibid.
58 Nakanishi and Special Reporting Group, 'Suzuki Muneo, Matsuoka Toshikatsu', pp. 100–1. Matsuoka's motives were attributed to his desire to repay favours back to Nokyo and to sell old rice at retail prices. Visit: http://piza.2ch.net/giin/kako/987/987905181.html
59 Absent was Matsuoka's old arch-enemy, Uozumi Hirohide, a member of the House of Councillors at the time.
60 Nakanishi and Special Reporting Group, 'Suzuki Muneo, Matsuoka Toshikatsu', p. 99.
61 Itô Hirotoshi, 'Shinbun ga Zettai ni Hojinai "Gyuniku Giso" no Anbu' ['The Black Spots of the "Beef Camouflage" that the Newspapers Absolutely Don't Report'], *Gendai*, October 2002, p. 82.
62 'Suzuki Muneo Giin no Ritô to Sono Yoha' ['Suzuki Muneo Diet Member's Split From the Party and its Aftermath'], *Nôsei Undô Jyânaru*, No. 42, April 2002, p. 1.
63 Nakanishi and Special Reporting Group, 'Suzuki Muneo, Matsuoka Toshikatsu', p. 101.
64 Satô Masaru, 2005. *Kokka no Wana: Gaimushô no Rasupuchi to Yobarete* [*National Trap: The So-Called Rasputin of the Foreign Ministry*], Shinchôsha, Tokyo.
65 Reed S. R.,'Revelations About Suzuki Muneo', 13 March 2002. Available from ssj-forum@iss.u-tokyo.ac.jp
66 Hasegawa, 'Nôsuishô o Haishi seyo', p. 36.
67 ibid.
68 ibid.
69 Nakanishi and Journal Reporter Group, 'Matsuoka Toshikatsu to Iu Giwaku Nin', p. 186.
70 Nakanishi, 'Matsuoka Toshikatsu', p. 29.
71 ibid.
72 'Matsuoka Daigishi ni Hisho no Taishokukin', p. 15.
73 Kitamatsu, *et al.*, 'Matsuoka Toshikatsu Daigishi Tettei Bunseki', p. 46.
74 'Hini Kaku "Matsuoka Toshikatsu Daigishi" no Patoron', p. 59.
75 Kitamatsu, *et al.*, 'Matsuoka Toshikatsu Daigishi Tettei Bunseki', p. 47.
76 ibid.
77 'Seiji Shikin Zenkoku Chôsa Kekka', available from: http://www.asahi.com/paper/special/shikin
78 ibid.
79 Reed, 'Revelations'. Available from: ssj-forum@iss.u-tokyo.ac.jp
80 'Seiji Shikin Pâtî ga Dai Seikyô' ['Great Success of Political Funds Parties']. Available from http://www.kenkin.com/etcetra/sikinparty.html
81 Nakanishi and Journal Reporter Group, 'Matsuoka Toshikatsu to Iu Giwaku Nin', p. 186.
82 Nakanishi, 'Matsuoka Toshikatsu', p. 29.
83 Ishii, 'Nôsuishô Osen', p. 194.
84 Kitamatsu, *et al.*, 'Matsuoka Toshikatsu Daigishi Tettei Bunseki', p. 48. The dates are not given but it is assumed that the years were 1996–2001.
85 ibid.
86 'Nishi no "Muneo"', p. 39.
87 For example, the association received subsidies of ¥35.4 billion to 'contribute technological development and data necessary for drawing up plans for forestry management in tropical rainforests'. Ishii, 'Nôsuishô Osen', p. 195.
88 Ishii, 'Nôsuishô Osen', p. 194.
89 Visit: http://www.nouminren.ne.jp/dat/200208/2002081202.htm. See also Nakanishi and Journal Reporter Group, 'Matsuoka Toshikatsu to Iu Giwaku Nin', p. 179.
90 This organisation was established in 1946 as a *gaikaku dantai* of the Forestry Agency for the purpose of developing the forestry industry and undertaking welfare works for the staff and retired officials of the Forestry Agency. See also Table 6.1.
91 Itô, 'Heisei Jiken Fuairu: Nôrin Jigyô Hojokin o Dokusen Suru Matsuoka Toshikatsu', p. 65.
92 ibid.
93 ibid.

94 Ishii, 'Nôsuishô Osen', p. 193.
95 *ibid.*
96 Nishikawa, 'Tako Tsubo', p. 48.
97 Ishii, *op.cit.*, p. 196.
98 *ibid.*, p. 197.
99 *ibid.*
100 *ibid.*, p. 194.
101 ibid., p. 195.
102 Kitamatsu, *et al.*, 'Matsuoka Toshikatsu Daigishi Tettei Bunseki', p. 47.
103 Politicians must submit such a report to MIAC at the end of September each year.
104 Nakanishi and Journal Reporter Group, 'Matsuoka Toshikatsu to Iu Giwaku Nin', p. 186.
105 Nakanishi and Special Reporting Group, 'Suzuki Muneo, Matsuoka Toshikatsu', p. 104.
106 'Hini Kaku "Matsuoka Toshikatsu Daigishi" no Patoron', p. 58.
107 *ibid.*
108 *ibid.*, p. 59.
109 *ibid.*, pp. 58-9.
110 *ibid.*, p. 59.
111 JCP Lower House member Ozawa Kazuaki survey. Available from http://www.nouminren.ne.jp/dat/200208/2002081202.htm
112 'Hini Kaku "Matsuoka Toshikatsu Daigishi" no Patoron', p. 59.
113 Under the Political Funds Control Law, the donation threshold at which the name of the donor must be disclosed is ¥50,000.
114 Kitamatsu, *et al.*, 'Matsuoka Toshikatsu Daigishi Tettei Bunseki', p. 47.
115 The table was obtained from Kitamatsu, *et al.*, 'Matsuoka Toshikatsu Daigishi Tettei Bunseki', p. 48.
116 The Forestry Benefit Association, also on the above list, claimed that it did not associate with Matsuoka after 1999. Kitamatsu, *et al.*, 'Matsuoka Toshikatsu Daigishi Tettei Bunseki', p. 48.
117 'Matsuoka Daigishi ni Hisho no Taishokukin', p. 15. See also below.
118 Nakanishi, 'Matsuoka Toshikatsu', p. 29.
119 Kitamatsu, *et al.*, 'Matsuoka Toshikatsu Daigishi Tettei Bunseki', p. 48.
120 The table was obtained from Kitamatsu, *et al.*, 'Matsuoka Toshikatsu Daigishi Tettei Bunseki', p. 48.
121 This information was revealed in the LDP's Kumamoto No. 3 Electoral District Branch revenue and expenditure report. Nakanishi, 'Matsuoka Toshikatsu', p. 29.
122 Nakanishi, 'Matsuoka Toshikatsu', p. 29.
123 *ibid.*
124 The table was obtained from Kitamatsu, *et al.*, 'Matsuoka Toshikatsu Daigishi Tettei Bunseki', p. 49.
125 Kitamatsu, *et al.*, 'Matsuoka Toshikatsu Daigishi Tettei Bunseki', p. 49.
126 *ibid.*
127 *Sankei Shinbun*, 30 September 2005.
128 'Matsuoka Daigishi ni Hisho no Taishokukin', p. 15.
129 *ibid.* All secretaries of Lower House members are automatically relieved of their posts when the house is dissolved, but are automatically rehired within 40 days of the election, if a form is submitted. Beyond that deadline, they are deemed automatically retired, and their retirement allowance is paid.
130 'Matsuoka Daigishi ni Hisho no Taishokukin', p. 15.
131 *ibid.*
132 *ibid.*
133 *ibid.*
134 *ibid.*
135 *ibid.*
136 The state subsidises three secretaries for each Diet member.
137 'Matsuoka Daigishi ni Hisho no Taishokukin', p. 15.
138 Nakanishi, 'Matsuoka Toshikatsu', p. 28.

139 *Yûkan Fuji News*. Available from: http://www.fujinews.com/today/2000-08/20000809/0809-08,htm>.
140 *ibid.*
141 *ibid.*
142 'Seijika o Katte ni Kenkyû Suru'. Available from: http://ww.fujinews.com/today/2000-08/200000809/0809-08.htm
143 *Yûkan Fuji News*. Available from: http://www.fujinews.com/today/2000-08/20000809/0809-08,htm
144 *ibid.*
145 Nakanishi, 'Matsuoka Toshikatsu', p. 29.
146 *ibid.*, p. 28.
147 Editorial Group, Ushiroda Ryôe, 'Okinawa de Mitsuketa Umami' ['The Allure Found in Okinawa'], *Aera*, 11 March 2002, p. 16.
148 '"Nishi no Muneo"', p. 38.
149 Nakanishi and Journal Reporter Group, 'Matsuoka Toshikatsu to Iu Giwaku Nin', p. 186.
150 'Matsuoka Daigishi ni Hisho no Taishokukin', p. 15.
151 Hasegawa, 'Kanjûdanomi no Hazama de Shundô', p. 24.
152 Kitamatsu, *et al.*, 'Matsuoka Toshikatsu Daigishi Tettei Bunseki', p. 46.
153 *ibid.*, pp. 46 and 48.
154 It was ¥2 million. Kitamatsu *et al.*, 'Matsuoka Toshikatsu Daigishi Tettei Bunseki, p. 48.
155 'Nishi no Muneo: Matsuoka Toshikatsu o Torimaku amari ni Kuroi Jinmyaku', p. 167.
156 Kitamatsu *et al.*, 'Matsuoka Toshikatsu Daigishi Tettei Bunseki, p. 49.
157 *ibid.*, p. 48.
158 Reed S. R., 'Revelations'. Available from: ssj-forum@iss.u-tokyo.ac.jp
159 Reed S. R., 'More Muneo', 20 March 2002. Available from: ssj-forum@iss.u-tokyo.ac.jp
160 Nakanishi and Journal Reporter Group, 'Matsuoka Toshikatsu to Iu Giwaku Nin', p. 186.
161 Visit: http://www.nouminren.ne.jp/dat/200208/2002081202.htm
162 Nakanishi and Journal Reporter Group, 'Matsuoka Toshikatsu to Iu Giwaku Nin', p. 180.
163 *ibid.*
164 See also Nakanishi and Journal Reporter Group, 'Matsuoka Toshikatsu to Iu Giwaku Nin', pp. 180–2.
165 '"Nishi no Muneo"', p. 39.
166 Nakanishi and Journal Reporter Group, 'Matsuoka Toshikatsu to Iu Giwaku Nin', p. 180.
167 Satô Saburô was Katô Kôichi's secretary, to whom he was extremely close, and who was prosecuted for evading taxes amounting to ¥100 million. Katô put a lot of trust in Satô, who was like his double. Katô was reported as saying, 'what Sato says, you can think of as being said by me', although he denied being involved in Satô's tax evasion scandal. 'Katô Kôichi yo Semete Muneo yori Hayaku Yamenasai' ['Katô Kôichi, You Need At Least to Resign Earlier Than Muneo'], *Shûkan Bunshun*, 21 March 2002, p. 166.
168 Nakanishi and Journal Reporter Group, 'Matsuoka Toshikatsu to Iu Giwaku Nin', p. 186.
169 '"Nishi no Muneo"', p. 39.
170 Nakanishi and Journal Reporter Group, 'Matsuoka Toshikatsu to Iu Giwaku Nin', p. 182.
171 '"Nishi no Muneo"', p. 39.
172 'Matsuoka Toshikatsu to Iu Giwaku Nin', p. 182.
173 *ibid.*
174 *ibid.*
175 'Sôsa Shinsa Repôto: Suzuki Muneo, Matsuoka Toshikatsu—Kesareta Shokuniku Rûto' ['Report from the Depths of the Criminal Investigation: Suzuki Muneo, Matsuoka Toshikatsu—The Meat Route that Has Been Wiped Out'], *Shûkan Bunshun*, 19 September 2002, p. 150.
176 Hasegawa, 'Nôsuishô o Haishi seyo', p. 38.
177 *ibid.* For a similar account of institutional factors within the MAFF, which were responsible for the BSE outbreak in Japan, see 'Nôsuishô "Chikusan Riken": Inamikitta "Niku" to "Uma" Gyôsei' ['The

Ministry of Agriculture and Fisheries' "Livestock Concessions"" The Totally Denied "Meat" and "Horse" Administration'], *Sentaku*, May 2002, pp. 126-129.
178 Ayukawa, 'Jimintô', p. 20.
179 *ibid*.
180 *ibid*.
181 Itô, 'Shinbun ga Zettai ni Hojinai', p. 84.
182 Nakanishi and Special Reporting Group, 'Suzuki Muneo, Matsuoka Toshikatsu', p. 101.
183 'Sôsa Shinsa Repôto', p. 150.
184 The Meat and Egg Division of the Livestock Department of the Production Bureau had official responsibility for administering the beef industry.
185 Nakanishi and Special Reporting Group, 'Suzuki Muneo, Matsuoka Toshikatsu', p. 101.
186 Ayukawa, 'Jimintô', p. 20.
187 'Sôsa Shinsa Repôto', p. 150.
188 'Nôsuishô "Chikusan Riken"', p. 127.
189 'Sôsa Shinsa Repôto', p. 150.
190 *ibid.*, p. 150.
191 'Nôsuishô "Chikusan Riken"', p. 127.
192 Nakanishi and Special Reporting Group, 'Suzuki Muneo, Matsuoka Toshikatsu', p. 101.
193 *ibid.*, pp. 101–2.
194 *ibid.*, p. 102.
195 Ayukawa, 'Jimintô', p. 20.
196 Nakanishi and Special Reporting Group, 'Suzuki Muneo, Matsuoka Toshikatsu', p. 102.
197 Ayukawa, 'Jimintô', p. 20.
198 Itô, 'Shinbun ga Zettai ni Hojinai', p. 82.
199 'Nôsuishô "Chikusan Riken"', p. 127.
200 Itô, 'Shinbun ga Zettai ni Hojinai', p. 83.
201 'Sôsa Shinsa Repôto', p. 151.
202 Ayukawa, 'Jimintô', p. 20.
203 'Sôsa Shinsa Repôto', p. 151.
204 *ibid.*, p. 150; 'Nôsuishô "Chikusan Riken"', p. 127.
205 'Nôgyô Kankei Seisaku Kettei no Ashidori', *Nôsei Undô Jyânaru*, No. 42, April 2002, p. 29.
206 'Sôsa Shinsa Repôto', p. 150.
207 'Nôsuishô "Chikusan Riken"', p. 127.
208 Nakanishi and Special Reporting Group, 'Suzuki Muneo, Matsuoka Toshikatsu', p. 102.
209 'Sôsa Shinsa Repôto', p. 150.
210 'Nôsuishô "Chikusan Riken"', p. 127.
211 'Sôsa Shinsa Repôto', p. 152.
212 Asada was implicated in a corruption scandal involving the Livestock Industry Promotion Corporation in 1987, and was arrested on suspicion of bribery. Itô, 'Shinbun ga Zettai ni Hojinai', p. 76.
213 *ibid.*, p. 82.
214 'Sôsa Shinsa Repôto', p. 152.
215 'Nôsuishô "Chikusan Riken"', p. 1278.
216 Itô, 'Shinbun ga Zettai ni Hojinai', p. 82.
217 'Sôsa Shinsa Repôto', p. 152.
218 Hasegawa Hiroshi, 'Kokusan Gyûniku Kaiage no Nazo' ['The Mystery of the Buy-up of Domestic Beef'], *Aera*, 8 July 2002, p. 22.
219 Itô, 'Shinbun ga Zettai ni Hojinai', p. 77. See also 'Nôsuishô "Chikusan Riken"', p. 127.
220 Itô, 'Shinbun ga Zettai ni Hojinai', p. 77.
221 *ibid.*, p. 77.
222 'Nôsuishô "Chikusan Riken"', p. 127.
223 Itô, 'Shinbun ga Zettai ni Hojinai', p. 77.

224 Hasegawa, 'Kokusan Gyuniku Kaiage no Nazo', p. 21.
225 Itô, 'Shinbun ga Zettai ni Hojinai', p. 77.
226 'Sôsa Shinsa Repôto', p. 151.
227 *ibid.*
228 *ibid.*, p. 152.
229 *ibid.*
230 'Matsuoka Toshikatsu Daigishi Hannan nado Giwaku Sanseki' ['Suspicions Accumulate About Diet Member Matsuoka Toshikatsu Such As Hannan'], *Kokumin Shinbun*, July 2002. Available from http://www5f.biglobe.ne.jp/~kokumin-shinbun/H14/1407/140769matsuoka.html
231 He was later sentenced to seven years in prison.
232 *The Japan Times*, 27 June 2004; 28 August 2004.
233 *ibid.*, 21 August 2004.
234 *ibid.*, 27 June 2004.
235 *ibid.*, 21 August 2004.
236 *ibid.*, 24 April 2004.
237 *ibid.* 28 May, 2005.
238 *ibid.*, 23 April 2004.
239 'Posuto "Muneo Sôsa" no Shôten e: Tokusôbu ga Kanshin o Motsu Matsuoka Toshikatsu no "Kôdô"' ['The Focus of the Post "Muneo Investigation": The "Action" of Matsuoka Toshikatsu in Which the Special Investigation Department Has an Interest'], *Themis*, July 2002, p. 35.
240 *ibid.*
241 *ibid.*
242 *ibid.*
243 Ayukawa, 'Jimintô', p. 20.
244 *ibid.*
245 *ibid.*
246 According to another source, this was the Supreme Court, not the High Court. *Tokyo Shinbun*, 20 February 2002.
247 These details were obtained from 'Nishi no Muneo: Matsuoka Toshikatsu o Torimaku Amari ni Kuroi Jinmyaku', p. 167.
248 'Muneo "Waido": "Usotsuki Toshu Otoko" to Kakarete Honshi o Utaeta "Suzuki Muneo" no Sente Shisaku' ['Muneo "Wide Show": Having Written That He Was an Habitual Liar', The Failed First Move of "Suzuki Muneo" in Suing Our Magazine'], *Shûkan Shinchô*, 14 March 2002, p. 48.
249 Tanikawa retired from the Lower House in 2003.
250 'Nishi no Muneo: Matsuoka Toshikatsu o Torimaku Amari ni Kuroi Jinmyaku' p.167. He also reportedly said: 'Keep your hands off because this property (building) has been settled politically, and the Foreign Ministry knows it.' *Tokyo Shinbun*, 20 February 2002.
251 *Tokyo Shinbun*, 20 February 2002.
252 'Nishi no Muneo: Matsuoka Toshikatsu o Torimaku Amari ni Kuroi Jinmyaku', p. 167.
253 'Muneo "Waido"', p. 48.
254 'Nishi no Muneo: Matsuoka Toshikatsu o Torimaku Amari ni Kuroi Jinmyaku', p. 167.
255 *ibid.*
256 'Muneo "Waido"', p. 48.
257 *ibid.*
258 *Tokyo Shinbun*, 20 February 2002.
259 The list of people attending the dinner party varies depending on the source. See also *Tokyo Shinbun*, 20 February 2002.
260 Nakanishi and Special Reporting Group, 'Suzuki Muneo, Matsuoka Toshikatsu', pp. 95–8.
261 'Muneo "Waido"', p. 48.
262 *ibid.*
263 Nakanishi and Special Reporting Group, 'Suzuki Muneo, Matsuoka Toshikatsu', pp. 95–8.

264 'Muneo "Waido"', p. 49.
265 *ibid.*
266 *Tokyo Shinbun*, 20 February 2002.
267 *ibid.*
268 '"Nishi no Muneo": Matsuoka Toshikatsu o Torimaku Amari ni Kuroi Jinmyaku', p. 167.
269 Nakanishi and Special Reporting Group, 'Suzuki Muneo, Matsuoka Toshikatsu', pp. 95–8.
270 '"Nishi no Muneo": Matsuoka Toshikatsu o Torimaku Amari ni Kuroi Jinmyaku', p. 167.
271 *ibid.*
272 *ibid.* See also below.
273 Yamarin's main office is in Obihiro City in Hokkaido. It is a leading timber company, built up over the lifetime of its president, Yamada Isao. Nakanishi and Journal Reporter Group, 'Matsuoka Toshikatsu to Iu Giwaku Nin', p. 179.
274 'Kinkyû Nyûin shita', p. 28.
275 Nakanishi and Journal Reporter Group, 'Matsuoka Toshikatsu to Iu Giwaku Nin', p. 180.
276 *ibid.*, p. 179.
277 *ibid.*, pp. 179–80.
278 'Posuto "Muneo Sôsa"', p. 34.
279 Nakanishi and Journal Reporter Group, 'Matsuoka Toshikatsu to Iu Giwaku Nin', p. 179.
280 *The Japan Times*, 27 June 2002.
281 Nakanishi and Journal Reporter Group, 'Matsuoka Toshikatsu to Iu Giwaku Nin', p. 179.
282 '"Muneo no Bôrei"', p. 28.
283 '"Nishi no Muneo"', p. 39.
284 *The Japan Times*, 27 June 2002.
285 *ibid.*
286 Nakanishi and Journal Reporter Group, 'Matsuoka Toshikatsu to Iu Giwaku Nin', p. 180.
287 Visit: http://www.nouminren.ne.jp/dat/200208/2002081202.htm
288 Visit: http://news.kyodo.co.jp/kyodonews/2002/suzuki/news/20020626-492.html
289 *The Japan Times*, 22 June 2002.
290 *ibid.*
291 'Posuto "Muneo Sôsa"', p. 34.
292 *The Japan Times*, 22 June 2002.
293 '"Nishi no Muneo"', p. 38.
294 'Kinkyû Nyûin shita', p. 28.
295 Itô, '"Muneo no Meiyû" no Arata na Taidô', p. 286.
296 'Sukûpu! Mitsui Bussan Kanbu Shain o Taiho: Chiken Tokusôbu ga Jimintô Daigishi Futari o Chôshu' ['Scoop! The Arrest of Mitsui Co. Executives: The Special Investigation Department of the Public Prosecutor's Office Listens to Two LDP Diet Members'], *Shûkan Gendai*, 20 July 2002, p. 55.
297 'Posuto "Muneo Sôsa"', p. 34.
298 Itô, '"Muneo no Meiyû" no Arata na Taidô', p. 288.
299 *ibid.*
300 Nakanishi and Journal Reporter Group, 'Matsuoka Toshikatsu to Iu Giwaku Nin', p. 179.
301 'Kinkyû Nyûin shita', p. 27.
302 *ibid.*
303 Kokita, 'Suzuki Muneo no Tsukurareta', p. 19.
304 'Kinkyû Nyûin shita', p. 28.
305 *ibid.*
306 *ibid.*
307 *ibid.*
308 *ibid.*
309 *ibid.*

310 Nakanishi and Journal Reporter Group, 'Matsuoka Toshikatsu to Iu Giwaku Nin', p. 178.
311 Itô , 'Shinbun ga Zettai ni Hojinai', p. 82.
312 'Posuto "Muneo Sôsa"', p. 34.
313 *The Japan Times*, 24 July 2002.
314 *ibid.*
315 *ibid.*
316 Suzuki was subsequently accused of four crimes, including accepting a bribe in return for services promised and accepting a bribe as a mediator. On 5 November 2004, he was convicted in the Tokyo District Court on all charges and sentenced to a two-year prison term and a fine of ¥11 million. The chief judge stated that 'although the defendant held an important post in the cabinet and was expected to show a high degree of morals and honesty, he took a bribe from his supporters in order to pursue his own interests and betrayed the citizens' trust. In spite of this, he dared to make false statements in the criminal investigation and in the trial. Therefore, he should face a goal sentence.' *Asahi Shinbun*, 5 November 2004. Suzuki stood unsuccessfully in the July 2004 Upper House election, having been forced to resign his Lower House Diet seat in 2002. He successfully regained a seat in the Lower House in September 2005. See below.
317 Itô, 'Shinbun ga Zettai ni Hojinai', p. 77.
318 'Nishi no Muneo: Matsuoka Toshikatsu o Torimaku amari ni Kuroi Jinmyaku', p. 167.
319 'Muneo no Meiyû "Matsuoka Toshikatsu" Jr. no Igai na Kekkon Aite' ['The Unexpected Marriage Partner of Muneo's Sworn Friend "Matsuoka Toshikatsu" Jr.'], *Shûkan Shinchô*, 28 November 2002, p. 32.
320 'Kinkyû Nyûin shita', p. 27.
321 Nakanishi Akihiko and the Journal Reporter Group, 'Matsuoka Toshikatsu to Iu Giwaku Nin', p. 178.
322 *Rondan, Kisha Kurabu* [*Discussion, Press Club*], visit: http://www.rondan.co.jp/html/kisha/0206/020625-3.html. Refer to the discussion concerning Mr 'A'.
323 *ibid.*
324 Nakanishi and Journal Reporter Group, 'Matsuoka Toshikatsu to Iu Giwaku Nin', p. 178.
325 *ibid.*
326 Kitamatsu, *et al.*, 'Matsuoka Toshikatsu Daigishi Tettei Bunseki', p. 46.
327 'Posuto "Muneo Sôsa"', p. 34.
328 Nakanishi and Journal Reporter Group, 'Matsuoka Toshikatsu to Iu Giwaku Nin', p. 178.
329 'Suzuki Muneo Giin no Ritô', p. 1.
330 *ibid.*
331 'Kinkyû Nyûin shita', p. 26.
332 *ibid.*, p. 27.
333 See Chapter 7 on 'Electoral Vicissitudes'.
334 Itô, '"Muneo no Meiyû" no Arata na Taidô', p. 288.
335 Nakanishi and Journal Reporter Group, 'Matsuoka Toshikatsu to Iu Giwaku Nin', p. 179.
336 *ibid.*
337 It is unclear whether Mr 'A' is the same as Mr 'K'. Certainly there are strong similarities. The investigative journal might have used Mr 'A' in the sense of Mr 'A' and Mr 'B', that is, in order to preserve his anonymity.
338 '"Nishi no Muneo"', p. 39.
339 Hasegawa, 'Kanjûdanomi no Hazama de Shundô', p. 24.
340 '"Nishi no Muneo"', p. 39.
341 *ibid.*
342 *ibid.*
343 *ibid.*
344 Hasegawa, 'Kanjûdanomi no Hazama de Shundô', p. 23.
345 *ibid.*

346 *ibid.*, p. 24.
347 *ibid.*, p. 23.
348 *ibid.*
349 *ibid.*
350 *Sankei Shinbun*, 24 September 2002.
351 *ibid.*
352 'Suzuki Muneo Giin no Ritô', p. 1.
353 *ibid.*
354 'Rieki Yûdô' ['Guiding Benefits'], Wikipedia. Available from: http://ja.wikipedia.org/wiki/%E5%88%A9%E7%9B%8A%E8%AA%98%E5%B0%8E
355 Its Japanese title is *Shintô Daichi*. The professed ideals of the party are 'to return to the land', 'to learn from the land', 'to respect nature and be thankful for it'. To achieve these goals 'reform of consciousness is necessary. Real reform is not from authority but from the people'. Visit: <http://www.muneo.gr.jp/>. Suzuki won 433,938 votes, slightly less than half the LDP's tally in the Hokkaido bloc. Muneo's party fielded two other candidates (both graduate students, one of which was an Ainu) in the Hokkaido regional bloc, but only Suzuki won a seat. It also fielded a candidate (a former Olympic ski-jumping athlete) in Hokkaido (1). He came last with 16,000 votes.
356 *Mainichi Shinbun*, 4 October 2005.
357 *ibid.*

7

ELECTORAL VICISSITUDES

The fallout from the Muneo affair and various scandals continued to swirl around Matsuoka in 2002 and 2003. He tried to rehabilitate his reputation in various ways, but in the end he paid a political price in the 2003 Lower House election. Paradoxically, this was also an election in which brakes were put on the popularity of Prime Minister Koizumi, whom Matsuoka openly opposed.[1]

THE SCARY 'RESISTANCE FORCE' POLITICIAN

Matsuoka became well known to the broader electorate in Japan as one of the most outspoken members of the *teikô seiryoku*. In a blatantly self-serving fashion, he went around preaching the errors of Koizumi's reforms.[2] He spoke out against these reforms both inside and outside the party.[3] In taking such a stand he 'demonstrated both his conservative side and his action side'.[4]

Matsuoka's first confrontation over Koizumi's economic reform policies went back to June 2001, only three months after Koizumi became prime minister. At the Executive Council of the LDP, of which he was a member at the time (see Table 4.1), Matsuoka, in his vocal role as a *nôrin zoku*, criticised Koizumi for 'pandering to the interests of urban voters'.[5] The council under Koizumi had become the main battlefield between the government and the ruling party in determining the nation's policies.[6] Many of Matsuoka's fellow rural stalwarts in the Executive Council were shouting and banging their desks in protest against Koizumi's proposals to benefit urban voters, such as diversifying the use of a special revenue source previously used exclusively for building roads,[7] cutting tax grants to local governments and giving more Diet seats to urban areas. If implemented, these proposals would have all hit rural regions hard.[8]

In the Executive Council meeting, Matsuoka openly opposed the proposal to expand the use of tax revenues earmarked specifically for road construction, claiming that 'earmarked taxes like the one for road construction should be used for their original purposes'.[9] He was quoted as saying, 'roads are the lifeblood of regional areas!' [*dôro wa chihô no mei da!*].[10] He had earlier argued against budget expenditure cutbacks 'without sanctuary' (*seiiki naki*) under the Hashimoto administration, arguing that 'there are not yet sufficient roads in local regions'.[11] He also opposed any increase in the rate of consumption tax.[12]

Matsuoka, as part of the bigger group of LDP resistance forces led by the road policy clique, successfully derailed Koizumi's plan to overhaul the tax revenues exclusively reserved for road construction. Koizumi aimed to turn such dedicated revenues into general revenues in the process of compiling the national budget for fiscal 2003.[13] However, Matsuoka always claimed that 'his objective [in opposing Koizumi's reforms] was not to guide benefits and concessions [to local areas]'.[14]

Matsuoka served on anti-reform, anti-Koizumi LDP committees whenever he got the chance. He became the mascot (*ojisan*) of the *teikô seiryoku* within the LDP[15] and the leader of a study group that rose in open revolt against Koizumi's reforms in November 2001. The name of Diet members' league to oppose Koizumi's 'structural reforms without sanctuary' (*seiiki naki kôzô kaikaku*) was the 'Diet Members' League to Save Japan From Crisis and Realize Real Reforms' (Nihon no Kiki o Sukui, Shin no Kaikaku o Jitsugen suru Giin Renmei), giving themselves the title 'Save the Nation League of Diet Members' (Kyûkoku Giren) for short. However, there was an immediate change of name to 'Diet Members' League for the Realization of Reforms and the Creation of a Bright Future' (Kaikaku o Jitsugen shi, Akarui Mirai o Sôzô suru Giin Renmei),[16] or 'Future-Creating Diet Members' League' (Mirai Sôzô Giren) for short. The reason for the change was because it was felt that the anti-Koizumi colour was too strong in the 'Save the Nation League of Diet Members', and so the words 'Creation of a Bright Future' were added, which also required a change in the abbreviated name.[17]

Matsuoka served as one of the representative executives of this league. In it were 13 Diet members from the Hashimoto faction (an anti-Koizumi faction) and 16 Diet members from the Etô-Kamei faction.[18] The Eto-Kamei faction was generally regarded as anti-Koizumi and anti-reform, like the Hashimoto faction. It represented the conservative 'old guard' of the LDP. Kamei, like Matsuoka, was an active leader of the anti-Koizumi force. He was very angry with Koizumi for including so few members of the Etô-Kamei faction in his first cabinet.

In December 2001, along with LDP Highway Investigation Committee Chairman and former party Secretary-General, Hiromu Nonaka, and Upper House Secretary-General, Aoki Mikio, Matsuoka opposed Koizumi's proposal to reform inefficient government-affiliated public corporations. These included the four hugely indebted public corporations related to highway construction, which the Koizumi administration wanted to amalgamate and privatise. At a meeting of the Mirai Sôzô Giren, Matsuoka stated that one of these public corporations, the Japan Highway Public Corporation (Nihon Dôrô Kodan), was highly regarded overseas, which was just the opposite of Koizumi's assessment.[19] Matsuoka appeared on the TV program, 'Takeshi's TV Tackle' hosted by Beat Takeshi. When 'he was asked the question: "Where do you want to build an expressway most?" he answered unashamedly, "everybody wants to build near themselves, such as Kumamoto, Kyushu"'.[20] The public was reportedly scandalised by his blatant sentiments in favour of guiding benefits to local areas (*rieki yûdô ishiki*).[21] An urban voter who undertook his own investigation of Matsuoka commented

> [h]is eyes look scary, his expression looks scary, I bet he speaks scarily. In any case, he looks scary. This is my honest impression when I first saw Matsuoka on TV...That's the kind of presence he has and that's the kind of force he exudes from his whole body. Because of how he looked in interviews as an opposition force when the Executive Council was reviewing the revenue source for road-building, the media portrayed him unequivocally as a 'baddie' (*akuyaku*).[22]

After Koizumi appointed Inose Naoki to head up a panel to look into reform of the Japan Highway Public Corporation and the three other road-building public entities, Matsuoka criticised Inose on the Asahi TV program, 'Sunday Project'. Matsuoka mentioned that he had quarrelled with Inose on the issue, and that Mr Inose was formerly a member of a radical student organisation (Zenkyôtô)[23] in Japanese universities in the 1960s. He later commented, 'I finished him'.[24] The discussion went on a bit and at one point, Matsuoka said: 'I will destroy Yamasaki',[25] in a reference to the secretary-general of the LDP at the time. Matsuoka reportedly did not show any sign of worrying about being called a 'resistance force' or being considered anti-Koizumi.[26] As one commentator observed, 'for Matsuoka, appearing on

TV saying such things is really a way of conveying a message to Koizumi that he opposed his policies.'[27]

By openly criticising Koizumi on television, Matsuoka became known as the new voice of opposition to the prime minister.[28] Initially the 'face' of *zoku* power and the conservatives and the advertising pillar of the *teikô seiryoku* had been Muneo. Matsuoka took over this role in the world of television,[29] becoming the 'face' of opposition to Koizumi's reforms. His stubbornness in rejecting Koizumi's programs was said to be similar to the old JSP's approach.[30] Occasionally, Matsuoka presented convincing arguments, such as the one he presented against the government's big-boned policy to cap the issuance of national bonds to ¥30 trillion (saying, '[d]on't just put the money together, make policies that deal with the causes of problems'),[31] but 'these were only fleeting'.[32]

When the resistance forces came under criticism, around the time the name of the Diet members' league was changed, Matsuoka began to make insincere statements, such as, 'our enemy is recession and the employment problem, not the Koizumi cabinet.'[33] Before long, he went as far as to say, 'I'm all for reforms, but if its done just on the mood of things, it will leave the root of evil.'[34] After that he dropped all pretence and emerged as an unequivocal resistance force. He claimed to be advocating what he called 'the spirit of the one hundred barrels of rice', which meant investing further than for what was just immediately ahead. On that basis, he argued, 'it would be wrong to stop [the construction of] highways. It would be like putting rice away in a warehouse if you don't build highways and just prioritise immediate debts'.[35] As one commentator observed, '[i]n Matsuoka's mind] there is no financial difficulty at all. Just build highway around the country by means of public works projects'.[36]

In December 2001, Matsuoka made several other public comments that were critical of Koizumi. He said that the Koizumi administration would not last long unless it shifted to a policy of aggressive public spending to shore up the economy: 'If the Koizumi Cabinet keeps running on the wrong track, we will have to have second thoughts [about supporting the prime minister]…That would be in the best interests of the public.'[37] He added that

> [t[he greatest achievement of the prime minister and leader of the LDP in his first eight months of office was that he had led the party to a resounding victory in the Upper House election in July 2001. I would give 150 points to Koizumi for winning a victory for the LDP in the Upper House election, but I would also give him minus 100 points for causing the economy to deteriorate.[38]

As far as the prime minister's high support rate was concerned, Matsuoka suggested that more people [who did not have a clear reason for supporting him] were becoming passive supporters of Koizumi: 'At first, the public was actively supporting Koizumi…But now people have no choice but to support him although he has achieved almost nothing to put the economy on a recovery path'.[39] Matsuoka argued: 'If the prime minister leaves the economic situation as is, further deflation and recession is inevitable'.[40] He suggested that Koizumi would lose public support, and thus the power to remain at the helm, if the economy deteriorated further in the following year.[41] In an interview with a journalist, he stated

> Koizumi's economic policy is wrong. When a country is in deflation, first it is necessary to adopt a policy to stimulate demand in order to grow out of deflation. Structural reform should focus first on bad debt management. The responsibility of executives should be clarified, and public funds should be thrown into banks in one hit.[42]

When asked whether he was sufficiently opposed to Koizumi to be called a resistance force, Matsuoka replied

> [i]t would be a lie if I said I don't have feelings of opposition. However, no matter how much people speak ill of me, I have the courage to keep going. Thanks to our Diet members' league, Mr Koizumi is increasing his popularity. We are caught in a dilemma… Although more than 90 per cent of LDP Diet members do not officially voice their objections, they are not thinking that Koizumi's reforms are right. If they listen to the context carefully, they understand that what I say on television is a sound argument.[43]

Matsuoka was critical of the Koizumi administration's budget for fiscal 2002, saying that it lacked sufficient measures to tackle the nation's serious unemployment problems. The real issue of the budget for a politician like Matsuoka, however, was that it promoted Koizumi's structural reform agenda, which aimed to slash spending on public works and therefore cut back on the projects Matsuoka could bring back to his own electorate. Matsuoka joined a number of other LDP Diet members in openly rejecting the Koizumi administration's cuts in public works spending because they shrank the pork barrel.

In Matsuoka's way of thinking, the problem with the Koizumi administration's reform policies was that they took direct aim at the type of politician that he was. They jeopardised the vote-winning contract he had with farmers and rural dwellers, as well as the financial supply contract he had with his company clients. The cuts in public works spending, including allocations to agricultural and rural public works, cramped Matsuoka's electoral style. Similarly, the 'trinity reform'—which aimed to devolve authority and spending-power from central to local governments and fix the debt-ridden finances of both⁴⁴—threatened to destroy the long-standing structure of vested interests tying together bureaucrats in the ministries, politicians (Diet members and local politicians), and industries (companies and individuals) benefiting from these subsidies. Matsuoka's primary support was based on a collection of special interests—local, sectional and client-based—rather than broad, policy-based programmatic appeals as a vote-collector for his party (the LDP) and its leader (Koizumi). If Matsuoka could not deliver on local projects, he was of little use to the particular clutch of vested interests that supported him.

Koizumi was also open in his advocacy of another crucial goal that took direct aim at the exercise of policy influence by individual Diet members such as Matsuoka—that of 'destroying the LDP' (*Jimintô o kowasu*). By this, the prime minister meant tackling the autonomous policymaking authority of PARC committees that undermined the power of the prime minister and cabinet, and which Matsuoka and the other members of the *teikô seiryoku* used to challenge Koizumi's reform initiatives. Destroying the LDP also entailed undermining the fundamental basis of the party's independent policymaking authority—the means by which LDP Diet politicians acquired this power—that is, their independent electoral support coalitions, which locked them into representing, promoting and protecting special interests as well as the interests of specific regional localities.

Matsuoka was identified with the LDP 'old guard', who were loath to change the established ways of doing things, who wanted to keep pork barrel politics alive, and who wanted to keep government subsidies flowing to rural-regional areas. For this type of LDP Diet member, politics was just 'distribution' (*haibun*). They functioned to distribute funds sucked up in the form of taxes to regional areas. For Matsuoka, '[t]he best part about being the LDP was the business of distributing resources through prior scrutiny'.⁴⁵

Koizumi, on the other hand, was aiming for the more equitable distribution of diminishing public resources.⁴⁶ He provided less room in the political system for politicians like Matsuoka, and less potential for guiding benefits from the central government to local areas because of his crackdown on public works spending and the indiscriminate scattering of subsidies (*hojokin no baramaki*) to rural-regional areas. This substantially weakened the appeal of politicians like Matsuoka who relied on this mechanism to win votes amongst locals.

Matsuoka, on the other hand, warned that Koizumi's top-down style of decision making—often ignoring traditional procedures within the party—might lead to 'self-righteousness or despotism'.[47] He criticised the LDP regime under Prime Minister Koizumi as the 'Taliban regime'.[48] He accused Koizumi of 'trying to establish cabinet decision-making as the way of doing things, thus bypassing the process of prior scrutiny by the LDP.'[49] Matsuoka stood for the old policymaking model, naturally enough, because under it, he exercised personal influence over policy through LDP committees. However, the 'traditional practice of having policies cleared by the LDP before submission to the cabinet for its approval... [was] incompatible with the sort of cabinet-led policymaking process that Koizumi...was trying to achieve'.[50]

In taking such a stand against Koizumi's top-down decision-making style, Matsuoka echoed the views of other members of the LDP's 'old guard'. Former LDP Secretary-General Nonaka, who had previously been chief secretary of the cabinet under Prime Minister Obuchi, said that Koizumi was a fascist in the way that he tried to ram his own policies through the party. Matsuoka's view was that '[a] real leader should be able to wrap up various opinions within the group. That is what democracy is all about'.[51] Matsuoka said that he did not care if Koizumi called him and other LDP politicians opposed to his reforms 'resistance forces'. He retorted, '[w]e will continue to make proposals to Koizumi even if we are labelled the "bad guys"'.[52]

For example, Matsuoka and his close associates confronted Koizumi directly over FTAs, which became one of the big 'reform' issues in agriculture during the Koizumi administration. The prime minister saw FTAs as an instrument of 'structural reform' of agriculture, which meant making the farm sector more internationally competitive by expanding the scale of farming. Koizumi kept on saying that reform was needed in agriculture and that it could not be allowed to hold up agreements on trade.

Matsuoka and others had a major tussle of wills with the prime minister over this issue. Koizumi set up all manner of policy groups essentially to bypass opposition to FTAs from the *nōrin zoku*. In 2002, he attempted to wrest control from the *nōrin zoku* by establishing a special LDP committee to study FTAs (FTA Tokumei Iinkai) chaired by his appointee as PARC Chairman, Nukaga Fukushirō, a commerce and industry *zoku*. The committee's task was to work with METI to formulate basic party policy on FTAs.[53] But the *nōrin zoku* were bitterly opposed to the formation of this special LDP committee to study

FTAs, saying, '[u]nder Japan's parliamentary cabinet system, the party decides policy, not the Kantei. The FTA Special Committee neglects the traditional way of deciding party policy, which gathers the opinions of each division such as the Agriculture and Forestry Division'.[54]

In December 2003, after the failure first of WTO negotiations at Cancun in September and then the bilateral Japan-Mexico FTA negotiations in October, the prime minister sought to avoid the possibility of Japan's being left on the sidelines of regional progress on FTAs. He set up an FTA Kankei Shôchô Kaigi (Council of Related Ministries and Agencies on FTAs) under Kantei leadership. For Koizumi, the *nôrin zoku*, who seemed to display blatant disregard for the national interest, had become the object of his irritation. He declared, '[a]fter this, I cannot leave it to the MAFF and the *norin zoku*'.[55]

However, Matsuoka and the other *norin zoku* bosses were not going to take Koizumi's moves lying down. Immediately after Koizumi's announcement of the new council on 11 December, they gatecrashed the Kantei for discussions with Chief Cabinet Secretary Fukuda Yasuo. They went with the aim of correcting the foreign policy position of the prime minister on FTA negotiations. CAPIC Chairman Norota Hôsei and Acting Chairmen Yatsu and Ôshima Tadamori, together with Chairman of the Agriculture, Forestry and Fishery Products Trade Investigation Committee, Sakurai, and Matsuoka as secretary-general as well as Chairman of the Agriculture and Forestry Division, Nakagawa Yoshio, assembled in full force and raised the stakes. They demanded to Fukuda: 'you should not progress FTAs over the heads of the party…to swallow all the demands of other countries is weak diplomacy…as we've said up to now, you should move forward in consultation with the party'.[56] They proposed to Fukuda that it was necessary for the government and the ruling parties to unite as one, hold the same opinion, and face negotiations in the future so as not to be taken unfair advantage of in the negotiations with the partner country.[57] In short, they demanded their rights of intervention in the matter. Fukuda tried to placate them by saying that of course the party would be consulted. He showed a certain understanding of their position with his comment that 'although the prime minister instructs in many ways when necessary, it is taken for granted that the prime minister discusses issues with others…The prime minister does not consider that only he leads on all points'.[58]

Details of the Japan-Mexico FTA that was signed in March 2004 revealed that a compromise on agricultural market access had been made, falling well short of endangering domestic agricultural producers in Japan or bringing about free trade in agriculture between the two countries. Not only did Japan commit itself to removing tariffs on just 86 per cent of all imports from Mexico, rather than the required 90 per cent, but politically sensitive commodities such as pork, orange juice, beef, chicken, and fresh oranges remained subject to special arrangements that would continue to protect the producers of these commodities. After the FTA agreement was signed with Mexico, Matsuoka stated: 'The agreement is balanced. We can stage FTA negotiations with other countries, based on this example'.[59] The precedent of only very limited concessions on agriculture in the Mexican agreement was very important. Not only did it introduce the notion that phased (i.e. incremental) liberalisation was compatible with bilateral FTAs, but it also allowed for liberalisation to occur over a long time period and/or be subject to quota limitations.

JUMPING ON THE BIOMASS BANDWAGON

During the 2002–2003 period, Matsuoka branched out by becoming the major political sponsor of the biomass industry in Japan, his 'new green revolution', which planned to generate energy from food, animal and timber waste. Matsuoka was hopeful of the promise of converting agricultural products such as sugar cane and corn and even timber into a source for ethanol gasoline and thus contributing to the development of agricultural and mountain village regions.[60]

Because biomass was relatively new and unexplored territory, pursuing this cause would make Matsuoka the first in the political world to be involved in it.[61] His primary reasons for jumping on the biomass bandwagon, however, were financial. From his own self-interested perspective, Matsuoka eyed the biomass industry not only as a potential source of budget outlays for public works projects in Kumamoto Prefecture, but also as a new business for companies on which he could prey for political funds.

The project followed the pattern that, whatever venture Matsuoka backed, there were usually advantageous corporate connections in the background. In this case, the executive of 'a particular company that could be said to be Matsuoka's "sworn friend" decided to make biomass into an industry'.[62] In fact, a group of companies with links to Matsuoka suddenly showed heightened interest in the biomass project. These companies had their main offices in Shinjuku in Tokyo and they all suddenly 'converted' to the business of the industrialisation of biomass.

One company was called Japan Geo-System Approach (as of 1 January 2003). Its previous company name was Japan Amusement System Incorporated (later Nazca), a company established to make pre-paid pin-ball (pachinko) cards.[63] Another company in the group was Caldean Integrate Incorporated that had provided information and communication services for computer users, and which changed its name to Green Energy Research Association Incorporated on 19 August 2002. With the change of name, it added 'research and development of biomass methanol and ethanol and its production/sale' to the details in its business purpose column.[64]

The person acting as the representative of the group of companies was Mitsuzuka Kôkichi, a nephew of retired former Minister of Finance and LDP faction leader, Mitsuzuka Hiroshi (Matsuoka's old faction leader). Mitsuzuka Kôkichi was three years younger than Matsuoka, but he was Matsuoka's influential sponsor at one time. He aggressively expanded his real estate business during the bubble period and became friendly with Matsuoka when Matsuoka left the Forestry Agency for the political world and achieved his first election victory in 1990. After that, the two remained close.[65] Mitsuzuka took pride in being a 'cheer squad' (*ôendan*) for Matsuoka.[66]

In addition to Japan Geo-System Approach and Caldean Integrate Incorporated, Mitsuzuka owned Japan Technoblast Incorporated for purposes of construction engineering and architectural contracts and consulting, and Mitsue Incorporated, a real estate business. Both Japan Technoblast Incorporated and Mitsue Incorporated also wrote 'research into biomass and its production and sales' in the company purpose column during September-October 2002. What is more, Japan Technoblast had amongst its employees, Ikeda Kazutaka, Matsuoka's policy secretary.[67]

According to a source in the real estate industry, Mitsuzuka, whose companies had not done well since 1990, seemed energised by his move into biomass saying, 'I'm going with biomass from now on'.[68] Mitsuzuka apparently did not keep company with anyone from the Mitsuzuka (now Mori) faction: 'After mixing with several Diet members, the person whom he thought "would be useful" was Matsuoka'.[69] Mitsuzuka was reportedly energetic at both work and play, and he and Matsuoka would bar-hop across four or five expensive clubs in Kabuki-chô, ending the evening with noodles (*ramen*).[70] The 'force of Mitsuzuka [reportedly] matched with the force of Matsuoka, known to make bureaucrats and local industries in Kumamoto flinch'.[71]

It was suspected that Mitsuzuka had advised Matsuoka to shake himself free of the 'concessions triangle' of money (*kane*), votes (*hyô*) and business (*shigoto*) between politicians and the construction industry.[72] It was also said that because Matsuoka's blatant guidance of benefits to his local Kumamoto Prefecture was becoming an issue, Mitsuzuka had advised, '[i]t is no longer the era of construction public works politics. Why don't you sponsor businesses in new areas like those related to biomass environment and energy equipment and organise that'?[73] Mitsuzuka himself was concentrating all his management resources and putting all his bets on biomass and was busy securing funds to invest in the industry.[74] He also tried to participate—in various forms—in biomass-related subsidised projects that had begun nationally.[75]

As far as Matsuoka was concerned, the biomass project killed a number of birds with one stone. First, it generated 'new works' in the MAFF budget to support the development of research and technology for utilising biomass, and the construction of action models across the whole country and the development of facilities. These facilities could be to Matsuoka's political advantage if they were located in Kumamoto.

The MAFF launched a 'Biomass Nippon Comprehensive Strategy Project Team' in June 2002. It received the green light from the Koizumi administration's 'big-boned policies number two' in the same month. Matsuoka had struggled to get his pet project into the big-boned policies by desperately badgering Prime Minister Koizumi.[76] The team consisted of seven MAFF officials under an office chief, with their own room on the first floor of the MAFF. The initial budget for the project allocated ¥22 billion, with most of it going to the MAFF under the government's so-called New Energy Strategy.[77] The team then asked for ¥29 billion in the draft budget for 2003 to promote the realisation of 'Biomass Nippon', a 'society that uses biomass to its fullest extent with a view to halting global warming and constructing a cyclic society'.[78] The total budget for 'Biomass Practical Use Frontier Infrastructure Works' in 2003 was ¥22.2 billion while ¥22.6 billion was allocated in 2004.[79] A MAFF official said, 'in the future, it will increase to many billions of yen'.[80]

Unsurprisingly, Kamoto Town (in Matsuoka's electorate) was chosen for the construction of a plant for processing livestock and food waste at a cost of ¥8.9 billion, under the budgetary rubric of 'Biomass Practical Use Frontier Infrastructure Works'.[81] Because Kumamoto had an extensive livestock industry, the biomass centre for processing livestock waste was ostensibly located there.

However, 'it was logical to think that the location was no accident—that it had been engineered by Matsuoka—although this was denied by the town mayor'.[82]

Second, from Matsuoka's perspective, the biomass project justified a new public value for both farming and forestry—as a source of energy. Not only Matsuoka but also other *nôrin giin* saw biomass as a means of revitalising agriculture and rural areas. In 2002, a new book on biomass entitled '21st Century: Escaping from Limitations and Chaos' under Matsuoka's authorship was about to be published. The subtitle was 'Environmental Regeneration and the New Energy Revolution'. The authors were Matsuoka and the Biomass Methanol Research Association (Biomasu-Metanôru Kenkyûkai). The timing of its intended publication matched the launch of the MAFF's 'Biomass Nippon' project.

Besides extolling the global environmental crisis in the afterword, Matsuoka wrote 'this book is not only a warning bell, it is also a book that suggests policies'.[83] He wrote in the epilogue (dated on an auspicious day in April 2002)

> [a]s you can see from reading this book, it has its starting point my feelings of 'crisis' in regards to limitations and chaos manifested in the 11 September terrorist attacks…in all circumstances politicians have the responsibility to present measures to resolve a problem. This book developed an argument for the potential of a second industrial revolution based on biomass methanol.[84]

The book was extremely hostile to the idea of a market principles and economic orthodoxy because of the way these principles treated 'losers' and those deemed 'unfit for the market'. In the book, Matsuoka called for economic principles that that were kind to the 'losers' in a system.[85]

For some reason, the book was not published, even though a publication party was planned. The binding had virtually finished but the publication was cancelled.[86] It was truly a phantom book.[87] Matsuoka's office denied any connection with it, although given that he wrote part of the book and that he was the most prominent politician-promoter of the biomass industry himself, this was incorrect. Matsuoka's prominence in pushing the issue was the reason why he was approached to participate.

It seems that the Muneo scandal was the reason why the book was not published. Muneo was arrested on suspicion of accepting bribes for mediation on 19 June 2002. His arrest was not simply a matter involving Matsuoka's sworn friend. There was danger for Matsuoka in it as well. The flurry of criticism extended to him as well,[88] encouraging him to keep a low profile. The publication date of the book was 16 July 2002, right in the middle of the maelstrom of the Muneo scandal, and the book clearly could not be published in the circumstances.[89] Despite having pushed so hard for the biomass project to get off the ground, Matsuoka refused all interviews on the subject. According to his policy secretary, Ikeda Kazutaka, he was so careful that he would not even issue a comment of 'No comment'.[90]

GREENWASH

Matsuoka took direct action to revamp his political reputation, which had been so sullied by his association with Muneo and by the BSE and Yamarin scandals, by becoming an outspoken advocate of interests that appeared directly to contradict his earlier, self-interested activities. He appeared on the program 'Jam the World' (J-WAVE FM radio) in Tokyo in February 2004. The topic of conversation was the import ban on American beef after the outbreak of BSE in the United States[91] and the disappearance of *gyûdon* (beef bowl) from popular restaurants. In his comments, Matsuoka took a hardline stance from a consumer perspective, which was not hard to do given the issue. Matsuoka explained that

> [a]lthough Japan gave notice to the United States to secure the safety of its beef and to implement cow inspections along the same lines as Japan and other countries are doing, the United States did not do so for various reasons. So for Japan, the essential problem was that consumer safety cannot be guaranteed. As long as safety and security cannot be demonstrated, Japan cannot comply with the U.S. unilateral demand to resume imports.[92]

In the wake of this appearance, Matsuoka reported that he received 'favorable comments from urban listeners, who were not usually familiar with agricultural issues, but who said that his comments were good and very easy to understand'.[93]

Matsuoka's main efforts, however, were directed to revamping his image in the area of environmental policy. He undertook a rather transparent effort to rehabilitate himself through a process of 'green-wash',[94] which involved the public relations 'greening' of his image. Matsuoka had always professed environmental credentials, showing his (opportunistically) environmentalist side in a number of policy activities, including, of course, biomass and the cause of agricultural protection.

Matsuoka's professed environmentalism was well publicised on his website. The headlines screamed: 'Protecting Water and Greenery and Food', and proclaimed that Matsuoka was tackling 'global-scale population, food and environmental problems in the twenty-first century with all his power (*zenryoku de)*'.⁹⁵ Matsuoka asserted that he was playing a major role in the 'Green Energy Revolution' (Midori no Enerugî Kaikaku) in order to protect the global environment. This stance he justified in terms of achieving the 'vitalisation' (*kasseika*) of regional people and agricultural and forestry industries that bring forth green resources.⁹⁶

It was the Yamarin scandal that was the spur to Matsuoka's apparent, full-blown conversion to 'green' environmentalism. Following the scandal, Matsuoka went to great lengths to strengthen his environmental credentials in the international campaign against illegal logging. The need for Matsuoka to promote such an environmental cause was blatant, given his association with Yamarin's illegal logging. Matsuoka tried to bury his past record by becoming a champion of the fight against illegal logging, not only in Japan but also around the world. He changed from someone who received political donations from a company engaged in illegal logging in Japan to someone who actively campaigned against it, particularly outside Japan.

In 2003, Matsuoka became chairman of the LDP's Forestry Illegal and Unlawful Logging Countermeasures Investigation Team (Shinrin Ihô, Fuhô Bassai Taisaku Kentô Chîmu). It discussed reports from the Forestry Agency and the Ministry of Environment about discussions in the WTO Trade and Environment Committee, and the results of the third meeting of the Asia Forestry Partnership and the Japan-Indonesia Illegal Logging Cooperation Action Plan.⁹⁷ In 2004, the committee changed its name to Illegal Logging Countermeasures Investigation Team to Protect the Global Environment (Chikyû Kankyô o Mamoru Fuhô Bassai Taisaku Kentô Chîmu), more in keeping with the times.

Matsuoka attended an International Symposium on Countermeasures for Illegal Logging in Tokyo in June 2003. The symposium was organised by the Japan Federation of World Timber Industry Associations (i.e. the main users of tropical timbers). At one point during the proceedings, Matsuoka made a rather vacuous speech, talking about his hopes for efforts to be taken against illegal logging, about Japan and Indonesia's efforts to stop illegal logging, and the fact that Diet members were prepared to form an international confederation of parliamentarians against illegal logging.⁹⁸

On domestic forestry policy, Matsuoka participated in a meeting of the Agriculture, Forestry and Fisheries Joint Council (Nôrinsuisan Gôdô Kaigi) in July 2003. This was a joint council of the two main agriculture committees in the PARC, the Agriculture and Forestry Division and CAPIC, as well as of the main LDP policy committee on forestry policy, the Forestry Policy Investigation

Committee, and the main party committee on fisheries policy, the Fisheries Comprehensive Investigation Committee (Suisan Sôgô Chôsakai). The joint council met to receive an interim report of the Forestry Agency's Research Association for Citizens' Support for Promoting the Absorption Source[99] Countermeasures to Prevent Global Warming (Chikyû Ondoka Soshi Kyûshûgen Taisaku no Suishin no tame no Kokumin Shien ni kansuru Kenkyûkai). The meeting acknowledged the need to work even more positively for the introduction of preferential measures for forest preservation through tax reform and securing revenue sources, because forests were a primary source of absorption of greenhouse gases.[100] In Matsuoka's view, Japan's forest maintenance program should play the biggest role in absorbing carbon dioxide. However, forest preservation had not been positively promoted. It was suffering from the long-term deterioration in the domestic forest industry, a shortage of forestry workers and the uniform cutback policy for public works etc.[101] When, in 2004 Matsuoka became chairman of the Forestry Basic Problems Subcommittee, one of its key tasks was to discuss the future development of 'Forestry Absorption Source Countermeasures' (*Shinrin Kyûshûgen Taisaku*), and to ensure that finance for the countermeasures was included in the Forestry Agency's draft budget for 2005.

Another issue of concern for the Forestry Basic Problems Subcommittee was that ministries and agencies should use domestic timber in the provision of their services and public works. Each year, the committee received a report from ministries and agencies on this matter, and its members energetically promoted the utilisation of regional timber products in government-sponsored public works.

Yet another task for the committee was ensuring that new production systems were budgeted for, which, according to Matsuoka, would trigger the regeneration of the forestry industry and green employment projects as well as projects for 'successors' (*kôkeisha*) to forest owners. A meeting of the committee was held about this in October 2005 in order to secure funding for such works.[102]

In 2003–5, Matsuoka served as chairman of the LDP's Countermeasures Investigation Team to Protect the Earth's Environment from Global Scale Illegal and Unlawful Deforestation, Import and Export and so on (Sekai Kibo no Shinrin no Ihô, Fuhô na Bassai oyobi Yushutsunyû tô kara Sekai Kanyô o Mamoru tame no Taisaku Kentô Chîmu). Its motto was: 'Stop the Destruction of the Environment, Solve the Problem by Using Domestic Timber!'. In May 2004, the investigation team held hearings with representatives of NGOs on ways to counter illegal deforestation in Southeast Asia, including a call for Japanese foreign aid to pay for the planting of nursery trees in areas that had suffered illegal deforestation as a result of demand from timber companies in Japan for wood.[103]

In August 2004, Matsuoka gave the keynote speech at a Regional Workshop on Strengthening the Asia Forest Partnership (AFP) organised by the Ministry of Forestry of Indonesia. Both Matsuoka and one of the other recipients of funds from Yamarin, Matsushita Tadahirô, attended, along with two other Lower House Diet members and officials from the Forestry Agency. Matsuoka's keynote speech described 'the current activities in Japan through the AFP for promoting sustainable forest management and controlling illegal logging and its associated trade'.[104] After the workshop, Matusoka led a delegation of 10 Japanese Diet members to East Kalimantan, promising local officials that they would help the local government combat illegal logging in the province. Matsuoka 'said the legislators were seeking information as to those areas to which they could contribute in the fight against illegal logging'.[105] On several earlier occasions Matsuoka had made public presentations on the decline in the world's forestry resources and its impact on water resources, and on illegal deforestation problems and related issues.[106]

In 2005, Matsuoka became chairman of the LDP's Illegal and Unlawful Logging Countermeasures Investigation Team (Ihô, Fuhô na Bassai Taisaku Kentô Chîmu). The team discussed putting effort into the positioning of countermeasures against illegal logging in the G-8 summit in England in 2005. At a meeting in March 2005, government spokespersons provided details of the United Kingdom and European Union's illegal logging countermeasures, and discussed proposals for the summit. In following month, Forestry Agency officials talked to the group about the timber trade in Japan and conditions of domestic distribution. A few days later, they held hearings where timber-importing companies made representations, followed by timber groups and NGO groups, and they discussed future action. They conferred on topics to be investigated concerning illegal logging and approved them. They agreed to firm up their standpoint, which would be transmitted to MoFA for the G-8 summit. In the team's view, illegal logging should be on the main agenda of the G-8 in Gleneagles, along with aid to Africa.

In June 2005, Matsuoka chaired a meeting of the investigation team, in which he formulated a system to remove illegally deforested timber from Japanese government procurement as a measure to protect the earth's environment.

One way to do this was through a traceability system for timber to prove its legality. Matsuoka claimed to have great responsibility for this issue since the illegal deforestation problem, he hoped, would be an important item on the agenda of the United Kingdom summit in July 2005.[107] The meeting was followed up later in the year with another, which discussed how timber logged illegally should be excluded under the government's Green Purchasing Law (*Gurîn Kônyûhô*), which made it a duty to consider the environment when the government procured goods.[108]

Matsuoka visited Britain at the invitation of the Environment, Food and Rural Affairs Minister Morley on how to put the deforestation problem on the agenda of the G8 Summit to be held in the United Kingdom later in the year. Matsuoka also held discussions with NGOs, interested groups and government officials concerned with supply policy and trade measures for obtaining timber from sustainable forests, and limiting timber supply for the central government to legitimate timber, a policy that the United Kingdom already had in place.[109]

During the same visit to the United Kingdom, Matsuoka attended a meeting sponsored by the Royal Institute of International Affairs on 'Forest Governance and Trade – Japan, the United Kingdom and Eureopean Union Initiatives'. The formal objectives of the meeting were 'to share information about efforts by United Kingdom and European Union governments and the private sector to combat illegal logging and associated trade and to discuss policy options available to the Japanese Government and Japanese private sector'.[110] In his speech, Matsuoka

> emphasised the vital importance of international cooperation and sharing of experience and best practice in tackling illegal logging....Leadership on the issue by the G8 was felt to have great potential and to fairly reflect the responsibilities of consumer nations [i.e. Japan]. Japan's engagement with the East Asia Forest Law Enforcement and Governance Conference and the Asia Forest Partnership were noted, as well as bilateral efforts to work with Indonesia through a Memorandum of Understanding.[111]

A few days prior to Prime Minister's Koizumi departure for the G8 summit, the investigation team briefed him on measures for protecting the earth's environment from illegal logging. It presented a series of recommendations, one of which was that the government should only procure timber that could prove that it was logged legally, and that support for exporting countries should be strengthened. Matsuoka said to the prime minister, 'I want you to assert [these policies] as the government-LDP draft at the summit',[112] to which Koizumi replied: 'you can rely on me to do this'.[113]

The team was very pleased to see that tackling illegal logging was part of the Action Plan coming out of the G8 summit. Item 37 of the Action Plan acknowledged that tackling illegal logging was an important step towards the sustainable management of forests, and that tackling this issue effectively required action from both timber producing and timber consuming countries.[114]

After the summit, Matsuoka attended a meeting with representatives of the timber industries of Canada, India, Indonesia and Norway and of Japanese groups. He also made an on-the-spot survey in Indonesia in his capacity as chairman of the investigation team, as well as attending a regional workshop of the Asia Forest Partnership. In Matsuoka's view, illegal deforestation was not only an important discussion item at the summit, but measures against it were an essential part of the solution to global-scale environmental deterioration and a means of reviving the domestic forestry industry. As he claimed

> [i]n 2001, I organised an investigation team in the LDP and have continued to appeal [for this cause] not only in Japan but also to countries around the world. Bit by bit, the problem has been recognised even in international conferences. Finally, illegal deforestation measures have become one of the main items at the summit this time.[115]

For Matsuoka, the answer was for 'countries not to use timber that was logged illegally. Such a system was already in place in the United Kingdom and in other countries that had removed illegally deforested timber from government-procured materials. He wanted to move the Japanese government finally to follow suit'.[116]

Matsuoka later reported on illegal logging countermeasures to the Forestry Management Activization Council (Shinrin Keiei Kasseika Kyôgikai), a group of Diet members in the LDP, of which he was the chairman. The council represented the forestry and the timber industry in Japan with a view to getting funding allocated in the Forestry Agency's draft budget for a forestry management revitalisation fund.

To further his international work on forestry, Matsuoka became the acting chairman of the supra-partisan Japan-China Tree-Planting Promotion Diet Members' League (Nicchû Ryokuka Suishin Giin Renmei). It promoted a tree-planting project in China using Japanese expertise on how to revive Chinese forests. Forest devastation was a leading cause of large-scale floods in China.[117] The league also aimed to assist the Japan-China Tree-Planting Fund (Nicchû Ryokuka Kikin) established by the late Prime Minister Obuchi. This fund was developing a tree-planting campaign in China, where land impoverishment had become a serious problem. The executive committee of the league decided to raise independent contributions focussing on Diet members and actively to support the campaign.[118] In discussions with the Chinese ambassador, Matsuoka took the opportunity to discuss subjects such as environmental problems, FTA negotiations, the Green Energy Revolution[119] and the idea for exporting farm products to China, which Matsuoka had proposed earlier.[120]

Quite apart from the blatantly self-serving nature of Matsuoka's leadership of the campaign against illegal logging, the sincerity of his environmental credentials can be questioned on a number of other grounds. First, for Matsuoka, environmentalism was really disguised agricultural protectionism. One of the main arguments that Matsuoka consistently advanced in opposition to agricultural trade liberalisation was the environmental one. Supporting and protecting the domestic farm sector was justified on environmental grounds insofar as agriculture was deemed to possess various environmental values. This fervently held position was one reason why Matsuoka pursued a position in environmental policy committees from an early stage in his Diet career (see Table 4.1). The other reason was his involvement in the forestry industry that had diverse, officially recognised environmental functions. There was a large area of 'protection forest' in Japan, intended to serve the interests of the public. 'Protection forests' included 'headwater conservation forest', 'soil run-off prevent forest', 'landslide prevention forest' and so on.

Second, Matsuoka had actively worked as a Forestry Agency bureaucrat, sponsoring the construction of unnecessary forest roads that required the cutting down of areas of forest.[121] While the Forestry Agency's mission was to protect mountains and forests, it specialised in the felling of trees to build forest roads.[122] Further, as a politician, Matsuoka worked as political broker for construction companies wanting to get involved in the construction of forest roads, and for developer-clients who sought to convert forestland to other uses. During the period of Matsuoka's close association with forestry administration and forest policy, Japan's domestic forestry industry steadily declined with mountain forests, in particular, falling into ruin.[123]

Matsuoka saw nothing contradictory about promoting higher prices for the timber sold by domestic forest owners on the Japanese market, while at the same time telling them that they and their organisations protected the green spaces of Japan.[124] Nor did he see any contradiction in utilising forests for timber while husbanding them as a national resource. He acknowledged that forests were diminishing around the world, but he sought to promote the use of Japanese timber by the Japanese government.[125] In this case, the economic self-interest of Japan's forest owners and Matsuoka's political self-interest took precedence over any environmental cause. Matsuoka's position as a chairman of the Forestry Policy Basic Problems Subcommittee was a venue in which he could push this line, arguing that as forests were a resource that were cultivated domestically, they should be positively utilised.[126]

From his vantage point as chairman of the subcommittee, Matsuoka made common cause with the head of Zenshinren about the need to expand the demand for the timber sold by the forest associations. At the same time, he was fond of motherhood statements such as 'making forests healthy protects our lives and the earth', by helping to protect against global warming. Those with a vested interest in Japan's domestic forestry industry (which, ironically meant felling rather than conserving trees)—such as Matsuoka and Zenshinren – latched on to the cause of 'global warming', and sought to harness it for their own political purposes, just as the agriculture lobby harnessed the issue of food self-sufficiency and food security.

Third, as Matsuoka's involvement with the biomass project demonstrated, some aspects of Matsuoka's environmentalism were more about boosting government spending in the agriculture and forestry sector, including for public works projects in Matsuoka's own electorate and thus gaining personal political advantage, than about supporting any particular environmental policy principle. Matsuoka became secretary-general of the Diet Members' League for Promoting the Green Energy Revolution (Midori no Enerugî Kaikaku Suishin Giin Renmei). The foundation general meeting of the league was held in the LDP headquarters on 24 January 2003. In Matsuoka's words, the league 'was established to promote the use of biomass and to aim to manage both the rehabilitation of the earth's environment and new energy production'.[127] Chairing the proceedings was Diet member Arai Hiroyuki, while the first person to offer greetings was Etô Takami.[128] Amongst those attending were members of both houses of the Diet as well as officials of the MAFF, the Natural Resources and Energy Agency and the Ministry of Environment, with the bureaucrats attending as observers. Matsuoka made a progress report to the gathering and read out the 'foundation purpose document' of the league, which stated

> [i]n the twenty-first century, the regeneration of the earth's environment, which is the foundation of existence, is the greatest theme for human kind....It is estimated that the scale of green energy-related industries is about ¥700 billion and the anticipated impact on industry is ¥1.3 trillion, and through this, the steady decline in our primary industry can be reversed into a large-growth industry.[129]

A total of 91 LDP Diet members participated in the foundation general meeting of the league.[130] The MAFF had high hopes for the league to become its main link to the political world.[131]

The meeting agreed to hold study groups to which academic experts were invited to speak about various aspects of the 'green revolution'. For example, in July 2003, they invited two academics to talk to them on the subject of biomass energy.[132] They hoped to make a case for supporting and protecting Japanese agriculture on the grounds that it would be a source of biomass energy, saying '[u]ntil now, local administration and farm products policy have always been on the defensive. However, adopting a policy to utilise biomass energy effectively will create an opportunity to be at the forefront of the times'.[133] The study group resolved to work hard with the ministries and agencies concerned to realise a concrete biomass policy.[134]

Matsuoka was the driving force behind the league, having the most knowledge and understanding of the issue amongst Diet members.[135] Even Etô Takami, who attended the general meeting as a representative promoter, said, 'biomass is Mr Matsuoka's endeavour. You need to ask Mr Matsuoka for details [not me]'.[136] Matsuoka, however, continued to refuse all interviews about biomass.[137]

Internationally, it was another matter. Matsuoka visited Brazil for agricultural trade diplomacy in June 2004, but he took the opportunity to discuss the Green Energy Revolution with the Brazilian ministers. The interest in the visit was Brazil's ethanol application policy.

Matsuoka was also a member of several Diet members' leagues focussing on the environment, including GLOBE Japan (Global Legislators Organization for a Balanced Environment, or Chikyû Kankyô Kokusai Giin Renmei), which held its first general meeting in June 2004.[138] Around 42 non-partisan Diet representatives with an interest in environmental problems participated in GLOBE Japan. The league was part of an international grouping consisting of parliamentarians hoping to build international cooperation for dealing with global environmental problems. It comprised volunteer legislators from parliaments in Japan and the European Union, as well as the United States Congress.

However, it is, perhaps, the project to reclaim Isahaya Bay that most reveals Matsuoka's shallow environmentalism. Local residents and fishermen have strenuously opposed the project, the fishermen arguing that it damaged the local fishing industry as well as the seaweed catch. In late 2000, an out-ofseason red tide occurred in several places in the Ariake Sea. The discolouration of cultured seaweed (*nori*) crops began to be noticeable, and this started to worry some MAFF bureaucrats. A bad *nori* harvest also ran the risk of influencing the support bases of Koga Makoto (secretary-general of the LDP) from Fukuoka, Matsuoka from Kumamoto, and Noda Takeshi, secretary-general of the Conservative Party.[139]

The MAFF set up a committee of specialists in order to review the project. Most of the members were sceptical about it, and most wanted the dyke gates opened and investigated.[140] There was a distinct split in the committee between MAFF *jimukan*, who were critical of the project, and MAFF *gikan*, who backed it with support from *nôrin zoku* including Matsuoka. The *gikan* on the committee resisted proposals to freeze the reclamation works that were affecting the water quality and an investigation into the dyke opening. They said that a huge budget would be needed to remove 4 million cubic metres of mud before the dyke could be opened.[141]

In August 2001, the *jimukan* officials proposed a large review of the project, which would mean cancelling it and leaving the dyke gates open for the time being. However, because this would require new expenditure, the plan failed to get MOF approval, and a decision was taken to leave the dyke gate unopened and to reduce the area of reclamation by half.

THE 2003 ELECTION

As a member of the LDP resistance forces, Matsuoka tried to exercise denial rights over Koizumi's reform proposals. Because of the lack of policy cohesion within the LDP caused by the party's decentralised policymaking process, it was not only possible but also acceptable for individual LDP Diet politicians like Matsuoka openly to oppose the prime minister and the Kantei. However, Matsuoka's individual power base in his own electoral district, from which he mounted his attacks on Koizumi, turned out to be not quite so secure after all, as the results of the 2003 election showed. As one political journalist pointed out

> Scary resistance politicians, such as Suzuki and Matsuoka, are completely disliked by the population. In particular, since the establishment of the Koizumi government, their role as the baddies has stuck, so they must be fairly anxious.... [Their attacks on pro-Koizumi politicians] probably come from this sense of anxiety, but in the end, they are just strangling themselves with their own hands.[142]

Matsuoka initially capitalised on his opposition to Koizumi (by gaining public prominence) but then suffered as a result. He lost the seat of Kumamoto (3) in the November 2003 election, in spite of the fact that the Komeitô once again endorsed his candidacy[143] and organisations connected to agriculture campaigned vigorously in support of him. One election rally in Kikuyo Town in Kikuchi County drew 3000 people, with the banners of the Kômeitô, the Kumamoto Prefecture Farmers' Political League and local government organisations all visible in the throng. The rally was to hear LDP Secretary-General, Abe Shinzô, give a speech in support of Matsuoka. Other people also spoke, including the heads of the LDP's federation of Kumamoto Prefecture party branches and Nokyo-related groups. They unanimously praised Matsuoka.[144] However, one construction company owner lamented on the eve of Abe's visit to Kikuyo Town to support Matsuoka's campaign, '[w]hy must Abe support such a human being'?[145] Nevertheless, Abe's appearance on the hustings helped to cement Matsuoka's loyalty and eventual backing for Abe's bid to succeed Koizumi, which was rewarded, in turn, with Matsuoka's elevation to the position of MAFF minister in the first Abe cabinet.

In line with the growing custom amongst the competing parties in the election, Matsuoka drew up his own policy manifesto, which was broadly publicised on his website, in order to convince voters to vote for him rather than for one of the other candidates. Not only did Matsuoka's advocacy of a policy manifesto suggest that he was still operating as an independent political entrepreneur with his own political marketing strategy, 'combining "position statements" on the big issues of the day...with special favours to local interests',[146] but it also indicated that he was, or assumed himself to be, in a position to deliver on the promises contained in his manifesto.[147]

The results of the election in Kumamoto (3) were very close: Matsuoka lost his seat by only about 3000 votes. He won 76,469 votes, or 40.9 per cent of the total cast vote of 186,857 (see Table 7.1), whilst his main rival, standing as an Independent, won 79,500 votes (42.5 per cent of the total). On the other hand, Matsuoka's vote tally was a precipitous decline of 32,658 votes on his 2000 win (see Appendix), with the loss interrupting five consecutive election victories in his local Kumamoto electoral district since 1990. Even in his *jiban* of Aso County, which, although it remained a rock solid base of support supplying about one third of his total vote, Matsuoka won only 51.9 per cent of votes (an overall decline of around 10,000 votes), compared with just under 74.4 per cent in 2000 (see Table 3.2 and Table 7.1).

Matsuoka's total county vote dropped by a third (see Table 3.2 and Table 7.1), representing less than half of the total cast vote (40.6 per cent) compared with almost two-thirds in 2000 (see Table 3.2 and Table 7.1). It was in the counties as much as in the cities where Matsuoka failed to gain his customary levels of support. The most telling decline was in Matsuoka's proportion of the total vote won, which fell from 63.6 per cent in 2000 to 40.9 per cent in 2003 (see Table 3.2 and Table 7.1).

Even Matsuoka's most fervent supporters seemed to catch a whiff of impending disaster. In the early morning of polling day, a prominent member of the Dôshikai in Aso Town visited someone who had strongly criticised the Matsuoka-prefectural assembly member-Kawasaki regime, notifying him of Matsuoka's impending defeat.[148]

Fortunately for Matsuoka, he was saved from electoral oblivion by the PR district system, scraping in at the bottom of the party list as one of three LDP SMD candidates in Kyushu who lost their seats but who were 'revived' (*fukkatsu*) by the party list in the Kyushu regional bloc constituency. The 'best loser' provision of the Public Office Election Law allowed losers in SMDs such as Matsuoka to be elected under PR if they received more than a legal minimum of votes (which was at least 10 per cent of the total vote in the SMD in which they stood). The LDP's overall vote tally in the Kyushu bloc was 36 per cent, which entitled it to eight PR seats. Matsuoka was ranked eighth.[149] He only managed to retain a Diet seat because he won 96.1 per cent of the victor's vote, which placed him third on the list of SMD losers. The party rewarded only those SMD losers 'who came closest to winning in their local district races'.[150] The top five on the party list were ranked by officials of the party executive and were not standing in SMD seats. They were given priority over Matsuoka and others who were simultaneously running in Kyushu SMDs.

Given his low ranking on the winners' list, the 2003 Lower House election was hardly a resounding victory for Matsuoka. This did not stop him and his followers letting off loud fireworks in Aso Town in the early dawn hours of the day after the election, to the anger and disgust of some of the residents.[151] During the campaign, Matsuoka's supporters had also put up posters of Matsuoka and Abe in each of the polling stations in Aso Town, which some residents argued violated the Public Office Election Law. They had to go through the town office to get them removed, resentful that Matsuoka and his followers acted as if they owned the town.[152]

The 2003 election thus made Matsuoka into a PR bloc politician rather than a local constituency politician. It meant that he was held in lower regard compared to his standing as a representative of Kumamoto (3). His position in the Diet and in the party was not as strong as it had been previously. He joined the group of 'zombie' candidates (who had risen from the dead),[153] winning only a bronze medal compared with the silver medalists (purely PR candidates with strong party endorsement) and gold medalists (those successful in the SMDs in their own right).

The loss of Matsuoka's SMD seat was a big shock, not only to Matsuoka himself but also to his supporters and to other *nôrin giin* in the party. Just past 1am on 10 October (the day after the election), a haggard Matsuoka emerged in his electoral office and according to one source in his *kôenkai* said, 'he reflected that "what was misunderstood was my own lack of power"'.[154]

Some observers interpreted it as a 'tectonic shift' in an LDP 'stronghold' portending that 'in the undercurrents in Japan, something was trying to change'.[155] The election results seemed to suggest that 'the need for construction companies to engage and invite Matsuoka's attention appears to be diminishing. Matsuoka is becoming a "has-been" for a wide range of social classes'.[156] One old timer in Aso Town commented, '[a]n unusual and big change has occurred…it is a change that I haven't witnessed before in my lifetime'.[157]

Matsuoka's loss was symptomatic, amongst other things, of the punishment that Japanese voters frequently mete out to notoriously corrupt and tainted politicians. Matsuoka was seen as a typical conservative reactionary by many non-rural and non-farm voters, who disapproved of his brush with the political corruption scandal involving Muneo and Yamarin, and hints of others. After Muneo was arrested and had to give up contesting his seat in the 2003 Lower House elections, Matsuoka had gone quiet

> [and] it seemed that the unfavorable wind against Matsuoka had stopped. However, as if to be cursed by Muneo's ghost, he tragically lost his seat in the Kumamoto (3)…For Matsuoka, he was fighting the election amidst unfavourable winds….[In the campaign], Matsuoka appealed to his past record over four terms, but the topic of Muneo was brought up again and again, underlining Matsuoka's strong image as 'Muneo of the West', which was a negative image that worked against him. Matsuoka was fairly annoyed by this whole scenario.[158]

Matsuoka's electoral record between 1990 and 2000 suggested that he had never been very popular personally and that he had never really established an impregnable electoral position in either Kumamoto (1) or in Kumamoto (3). For example, in 1993, Hosokawa garnered over 200,000 votes, while Matsuoka only secured just over 80,000 (see Appendix). It would seem that Matsuoka's supporters had only voted for him out of self-interest, which was not a sufficient basis for sustained electoral popularity.

What was particularly galling about the 2003 election result for Matsuoka was that he had not lost to someone from the main opposition DPJ whose candidate garnered only 26,317 votes (just over a third of Matsuoka's vote tally) but to an Independent candidate called Sakamoto Tetsushi, a man who was little known outside his local district of Kikuchi County.[159] Sakamoto was a former journalist for the *Kumamoto Nichinichi Shinbun*, and a former four-term LDP/Independent member of the Kumamoto prefectural assembly, endorsed by the prefectural *nôseiren*. He had to split from the LDP prior to the election because the party's endorsement went to Matsuoka. LDP supporters in the local area were reportedly in the habit of neatly dividing the political world into two halves: 'national politics = Matsuoka, and prefectural politics = Sakamoto'.[160] However, because Sakamoto stood as a candidate for a Diet seat (thereby breaking the unspoken contract), the mud-slinging began.[161] As Sakamoto was 53 years old, Matsuoka pilloried him saying, 'you are a betrayer. What can you do becoming a national Diet member past 50?'.[162] The Sakamoto camp retorted, 'Matsuoka's method is just consistently to throw mud'.[163] It also alleged: 'Mysterious documents [libelling Sakamoto] were distributed and there were as many as seven versions since the opening of the electoral office. They were distributed over the entire electorate, so they would amount to a few hundred thousand copies'.[164]

The campaign turned out to be a fierce contest between Matsuoka and Sakamoto. Matsuoka reputedly carried about 50 per cent of the LDP vote, and 60 per cent of the Kômeitô vote. In contrast, Sakamoto's support was a mixture of about 20 per cent of the LDP vote, 30 per cent of DPJ supporters, and 50 per cent of Social Democratic Party (SDP) supporters. According to the *Asahi*, Sakamoto gained great strength from criticising Matsuoka and made inroads into unaffiliated voters and supporters of the DPJ.[165] The DPJ candidate (like Matsuoka, also from the Aso region) was winning only about 50 per cent of the DPJ vote.[166]

Table 7.1 Farm household composition/votes cast for Matsuoka by municipality in Kumamoto (1) in 2003 Lower House election

Name of municipality	No. of farm households[a]	Farm households as % of total in municipality/ies	Votes cast for Matsuoka	% of total cast vote	% of Matsuoka's total vote	Placing among 4 candidates
Cities	3,224	16.1	14,019	42.3	18.3	1st
Yamaga City	1,490	13.2	7,414	41.7	9.7	1st
Kikuchi City	1,734	20.0	6,605	43.0	8.6	1st
Counties	17,293	20.7	62,450	40.6	81.7	2nd
Kamoto County	5,052	29.5	15,993	48.1	20.9	1st
Kahoku Town	795	54.3	2,173	60.9	2.8	1st
Kikuka Town	1,158	55.6	2,579	52.7	3.4	1st
Kamoto Town	603	22.8	2,492	48.6	3.3	1st
Kao Town	695	47.2	1,873	52.9	2.4	1st
Ueki Town	1,801	19.1	6,876	42.7	9.0	1st
Kikuchi County	4,570	11.1	21,706	29.8	28.4	2nd
Shichijo Town	795	39.4	1,658	45.9	2.2	1st
Kyokushi Village	1,158	36.8	1,607	47.1	2.1	1st
Ozu Town	603	12.6	3,586	22.6	4.7	2nd
Kikuyo Town	695	8.0	3,936	26.4	5.1	2nd
Koshi Town	1,801	7.4	3,604	29.7	4.7	2nd
Shisui Town	795	14.8	3,094	38.9	4.0	2nd
Nishigoshi Town	1,158	4.7	4,221	28.3	5.5	2nd
Aso County	7,671	31.1	24,751	51.9	32.4	1st
Ichinomiya Town	727	22.4	2,964	50.9	3.9	1st
Aso Town	1,643	27.6	7,510	62.7	9.8	1st

Minamioguni Town	578	38.3	1,287	41.9	1.7	2nd
Oguni Town	854	28.5	1,803	31.7	2.4	2nd
Ubuyama Village	286	47.2	784	67.1	1.0	1st
Namino Village	264	50.9	735	63.0	1.0	1st
Soyo Town	733	49.8	1,544	51.3	2.0	1st
Takamori Town	616	25.1	2,370	52.7	3.1	1st
Hakusui Village	584	44.6	1,409	48.6	1.8	1st
Kugino Village	420	57.1	1,015	57.3	1.3	1st
Choyo Village	401	18.7	1,444	49.0	1.9	1st
Nishihara Village	565	33.0	1,886	51.6	2.5	2nd
Total	20,517	19.9	76,469	40.9	100.0	

Notes: [a] Farm household data are for 2000.

Sources: Sōmuchō, Tōkei Kyoku, *Heisei 12-nen Kokusei Chōsa Hōkoku Dai 2-kan Dai 1-ji Kihon Shūkei Kekka Sono 2 Todōfuken, Shichōson Hen-43 Kumamoto-ken*, pp. 288-91 and 294-303; Kumamoto-ken Hōmu Pēji/Senkyo Kanri Iinkai, *(Dai 43-kai) Shūgiingiin Sōsenkyo (Shōsenkyoku) Kaihyō Kekka: Heisei 15-nen 11-gatsu 9-ka* [(The 43rd) *House of Representatives General Election (Single-Member Districts), The Results of the Vote Count: 9 November 2003*], p. 2; <http://www.pref.kumamoto.jp/gyousei/senkan/osirase/h15kekka/pdf/no43_03.pdf>.

On the other hand, in strong contrast to Matsuoka, Sakamoto, as vice-president of a body called 'Group Reform' (Gurûpu Kaikaku), was the only candidate who campaigned on a platform of open support for Prime Minister Koizumi's reform program. Sakamoto's policy platform advocated the decentralisation of taxing and subsidy powers from central to local governments, which was one of the main planks in Prime Minister Koizumi's reform program. Sakamoto claimed that he was running for the Diet to change politics and also enthusiastically represented causes such as cleaner politics and 'politics for the people'.[167] His successful election was an implicit criticism of Matsuoka.[168]

An LDP prefectural assembly member and supporter of Matsuoka explained his defeat in the following terms

> [w]e were defeated by our opponent's strategies. We were accused of benefit and concession politics, and I suppose this mudslinging confused the influential people. Because of the Yamarin affair, we were predicting a difficult election, but we didn't ever think that we'd lose. The LDP also was not a monolithic union. While saying that they would support Matsuoka, about half of the assembly members supported Sakamoto. There was also some rebellion as Matsuoka was only favoring a few specific industries. The Sakamoto camp also distributed three or four (libellous) documents. Sakamoto was elected, but the influential people didn't care who got elected. It means that they gathered the anti-Matsuoka vote.[169]

A journalist concurred, saying 'the anti-Matsuoka vote that flowed to Sakamoto is most likely the reason for his win'.[170]

Although Sakamoto subsequently joined the LDP's parliamentary caucus (*kaiha*) in the Lower House, meaning that he was considered an Independent member of the ruling camp, and even though he joined the Yamasaki faction, he remained an Independent for electoral purposes at the local constituency level. From the perspective of the local LDP organisation in Kumamoto (3), he was, therefore, not recognised as the LDP member for that constituency. Sakamoto himself acknowledged that, although he was in the LDP *kaiha* in the Diet, he was not recognised as a member of the LDP. Because of that, he received no subsidy from the party.[171] In fact, Matsuoka continued to give the address of the LDP's party branch in Kumamoto (3) as an address of his *kôenkai*. This alone suggested that, from Matsuoka's perspective, Sakamoto was merely keeping the seat warm until Matsuoka won it back at the next election. It also meant that Matsuoka would receive the party subsidy for Kumamoto (3) in any subsequent campaign. On top of this, Matsuoka remained secretary-general of the association of LDP Kyushu Diet members.

Even after his defeat, Matsuoka continued to cultivate the Kumamoto LDP federation of branches assiduously. A photograph on his website showed him addressing the women's division of the federation.[172] Clearly, the next Lower House election would be a test to see whether Sakamoto could withstand a renewed onslaught from Matsuoka.

THE SIGNIFICANCE OF THE MATSUOKA-SAKAMOTO CONTEST

The significance of the electoral battle between Matsuoka and Sakamoto in the 2003 election lay in the fact that it was a contest between the old and new style of politics, a microcosm of the contest that was being played out on the national political stage under Prime Minister Koizumi. The new style of politician relied primarily on programmatic appeals, with a focus on unaffiliated voters. Such politicians were not so reliant on the traditional electoral ingredients that LDP candidates had so often drawn on for success —a strong local *jiban* and *kôenkai*, the backing of various interest groups, plentiful political funding as well as the advantages of incumbency, which enabled them to guide benefits to local areas (*rieki yûdô*), to act as political brokers for individual clients and to represent organised interest groups. As one journalist commented, 'the Matsuoka camp fortified itself through these unchanging traditional, stable organisational votes centering on groups such as the construction world and the agricultural political league. The Sakamoto camp launched an election campaign that targeted not only these industries but also more widely'.[173]

Matsuoka's defeat was interpreted as a rejection of his political methods and style. Those voting against him were appealing for another kind of politics, condemning 'the kind of political methods that relied on state power politics and guiding benefits to local interests as bad'.[174] Matsuoka's defeat showed that a political rival making programmatic appeals could win out over an old-style LDP politician. Undoubtedly, Sakamoto's victory was testimony to voter support for Koizumi and his policy program. Sakamoto backed Koizumi's reforms, knowing that Matsuoka did not. In voting for Sakamoto, voters were voting for a reform program and the prime minister. A politician like Matsuoka could stand on a platform that opposed his party leader if the electoral coalition supporting him were strong enough. Such independence reflected the lack of policy cohesiveness in the LDP, its decentralised organisation, the weak link between the party and its own Diet members, and the extent to which LDP candidates relied on their own individual power bases.

Fundamentally, the loss of Matsuoka's SMD seat could be explained by his being exposed on two fronts. He was squeezed between the preferences of non-rural and non-farm voters in Kumamoto (3), who were plumping for change by voting for a candidate who had a strong reform image, and dissatisfied farmers and rural dwellers opposed to the Koizumi administration's structural reform orientation and its cuts in public works spending - policies that were contributing to the widespread perception in regional areas that the administration's economic policies, particularly cuts in subsidies and public works, had caused economic recession to deepen in regional areas.[175]

An investigative journalist from the *Asahi* journal *Aera*, visiting Kumamoto in 2002, encountered several people who supported Matsuoka's opposition to the structural reforms in his electorate and neighbouring electorates.[176] Even former Chairman Araki of the Araki Construction Group in Kumamoto, who was aware of Matsuoka's bad reputation, made his support for Matsuoka clear in his capacity as a representative of local construction businesses: 'I am opposed to the Koizumi government's structural reforms. When there is a change of government I want Matsuoka to be prime minister. I respect his anti-reformist position very much'.[177]

However, not all local opinions were positive about Matsuoka. Yamaguchi Rikio (54), who was born and lived in Aso Town, was a well known farmer who ran a private facility called 'Farmer Village' that took in people who wanted to experience farming life. He said

> [e]ven people who live in this area do not really think that character [Matsuoka] is good. Even the locals know that there are all sorts of rumors. These kinds of people thrive because our choices are being eliminated by the small electorate system. We're stressed about that. He's only riding on a system where the LDP and central government control the regional areas. He's only a pawn being used by the central government, and is nothing but an insignificant member sacrificed so that the larger organisation can survive.[178]

Yamaguchi's real gripe was the fact that he had to vote for Matsuoka, whom he really disliked, if he wanted the LDP to win the seat of Kumamoto (3) because the SMD system gave him no other choice.[179] Such a view reflected a perception of one commentator that

Matsuoka was a person who was immersed in the negative political structures of the country where the bureaucracy, LDP and *nôrin zoku* teamed together to organise policy outcomes in their own interests and were doing nothing but hanging on. If you walk around Kumamoto (3) electorate, criticism of Matsuoka's crawling around and participating in the execution of projects based on bureaucratic demand echo everywhere.[180]

Another stated 'Matsuoka isn't even liked in his local area, but there's an atmosphere of being unable to oppose him'.[181] His 'presence is like a local mafia boss. If someone humiliates Matsuoka on television, that person really might get stabbed'.[182]

TOEING THE KOIZUMI LINE

The biggest lesson for Matsuoka from his electoral loss and the main message that he took from it was that if he wanted his old seat of Kumamoto (3) back, he had to relinquish his anti-Koizumi, anti-reform position, and present a reformist face to the greatest possible extent that was consistent with his fervently held policy standpoints. This meant trading away everything but the core positions that retained the backing of his core supporters such as farmers.

The implications for Matsuoka's *giin katsudô* of his status as a PR member were also significant. His switch to the Kyushu regional bloc constituency produced a Kumamoto-wide policy focus. Matsuoka threw himself into a number of causes that were attractive to a wider range of voters.[183] However, because Matsuoka had an eye on winning back his old seat, he kept his main focus on his old electorate.

On issues where his local supporters opposed Koizumi, he chose the role of coordinator and mediator of the two sides, acknowledging their differing positions but also emphasising the need for compromise. In May 2004, Matsuoka travelled to Kumamoto along with all the other LDP Diet representatives from the prefecture to attend a conference. The meeting was held to discuss local issues with municipal mayors, the chairman of municipal assemblies and town and village associations, and the chief of the secretariat of the association of assembly chairmen. Topics included administrative problems in the prefecture such as deteriorating fiscal conditions, the prolonged deflation-recession, anxiety about employment and the impoverished regional economy.

The town and village representatives heartily complained that, if the Koizumi administration's 'trinity reform' were forcibly imposed without proper consideration of its impact on local government budgets, which were heavily dependent on subsidies and tax revenue allocated by the central government, even minimal residents' services could not be guaranteed.[184] Matsuoka's solution was for the national and local governments to acknowledge the problem and work together strongly to solve it.[185]

Matsuoka went on gently questioning the wisdom of fiscal reforms that would give rise to inequities between regions.[186] His manner was a far cry from the strident criticism that he had levelled at the Koizumi government's policies prior to his defeat in 2003. Matsuoka admitted that the 'trinity reform' was designed to restore fiscal health to the regions, but he also expressed real concerns for the agricultural and mountain village regions of Kumamoto where fiscal resources would become scarcer. It was, therefore, necessary to produce a reform plan that would make both cities and regional areas better, not worse.[187]

The loss of his 2003 seat reminded Matsuoka of his vulnerability in the SMD poll, and of the fact that he was beaten by someone who was a Koizumi supporter. He might also need LDP endorsement as a safety net in case he failed to win a plurality in Kumamoto (3) in the next Lower House election and put himself up again as a dual candidate on the party list.

Accordingly, Matsuoka abandoned his membership of the resistance forces and began toeing the Koizumi line. The big test was his position on postal privatisation, which ended up being Koizumi's test of LDP endorsement in the 2005 Lower House election. In 2005, Matsuoka was made a director of the Lower House Special Committee Relating to Postal Privatization (Yûsei Mineika ni kansuru Tokubetsu Iinkai). He claimed to be surprised at his nomination, asserting that it was a complete bolt from the blue.[188] Once he became a director of the committee, he said that his responsibility was to make sure that deliberations on the bill proceeded smoothly.[189] However, because some members of the LDP opposed the bill, committee deliberations did not proceed smoothly. Matsuoka said, 'my heart is full of anxiety in accepting such a difficult duty. However, since postal privatisation is a campaign pledge of Prime Minister Koizumi, I will do my best to manage the committee smoothly by making good use of my experience in various Diet deliberations up till now.'[190]

This was an important committee in discussing and progressing the postal privatisation bills, not only between the ruling and opposition parties but also amongst the LDP Diet members themselves, some of whom were opposed to Koizumi's pet project. In his position as director, Matsuoka was in frequent contact with Koizumi and became an ostensible convert to the cause of postal privatisation. He admitted that being a director, and thus being involved in managing the committee proceedings was hard, 'I am stuck all day in the Diet, coordinating etc. with the opposition, from briefing before the committee meetings to the meeting of directors after the committee meetings'.[191]

The committee kept him exceedingly busy through 2005, leading to the passage of the postal privatisation bills in the Lower House in October 2005. Matsuoka was in the front line of all the discussions, negotiations and coordination on the issue, describing himself as representing the LDP as a 'top batter' in asking questions on the issue.[192] His support for postal privatisation would have put him in a very difficult position, given that his faction boss, Kamei, was one of the chief hold-outs on the issue. In fact, as soon as there was agreement that a dissolution of the Lower House was inevitable, Matsuoka came out more strongly in favor of postal privatisation and parted company from Kamei. Most anti-postal privatisation Diet members were either from the Kamei or Hashimoto factions, both of which were generally critical of Koizumi's structural reform drive. A total of 12 out of the 37 Lower House members who had voted against the postal privatisation bills were from the Kamei faction.[193] Because of the speed of Matsuoka's departure from the Kamei faction, he was labelled a 'betrayer' by Kamei faction insiders.[194] Matsuoka voted for the postal privatisation bills along with 197 other LDP Diet members.

Blatant self-interest was behind Matsuoka's support for Koizumi's postal privatisation project. He was hoping for the position of MAFF minister in the next Koizumi Cabinet, which would have realised a long-held ambition.[195] During the subsequent September 2005 election, Matsuoka confidently asserted to those around him that he wanted to 'win votes fit for the position of a minister. The post I want is Minister of Agriculture, Forestry and Fisheries'.[196] In Nagata-chô, however, the probability of Matsuoka's achieving his ambition was put at only about 30 per cent. It was generally thought that MAFF Minister, Iwanaga Mineichi (who had been elevated to the position from deputy minister, following MAFF Minister Shimamura's sacking in August for opposing the Lower House dissolution) would naturally accede to the position. The fact that Matsuoka's name came up in connection with the Yamarin scandal caused by his sworn friend Muneo was generally thought to be Matsuoka's 'Achilles heel'. The majority view was that '[the scandal] is too fresh [for Matsuoka] to become Minister of Agriculture, Forestry and Fisheries'.[197] A Kasumigaseki bureaucrat observed that even if Prime Minister Koizumi gave Matsuoka a position based on merit, the best he could hope for would be 'about Minister of Environment'.[198] As it turned out, the MAFF minister's position went to ex-METI Minister, Nakagawa Shôichi, Matsuoka's long-term rival and Muneo-hater. Matsuoka was totally passed over for ministerial preferment.

THE 2005 ELECTION

In the September 2005 election, Matsuoka was again competing against the incumbent, Sakamoto, and a newcomer from the DPJ, Nakagawa Kôichirô. At a meeting in Koshi Town, the night after the dissolution of the House of Representatives, a Bon dance festival was held. Both Sakamoto and Matsuoka fronted up, and Matsuoka greeted the public by saying, '[f]or 16 years, it has only been me in the LDP'.[199] On the podium, Matsuoka and Sakamoto's eyes barely met. Matsuoka kept on repeating: 'This time round, I am the challenger'.[200] After the Diet's dissolution, Matsuoka spent most of his time in his local district, frequently making appearances at gatherings of his supporters. At his electoral office, his secretaries made the rounds of various areas in order to give out invitations to regional mayors and prefectural assembly members requesting cooperation with Matsuoka's re-election. There were also phone calls from supporters saying 'do your best'![201]

The head of Matsuoka's election office, Murata Kazuyoshi said, '[w]e want to dig up support in an honest way'.[202] At a sumô competition in Koshi Town, there was spirit and determination in Matsuoka's expression as he shook hands with each person present. He reflected, '[l]ast time, Sakamoto stood; and from my local area Aso, there was a Democrat candidate. I was in an adverse situation'.[203] This time, Matsuoka emphasised his achievements as the chairman of the LDP's Special Committee Relating to Postal Privatization. He campaigned as 'a helper in Koizumi's reform' and strove to revamp his image.[204] He began his campaign outside the Aso Shrine, saying '[t]he catchphrase "reform leader" indicates my will to see through reforms and not to give in to difficulties, no matter what. I want to develop agriculture into a No. 1 export industry and to lead and carry out reform of agriculture so that it will no longer need subsidies'.[205]

A week before election day, Matsuoka went down on his bended knees at an individual speech event for the LDP candidate for Kumamoto (3). It was held in the Aso City Gymnasium.²⁰⁶ The supporters who flocked to the gathering could not believe their eyes when they saw the behaviour of the candidate who made an impassioned speech, saying 'this is a once-in-a-lifetime request'.²⁰⁷ Matsuoka went down on both hands and knees and lowered his head deeply. The hall went silent for a moment, and then there was big applause.²⁰⁸ Matsuoka conducted his campaign as 'a fight in which he risked his political life and which he could not lose'.²⁰⁹

In an exceptional case of 'burning his bridges', Matsuoka withdrew his joint candidacy in the PR Kyushu bloc saying that 'last time was a 50 per cent victory; this time I will aim for a 100 per cent victory [meaning, winning back his SMD seat]'.²¹⁰ His withdrawal from the bloc seat was the trigger for Prime Minister Koizumi to show his support for Matsuoka, which enabled Matsuoka to secure votes from unaffiliated voters.²¹¹ It also put even greater pressure on Sakamoto.²¹² The head of Sakamoto's election office said: 'We'd like to fight it out on policy issues this time. We will strive for an election that doesn't use money'.²¹³ Sakamoto pushed his reform message again, saying '[l]et's carry out real reforms, improvements and politics'.²¹⁴

Matsuoka's loyalty on the postal privatisation issue was rewarded with LDP (and Kômeitô's) endorsement in the 2005 election. The latter's endorsement was reputedly worth about 2000 votes.²¹⁵ The organisation representing retired special postmasters and their families (Taiju) in Kumamoto issued a recommendation for a 'free vote' because all the LDP candidates from the prefecture had supported postal privatisation.²¹⁶ In a public meeting in his electorate, Matsuoka stated that he was 'in favour of privatisation', as did Sakamoto.²¹⁷ This made the competition between them even more severe as the differences in their policies were not clear. Matsuoka was described in the press as having infiltrated Kômeitô supporters, while Sakamoto reputedly broadened his support amongst unaffiliated voters and made inroads into supporters of the DPJ.²¹⁸ Sakamoto sent his election car out into the rural areas, travelling in search of houses on narrow roads and steep mountain paths. When he found someone, 'he would run towards them and shake their hands'.²¹⁹ In fact, he had 'continued to do the rounds of the local area every week even after he was elected last time, greeting his supporters. He claimed that "support is slowly being established"'.²²⁰

Matsuoka was able to capitalise on the fact that Sakamoto's attitude to postal privatisation was somewhat inconsistent, opposing it right up until just before the Lower House vote. Sakamoto acknowledged that this would have an impact on his support, but claimed that, as the son of the chief of a privately owned post office, he knew the post office best. Sakamoto also reportedly gained strength from criticising Matsuoka. Kumamoto (3) was widely portrayed in the press as an electorate where a fierce battle for victory was being fought between Matsuoka and Sakamoto, and where the new DPJ candidate could hardly get a look in.[221]

Sakamoto, however, had difficulty in raising sufficient funds, and put out a call on his website for financial backing for his support association. He explained that he received no subsidy from the LDP, in spite of the fact that he was a member of its parliamentary caucus. He said that he was not a member of the party itself (*seitôin*), and, as an Independent, he could not receive contributions from corporations. He had to rely totally on donations from individual persons.[222] Sakamoto also objected in principle to corporate donations in line with his promotion of clean politics and breaking up the adhesion between politicians and corporations.[223]

All the LDP candidates standing for SMDs and the Kyushu bloc from Kumamoto received the recommendation from the prefectural Chamber of Commerce and Industry. They exchanged policy agreements that committed them to wide-ranging coordination in the event that large stores would be built in local areas and to making efforts to secure budgetary funds and implementing countermeasures for the vitalisation of small and medium-sized businesses.[224]

As was customary, Matsuoka also received the backing of the local agricultural cooperatives. The *nôseiren* recommended all the candidates in Kyushu except for one in Kagoshima, an Independent standing in opposition to the postal privatisation bills. The reason given was that the candidates had strong ties with the *nôseiren*, and 'importance was placed on already established pipelines'.[225] Many Nokyo organisations in 2005 made opposition to a plan to break up agricultural cooperatives as a condition for their support of LDP candidates running in their constituencies. Although in Kumamoto there were no LDP incumbents who had voted against the postal services bill, the prefectural *nôseiren*, which feared that 'after the postal service the agricultural cooperatives will be targeted for reform',[226] presented a memo signed by the chairman of the *nôseiren*, Sonoda Toshiyuki, and the six LDP-endorsed representatives from the prefecture who were recommended by the organisation. The memo pointed out that the government's Deregulation and Privatisation Promotion Council (Kitei Kaikaku/Minkan Kaihô Suishin Kaigi) was attempting to announce the separation and division of the agricultural cooperatives, and on this basis they claimed that 'it is unwarranted intervention and should be withdrawn and reconsidered'.[227] Sonoda stated

> Koizumi is out to crush vested interests by saying he is going to 'destroy the old LDP'. Normally where the brakes would work, he puts forward an argument, and pushes [changes] through, and does not hide his feelings of caution. If he proposes it in the Diet, he'll face opposition that will far and away exceed the levels of the postal service.[228]

However, the prefectural *nôseiren* rationalised its support for the LDP by saying that 'the stability of the political situation is essential'.[229] The memo was a desperate measure taken under the pressure of necessity.[230]

Matsuoka won 86,688 votes or 43.7 per cent of the total cast vote (see Appendix). This was a little over 10,000 more votes than he received in 2003, but it made all the difference between victory and failure by putting him well ahead of Sakamoto. It was this surge in support that won Matsuoka the seat, because Sakamoto's vote tally changed very little (78,796 votes compared with 79,500 votes in 2003).[231] The Sakamoto camp bemoaned the fact that 'most of the Kômeitô's votes went to Matsuoka as the cause of their defeat'.[232] In defeating Sakamoto, Matsuoka reputedly 'vindicated his honor', while in Kumamoto as a whole, LDP candidates were so successful that a new 'conservative kingdom' appeared in the offing.[233]

Following his victory, Matsuoka bowed his head deeply saying, 'I am full of thanks and appreciation that cannot be expressed in words. I would like to repay everyone's kindness through my political activities'.[234] He declared that he would 'faithfully carry out the judgement of the people on administrative, fiscal and political reform, starting with postal privatisation'.[235] His post 2005 election victory statement stressed his pairing of environmental and regional economic objectives

> Amidst the stagnation of regional economies, by practising the 'Green Energy Revolution' that uses regional greenery as an energy resource, I will make efforts to make possible the combination of environmental preservation and the stimulation of the regional economy.[236]

However, Matsuoka won his seat back primarily because he was now seen publicly to be allied with Koizumi's reform program. Koizumi himself came to Kumamoto to publicly campaign with Matsuoka. On his website, Matsuoka proudly displayed a photograph of himself and Koizumi holding up their arms together on top of a campaign platform. Moreover, photos of Matsuoka with Koizumi were used extensively for Matsuoka's election posters. In Matsuoka's election speeches, he stressed his 'closeness of distance with Prime Minister Koizumi' by frequently raising his name.[237] Election analysts commented that in Matsuoka's victory, there was certainly a 'Koizumi effect. Unaffiliated voters were mobilised by the national surge of Koizumi's LDP superiority'.[238]

Some of the gloss on Matsuoka's victory was subsequently tarnished by the arrest of one of his campaign workers, Ieiri Katsukichi, a resident of Nishimachi in Aso City, for vote buying and thus violating the Public Office Election Law. According to the investigations that led to the charge, Ieiri plotted with Ichihara Shiegyuki, unemployed of Aso City, along with others from Aso City at the beginning of September to ask several voters to gather support for Matsuoka. They were suspected of giving voters several tens of thousands of yen each.[239]

The election result meant that Matsuoka's closest mates were now out of the party, including his old faction boss, Kamei, who stood and won for the People's New Party (Kokumin Shintô) in his constituency of Hiroshima (6). Kamei described Koizumi as 'worse than Hitler' for having sent the rebels 'to a gas chamber'.[240] Ibuki Bunmei took over Kamei's faction (the Shisuikai) and Matsuoka joined it, along with other *nôrin zoku* such as Yatsu Yoshio, Nakagawa Shôichi and Kawamura Takeo.

Following the election Matsuoka had to find a place for himself in the new, much more unified and policy-cohesive LDP under Koizumi, in which the dual structure of LDP—bureaucracy policymaking continued to be undermined by the shift towards a more prime minister-centred policymaking system. On the other hand, even though the LDP as a party became less dependent on organised interests in the election and although Matsuoka picked up the pro-Koizumi LDP vote, he was still dependent to a large degree on his customary supporters. In spite of his apparent conversion, Matsuoka remained very much a traditional LDP member, dependent on agricultural cooperatives, construction industry groups and other special-interest groups as well as his own supporters' association for his political base, with only a little help from unaffiliated voters. That help, however, could well have made the difference between success and failure, a lesson that would not be lost on Matsuoka.

In the wake of the 2005 election, and with his ambition to secure the post of MAFF minister still unrealised, Matsuoka tried to demonstrate his leadership potential and ability to act as a policy coordinator on issues such as agricultural trade. Appointed as chairman of the LDP's Agriculture, Forestry and Fishery Products Trade Investigation Committee, and in the newly created position of chairman of the LDP's Committee to Rapidly Promote Exports of Agricultural Products, Matsuoka endeavoured to find a compromise between maintaining agricultural protection and responding positively to a policy environment that was increasingly favorable to bilateral trade deals. The incoming prime minister, Abe Shinzô, had already indicated his strong support for FTAs with other countries in the Asia Pacific. Matsuoka, wearing his hat as a special-interest farm politician, demanded an expansion in the MAFF's agricultural export promotion budget to more than ¥2 billion.[241] However, wearing his hat as an agricultural trade policy leader, he sought to exploit the opportunities presented by further liberalisation, proposing that Japanese agricultural processors add value to foreign farm imports and then export them to other countries. His public acceptance of trade bilateralism, together with his long-standing loyalty to Abe were finally rewarded with appointment to the position of Minister of Agriculture, Forestry and Fisheries in Abe's first cabinet. However, Matsuoka remained a very traditional politician pretending to be a new style of politician in order to secure the position of minister.[242]

In an interview with the press shortly after his appointment, Matsuoka reiterated his trademark themes of aggressively promoting Japanese agricultural exports and expanding biomass energy-based production.[243] On trade matters, he blamed the United States for the failure of the WTO Doha Round whilst declaring that he was committed to the defence of Japan's position[244] and to adopting a stance of 'taking whatever we can and accepting whatever we should'.[245] On FTAs, he professed a 'give and take' approach, admitting to being less than enthusiastic about trade agreements with countries that would not reciprocate by taking Japanese agricultural exports.[246]

NOTES

1 *Yomiuri Shinbun*, 10 November 2003.
2 'Hini Kaku "Matsuoka Toshikatsu Daigishi" no Patoron', p. 58.
3 Hasegawa, 'Jimin "Gajô" no Chikaku Hendô', p. 25.
4 Nakanishi, 'Matsuoka Toshikatsu', p. 28.
5 *The Japan Times*, 23 May 2001.
6 *Nihon Keizai Shinbun*, 19 May 2005.
7 The 'special road revenue comes mainly from gasoline and vehicle weight taxes. It has been instrumental in improving the nation's roads and highway networks since WWII. However, critics argue that for many years, it has been used by LDP elements for pork-barrel politics.' *The Japan Times*, 23 May 2001.
8 *The Japan Times*, 12 June 2001.
9 *ibid.*, 23 May 2001.
10 Tamagawa Tôru, 'Matsuoka Toshikatsu Giin no "Senryaku"' ['Matsuoka Toshikatsu Diet Member's "Strategy"']. Available from http://www.tv-asahi.co.jp/scoop/update/director/20011201_010.html. This policy objective emerged as a Koizumi administration priority policy task following postal privatisation after the September 2005 Lower House election.
11 Nakanishi, 'Matsuoka Toshikatsu', p. 28.
12 *ibid.*
13 *Yomiuri Shinbun*, 30 September 2005.

14 'Hini Kaku "Matsuoka Toshikatsu Daigishi" no Patoron', p. 58.
15 Nakanishi, 'Matsuoka Toshikatsu', p. 28.
16 'Hini Kaku "Matsuoka Toshikatsu Daigishi" no Patoron', p. 58.
17 ibid.
18 'Nihon no Kiki o Sukui Shin no Kaikaku o Jitsugen shi Akarui Mirai o Sôzô suru Giin Renmei (Mirai Sôzô Giren) Sankasha Risuto' ['Participants List of Diet Members' League to Create a Bright Future by Saving Japan and Realising True Reform (Future Creating Giin League)']. Available from http://www.bea.hi-ho.ne.jp/naito38/mirailist.htm
19 'Nihon no Kiki o Sukui Shin no Kaikaku o Jitsugen shi Akarui Mirai o Sôzô suru Giin Renmei (Mirai Sôzô Giren) Sankasha Risuto'. Available from http://www.bea.hi-ho.ne.jp/naito38/mirailist.htm
20 Nakanishi, 'Matsuoka Toshikatsu', p. 28.
21 ibid.
22 'Seijika o Kattei ni Kenkyû suru' ['Researching a Politician Off My Own Bat']. Available from http://www.geocities.co.jp/WattStreet-Stock/4518/matsuoka_t.html
23 The formal English translation of the title of this group is National Joint Struggle. It was the leading group in the Japanese university students' movement in the 1960s.
24 Nakanishi and Special Reporting Group, 'Suzuki Muneo, Matsuoka Toshikatsu', p. 99.
25 This was a reference to Yamasaki Taku, who was secretary-general of the LDP at the time. Nakanishi and Special Reporting Group, 'Suzuki Muneo, Matsuoka Toshikatsu', p. 99. Yamasaki was Prime Minister Koizumi's right-hand man in pushing his structural reform program. After a sex scandal cost him his seat in the Lower House in 2003, the former LDP vice-president and secretary-general in the first Koizumi administration won back a seat in the Lower House in a Fukuoka (2) by-election in April 2005, although Yamasaki had remained special advisor to the prime minister after losing his seat.
26 Tamagawa, 'Matsuoka Toshikatsu Giin no "Senryaku"'. Available from http://www.tv-asahi.co.jp/scoop/update/director/20011201_010.html
27 ibid.
28 Nakanishi, 'Matsuoka Toshikatsu', p. 28.
29 ibid.
30 'Hini Kaku "Matsuoka Toshikatsu Daigishi" no Patoron', p. 58.
31 ibid.
32 ibid.
33 ibid.
34 ibid.
35 ibid.
36 ibid.
37 *The Japan Times*, 29 December 2001.
38 ibid., 29 December 2001.
39 ibid.
40 ibid.
41 ibid.
42 Nakanishi, 'Matsuoka Toshikatsu', p. 29.
43 ibid.
44 The trinity reform of central and local fiscal and taxation systems aims to cut national subsidies to local governments (used for public services, including works projects, education and welfare), to review tax allocations from the central government to municipalities (including a review of allocating taxes to all municipalities on an equal basis), and to cede some tax collection authority (.e. over income, consumption, corporate, liquor, petroleum and tobacco taxes) to municipalities. The reforms are intended to reduce the budget deficits of both central and local governments while promoting decentralisation. In the view of the Koizumi administration: 'Too much money in the form of subsidies and tax grants to municipalities pamper them…If each local authority has to raise its own revenue, this will reduce unnecessary projects for roads and infrastructure'. *The Japan Times*, 14 June 2003. In late 2004, the

government hammered out a trinity reform framework that called for cutting ¥4 trillion in state subsidies over the three years that ends in fiscal 2006 and transferring ¥3 trillion in tax revenue sources to local governments. *Asahi Shinbun*, 5 October 2005.
45 'Jimintô to iu Kettei Hôhô' ['The LDP Decision Method'], in *Director's Eye*. Available from http://www.tv-asahi.co.jp/scoop/update/director/20011201_010.html
46 *ibid.*
47 *The Japan Times*, 29 December 2001.
48 Tamagawa, 'Matsuoka Toshikatsu Giin no "Senryaku"'. Available from http://www.tv-asahi.co.jp/scoop/update/director/20011201_010.html
49 'Jimintô, to iu Kettei Hôhô'. Available from http://www.tv-asahi.co.jp/scoop/update/director/20011201_010.html
50 Okano Sadahiko, 'Progress Toward a New Policy-Making Process', *Japan Echo*, Vol. 10, October 2005, p.41.
51 *The Japan Times*, 29 December 2001.
52 *ibid.*
53 Several comments critical of the government's handling of this issue were made at a December 2003 meeting of the panel, such as 'the government lacks an FTA strategy' and 'its response is haphazard'. *Asahi Shinbun*, 4 December 2003.
54 'Kantei Shudô no FTA Suishin ni Hanpatsu shi Kessoku mo, Kyûshinryoku Kadai' ['The Prime Minister's Official Residence Leadership's FTA Promotion, Centripetal Force Topic'], *Nôsei Undô Jyânaru*, No. 53, February 2004, p.1.
55 'Za Sankuchuari', p. 58.
56 *ibid.*
57 Matsuoka Toshikatsu Official Site, '"FTA Kôshô no Kantei Dokusô" o Kensei' ['Restrain the "Unilateral Action of the Prime Minister's Official Residence on FTA Negotiations"'], in *Katsudô Hôkoku* [*Activity Report*]. Available from http://matsuokatoshikatsu.org/site002//public/048.html
58 '"FTA Kôshô no Kantei Dokusô" o Kensei'. Available from http://matsuokatoshikatsu.org/site002//public/048.html>.
59 *Nihon Keizai Shinbun*, 11 March 2004.
60 Matsuoka Toshikatsu Official Site, 'Midori no Enerugî Kakumei ga Zenshin' ['The Green Energy Revolution Advances'], in *Katsudô Hôkoku* [*Activity Report*]. Available from http://matsuokatoshikatsu.org/index1.html
61 Itô, '"Muneo no Meiyû" no Arata na Taidô', p. 287.
62 *ibid.*
63 *ibid.*, p. 289.
64 *ibid.*
65 *ibid.*
66 *ibid.*, p. 291.
67 *ibid.*, p. 290.
68 *ibid.*
69 *ibid.*
70 *ibid..*
71 *ibid.*
72 *ibid.*, p. 291.
73 *ibid.*
74 *ibid.*, p. 292.
75 *ibid.*
76 *ibid.*, p. 288.
77 *ibid.*, p. 286.
78 Itô, 'Matsuoka Toshikatsu Daigishi', p. 52.
79 Itô, '"Muneo no Meiyû" no Arata na Taidô', p. 292.

80 Itô, *op.cit*, p. 53.
81 Itô, *op.cit.*, pp. 292-293.
82 *ibid.*, p. 293.
83 Itô, 'Matsuoka Toshikatsu Daigishi', p. 53.
84 Itô, *op.cit.*, p. 287.
85 Itô, *op.cit.*, p. 53.
86 Itô, '"Muneo no Meiyû" no Arata na Taidô', p. 287.
87 Itô, 'Matsuoka Toshikatsu Daigishi', p. 53.
88 Itô, *op.cit.*, p. 288.
89 *ibid.*, p. 288.
90 *ibid.*, p. 287.
91 The first BSE case was discovered in the United States in December 2003.
92 Matsuoka Toshikatsu Official Site, 'Gyûdon Fukatsu no Hi wa Itsu?' ['When Will the Day Come When Beef Bowl Comes Back?'], in *Katsudô Hôkoku* [*Activity Report*]. Available from http://matsuokatoshikatsu.org/sit002//public/051.html
93 *ibid*.
94 Indonesia's timber king, for example, has been active in the same process of 'green-wash', as well as in the International Olympics. Personal communication, Paul Gellert, Visiting Fellow, Institute of Asian Cultures, Sophia University, 1 April 2005.
95 Visit: http://www.matsuokatoshikatsu.org/index1.html
96 Visit: http://www.jimin.jp/jimin/giindata/matsuoka-to.html
97 'Nôgyô Kankei Seisaku Kettei no Ashidori', *Nôsei Undô Jyânaru*, No. 53, February 2004, p.23.
98 '"Stop the Illegal Logging" Summary', International Symposium on Countermeasures for Illegal Logging. Available from http://www.zenmoku.jp/sinrin/symp2003_rep_e.html
99 The significance of the words 'absorption source' is that forests are a source of absorption of greenhouse gases, thus helping to prevent global warming.
100 Matsuoka Toshikatsu Official Site, 'Ondanka Bôshi no Kagi o Nigiru "Shinrin Seibi" no Kakujû o' ['Expansion of "Forestry Maintenance" Holds the Key to Preventing Global Warming'], in *Katsudô Hôkoku* [*Activity Report*]. Available from http://matsuokatoshikatsu.org/sit002//public/041.html
101 'Ondanka Bôshi no Kagi o Nigiru "Shinrin Seibi" no Kakujû o'. Available from http://matsuokatoshikatsu.org/sit002//public/041.html
102 Matsuoka Toshikatsu Official Site, 'Tô no Nôrin Kankei Kaigi ga Meijirooshi' ['The Party's Agriculture and Forestry-Related Council Stands Close Side by Side'], in *Katsudô Hôkoku* [*Activity Report*]. Available from http://matsuokatoshikatsu.org/index1.html
103 Matsuoka Toshikatsu Official Site, 'Sutoppu the Kankyô Hakai, Kokusan Zai o Tsukaeba Mondai Kaiketsu' ['Stop the Destruction of the Environment, Solve the Problem by Using Domestic Timber'], in *Katsudô Hôkoku* [*Activity Report*]. Available from http://matsuokatoshikatsu.org/sit002//public/054.html
104 'Results of the Regional Workshop on Strengthening the Asia Forest Partnership', *MAFF Update*, No. 563, 27 October 2004. Available from http://www.maff.go.jp/mud/563.html
105 Visit: http://www.illegal-logging.info/news.php?newsId=456
106 Matsuoka Toshikatsu Official Site, '"Mizu to Shinrin no Taisetsusa" Sekai e Uttaeru' ['An Appeal on the "Importance of Water and Forests" to the World'], in *Katsudô Hôkoku* [*Activity Report*]. Available from http://matsuokatoshikatsu.org/sit002//public/041.html
107 Matsuoka Toshikatsu Official Site, 'Mokuzai ni mo Torêsabiritei o' ['Traceability Even for Timber'], in *Katsudô Hôkoku* [*Activity Report*]. Available from http://matsuokatoshikatsu.org/index1.html
108 Matsuoka Toshikatsu Official Site, 'Gurîn Kônyûhô de mo Ihô Bassai Taisaku o' ['Illegal Logging Countermeasures Even Under the Green Purchasing Law'], in *Katsudô Hôkoku* [*Activity Report*]. Available from http://matsuokatoshikatsu.org/index1.html
109 Matsuoka Toshikatsu Official Site, 'Eikoku Môrî Kankyô, Shokuryô, Nôson Chiikishô Kakugai Daijin no Shôsei o Uke Hôei' ['A Visit to the United Kingdom at the Invitation of Minister Morley, Minister

of State for the Department of the Environment, Food and Rural Affairs'], in *Katsudô Hôkoku* [*Activity Report*]. Available from http://matsuokatoshikatsu.org/sit002//public/059.html
110 'Forest Governance and Trade – Japan, the UK and EU Initiatives', Chatham House, London. Available from http://www.illegal-logging.info/events/Japan_meeting_notes.doc
111 *ibid.*
112 Matsuoka Toshikatsu Official Site, 'Shinbun Kiji ni Miru Matsuoka Toshikatsu Daigishi no Seiji Katsudô' ['Matsuoka Toshikatsu Diet Members' Political Activities Seen in a Newspaper Article']. Available from http://www.matsuokatoshikatsu.org/site002//public/o69.html
113 'Shinbun Kiji ni Miru Matsuoka Toshikatsu Daigishi no Seiji Katsudô'. Available from http://www.matsuokatoshikatsu.org/site002//public/o69.html
114 Gleneagles Plan of Action, Climate Change, Clean Energy and Sustainable Development. Visit: www.g8.gc.ca/
115 Matsuoka Toshikatsu Official Site, 'Kinkyô Hôkoku' ['A Report on Recent Doings']. Available from http://matsuokatoshikatsu.org/index1.html
116 *ibid.*
117 Matsuoka Toshikatsu Official Site, 'Nitchû Ryokka Suishin Giin Renmei, Yakuinkai nite Konnendo no Shien Katsudô' ['The Diet Members' League to Promote Japan-China Tree-Planting Checks This Fiscal Year's Support Activities at an Executive Committee'], in *Katsudô Hôkoku* [*Activity Report*]. Available from http://www.matsuokatoshikatsu.org/sit002//public/003.html
118 *ibid.*
119 See below.
120 Matsuoka Toshikatsu Official Site, 'Chûnichi Chûgoku Taishi to no Iken Kôkan' ['Exchanging Opinions with the Chinese Ambassador'], in *Katsudô Hôkoku* [*Activity Report*]. Available from http://www.matsuokatoshikatsu.org/sit002//public/059.html
121 Even while the Forestry Agency carried a debt burden of ¥3.7 trillion, it continued to cut down forests all around Japan and build unnecessary roads. One of the 'forestry area development roads' running through the mountain forests of Fukushima Prefecture caught media attention. It lay unfinished under the snow. After 30 years it was still only 50 per cent complete. The total cost of the road was put at ¥46.5 billion, approximately ¥34 billion of which was provided by central government subsidies. 'Rinya Gyôsei ga Baramaki Tsuzukeru Dôtô Kensetsu Hojokin to Sugi Kafun' ['Forestry Administration Continues to Scatter Road Construction Subsidies and Cedar Pollen'], *Shûkan Daiyamondo*, 20 April 2002, pp. 66 and 69.
122 *ibid.*, p. 67.
123 *ibid.*, pp. 66-69.
124 'Aso Shinrin Kumiai Sôdaikai' ['A Representatives' Meeting of the Aso Forestry Association'], in *Katsudô Hôkoku* [*Activity Report*]. Available from http://matsuokatoshikatsu.org/index1.html
125 'Oguni-machi Gikai'. Available from http://matsuokatoshikatsu.org/index1.html
126 *ibid.*
127 Matsuoka Toshikatsu Official Site, 'Midori no Enerugî Kakumei Suishin Giin Renmei Hossoku!!' ['Launching the Diet Members' League to Promote the Green Energy Revolution!!'], in *Katsudô Hôkoku* [*Activity Report*]. Available from http://www.matsuokatoshikatsu.org/sit002//public/003.html
128 Itô, '"Muneo no Meiyû" no Arata na Taidô', p. 290.
129 *ibid.*
130 'Midori no Enerugî Kakumei'. Available from http://www.matsuokatoshikatsu.org/sit002//public/003.html. Another source put the figure at 93. Itô, '"Muneo no Meiyû" no Arata na Taidô', p. 290.
131 Itô, '"Muneo no Meiyû" no Arata na Taidô', p. 291.
132 Biomass energy is created from the fermentation of animal dung, which produces biogas—methane and carbon dioxide—for use as fuel. The researchers are professors belonging to the Society for the Study of Green Energy founded by Mitsuzuka. They, along with about eight other professors, are researching biomass energy. The society also accepts executives from Mitsuzuka's businesses and from general construction companies and engineering companies. Itô, '"Muneo no Meiyû" no Arata na Taidô', p. 292.

133 Matsuoka Toshikatsu Official Site, 'Midori no Enerugî Kaikaku Suishin Giren Dai-2 Kai Benkyôkai' ['The Second Study Group of the Diet Members' League for Promoting the Green Energy Revolution'], in *Katsudô Hôkoku* [*Activity Report*]. Available from http://matsuokatoshikatsu.org/site002//public/041.html
134 *ibid.*
135 Itô, '"Muneo no Meiyû" no Arata na Taidô', p. 291.
136 *ibid.*
137 *ibid.*
138 Matsuoka Toshikatsu Official Site, 'GLOBE Japan (Chikyû Kankyô Kokusai Giin Renmei) Sôkai Kaisai' ['General Meeting of GLOBE (Global Legislators Organization for a Balanced Environment') Held]', in *Katsudô Hôkoku* [*Activity Report*]. Available from http://matsuokatoshikatsu.org/sit002//public/054.html
139 asahi.com. Visit: http://mytown.asahi.com/saga/news01.asp?c=5&kiji=725
140 *ibid.*
141 *ibid.*
142 'Hirasawa Katsuei Vs Matsuoka Toshikatsu', p. 45.
143 In the 2003 elections, Kômeitô supported 198 out of 277 LDP candidates in SMDs. *Nikkei Weekly*, 24 November 2003.
144 Hasegawa, 'Jimin "Gajô" no Chikaku Hendô', p. 25.
145 *ibid.*
146 Self P., 1993. *Government by the Market? The Politics of Public Choice*, Macmillan, Houndmills, p. 31. This is how Peter Self describes the rational choice theory of Mayhew (*Congress: The Electoral Connection*, Yale University Press, 1974) about how Congressmen in the United States seek 'to get and stay elected'.
147 These promises included the following: realising the Green Energy Revolution; establishing welfare systems such as reliable pensions, medical and aged care, and nursing systems; establishing a new education system that would become the basis for building the country in the twenty-first century; establishing a society in which everyone could feel safe; promoting policies for an aged society and for the participation of women; establishing a base for economic activities and daily living; and establishing a base for the regional economy. Available from http://www.matsuokatoshikatsu.org/index1.html. A similar list of commitments can be found at http://www.jimin.jp/jimin/giindata/matsuoka-to.html
148 Hasegawa, 'Jimin "Gajô" no Chikaku Hendô', p. 26.
149 There were 10 more LDP SMD losers below him, six of whom lost to the DPJ.
150 Machidori, 'The 1990s Reforms Have Transformed Japanese Politics', p. 40.
151 Hasegawa, 'Jimin "Gajô" no Chikaku Hendô', p. 26.
152 *ibid.*
153 Krauss and Pekkanen, 'Explaining Party Adaptation to Electoral Reform', p. 7
154 '"Muneo no Bôrei"', p. 29.
155 Hasegawa, 'Jimin "Gajô" no Chikaku Hendô', pp. 25-27.
156 *ibid.*, p. 27.
157 *ibid.*, p. 25.
158 '"Muneo no Bôrei"', p. 28.
159 *Yomiuri Shinbun*, 15 September 2005.
160 '"Muneo no Bôrei"', p. 28.
161 *ibid.*
162 *ibid.*
163 *ibid.*
164 *ibid.*, pp. 28-29.
165 asahi.com. Visit: http://www2.asahi.com/senkyo2005/local_news/kumamoto/SEB2...
166 *Yomiuri Shinbun*, 5 November 2003.
167 'Sakamoto Tetsushi o Sasaerukai Nyûkai no Goannai' ['Information for Becoming a Member of Sakamoto Tetsushi's Support Association']. Available from http://www.tetusi.com/sasaeru/index.html

168 "'Muneo no Bôrei'", p. 28.
169 *ibid.*, p. 29.
170 *ibid.*, p. 28.
171 Visit: http://www.tetsusi.com/sasaeru/index.html
172 Visit: http://www.matsuokatoshikatsu.org/index1.html
173 "'Muneo no Bôrei'", p. 28.
174 *ibid.*
175 'Sôsaisen, Sôsenkyo, Tsugi no Rîdâ...' ['Election for the President, General Election, The Next Leaders...'], *Nôsei Undô Jyânaru*, No. 50, August 2003, p. 1.
176 Hasegawa, 'Kanjûdanomi no Hazama de Shundô', p. 24.
177 *ibid.*
178 *ibid.*, pp. 24-25.
179 This reflects wider frustration amongst farmers and rural dwellers, who, according to research done by Robin Le Blanc, are 'frustrated by the reduced selection of LDP-type candidates. Where two very different LDP members from two distinct factions had represented the area prior to the 1990s, voters now have to choose from a list typically including a single LDP candidate, a Communist, and a DPJ candidate'. 'The Small District System', visit: http://ssj-forum@iss.u-tokyo.ac.jp, 13 October 2005.
180 Hasegawa, 'Kanjûdanomi no Hazama de Shundô', p. 25.
181 Visit: http://piza.2ch.net/giin/kako/987/987905181.html
182 *ibid.*
183 See Chapter 4 on 'Exercising Power as a *Nôrin Giin*'.
184 Matsuoka Toshikatsu Official Site, 'Kumamoto Kenka no Zen Chôson to no Iken Kôkankai' ['Meeting to Exchange Opinions with All Towns and Villages in Kumamoto Prefecture'], in *Katsudô Hôkoku* [*Activity Report*]. Available from http://matsuokatoshikatsu.org/site002//public/053.html
185 'Kumamoto Kenka no Zen Chôson to no Iken Kôkankai'. Available from http://matsuokatoshikatsu.org/site002//public/053.html
186 Matsuoka Toshikatsu Official Site, 'Chiikikan Kakusa o Umu Zaisei Kaikaku de yoi ka?' ['Are Fiscal Reforms that Produce Differences between Regions Good?'], in *Katsudô Hôkoku* [*Activity Report*]. Available from http://matsuokatoshikatsu.org/index1.html
187 'Chiikikan Kakusa o Umu Zaisei Kaikaku de yoi ka?'. Available from http://matsuokatoshikatsu.org/index1.html
188 'Kinkyô Hôkoku'. Available from http://matsuokatoshikatsu.org/index1.html
189 *ibid.*
190 *ibid.*
191 Matsuoka Toshikatsu Official Site, 'Hakunetsu shita Yûsei Mineika Hôan Shingi' ['Deliberations on the Postal Privatization Bills Have Reached a Climax'], in *Katsudô Hôkoku* [*Activity Report*]. Available from http://matsuokatoshikatsu.org/index1.html
192 Matsuoka Toshikatsu Official Site, 'Yûsei Mineikahô Yoyatôan no Shingi' ['Deliberations of Government and Opposition Party Drafts of the Postal Privatization Laws'], in *Katsudô Hôkoku* [*Activity Report*]. Available from http://matsuokatoshikatsu.org/index1.html
193 *Sankei Shinbun*, 6 July 2005.
194 *Mainichi Ekonomisuto*, 18 October 2005.
195 *ibid.*
196 *ibid.*
197 *ibid.*
198 *ibid.*
199 asahi.com. Visit: http://www2.asahi.com/senkyo2005/local_news/kumamoto/SEB2...
200 *ibid.*
201 *ibid.*
202 *ibid.*
203 *ibid.*

204 *ibid.*
205 *Yomiuri Shinbun*, 31 August 2005.
206 On 11 February 2005, Aso Town became Aso City by merging with Ichinomiya Town and Namino Village.
207 *Yomiuri Shinbun*, 14 September 2005.
208 *Yomiuri Shinbun*, 14 September 2005.
209 *Yomiuri Shinbun*, 12 September 2005.
210 asahi.com. Visit: http://www2.asahi.com/senkyo2005/local_news/kumamoto/SEB2...
211 *Yomiuri Shinbun*, 12 September 2005.
212 asahi.com. Visit: http://www2.asahi.com/senkyo2005/local_news/kumamoto/SEB2...
213 *ibid.*
214 asahi.com. Visit: http://www2.asahi.com/senkyo2005/local_news/kumamoto/SEB2...
215 *Yomiuri Shinbun*, 15 September 2005.
216 asahi.com. Visit: http://www2.asahi.com/senkyo2005/local_news/kumamoto/SEB2...
217 *ibid.*
218 *ibid.*
219 *ibid.*
220 *ibid.*
221 '2005 Sôsenkyo' ['2005 General Election'], asahi.com. Visit: http://www2.asahi.com/senkyo2005/local_news/kumamoto/SEB2...
222 Visit: http://www.tetusi.com/sasaeru/index.html
223 *ibid.*
224 asahi.com. Visit: http://www2.asahi.com/senkyo2005/local_news/kumamoto/SEB2...>.
225 *ibid.*
226 *ibid.*
227 *ibid.*
228 *ibid.*
229 *ibid.*
230 *ibid.*
231 '2005 Sôsenkyo', asahi.com. Visit: http://www2.asahi.com/senkyo2005/kaihyo/A43003.html
232 *Yomiuri Shinbun*, 15 September 2005.
233 *Mainichi Shinbun*, 13 September 2005.
234 *Yomiuri Shinbun*, 12 September 2005.
235 'Matsuoka Toshikatsu Daigishi kara Minasama e'. Available from http://www.matsuokatoshikatsu.org/site003//public/077.html
236 *ibid.* See also below.
237 *Yomiuri Shinbun*, 15 September 2005.
238 *ibid.*
239 *Yomiuri Shinbun*, 6 October 2005.
240 *Los Angeles Times*. Visit: http://www.latimes.com/news/nationworld/world/la-fg-japan12sep...
241 *Nihon Nogyo Shinbun*, 2 August 2006.
242 Personal communication, Japanese government official, 4 October 2006.
243 *Yomuri Shinbun*, 29 September 2005.
244 *Mainichi Shinbun*, 4 October 2006.
245 *Yomuri Shinbun*, 29 September 2005.
246 *Mainichi Shinbun*, 4 October 2006.

8
CONCLUSION

What is it like to be a Japanese politician in the LDP? How do such politicians fill their days? What kinds of issues motivate them? How do they gather support? This book has tried to answer these and other related questions by personalising Japanese politics as a story of an individual politician. Its approach, although superficially similar to a political biography, is very different in purpose. While eschewing generalisation, it has aspired to yield the kind of understanding and insights about Japanese politics that have previously been derived from more general, orthodox studies. Japanese politics lends itself to such analysis because of the prominent role of individual LDP Diet members, which can be seen in several important contexts: in the electoral arena, in the arena of party politics, and in the arena of policymaking.

The book has told the 'inside story' of a Japanese politician, Matsuoka Toshikatsu. It is primarily based on what Matsuoka has said about himself and his activities, and what others have said about him. It shows how, as an LDP backbencher, Matsuoka has lived by the unspoken creed of 'power and pork'. The book details Matsuoka's early background, career progress, support structures, electoral fortunes, personal connections and policy activities. In doing so, it documents the very public side of Matsuoka's life as he has strutted the political stage, representing special interests, holding court in his Diet office like a feudal overlord, deliberating on policy measures with other LDP politicians in PARC committees, lobbying the government for certain causes, shaping policy outcomes and traversing the world like a salesman 'selling' the cause of Japanese agricultural protection.

At the same time, the book exposes more covert aspects of Matsuoka's political activities, uncovering some of the deals that have been struck and how Matsuoka has 'sold' his services as a political broker in order to secure political funding. Not surprisingly, Matsuoka has been called a 'concession-hunting politician to the marrow'.[1] As mediator and political 'fixer', Matsuoka has established a direct line of influence over public officials, particularly those in the MAFF, which has often skewed the distribution of public resources in favour of his own electorate.

Matsuoka has exemplified the political phenomena of localism, sectionalism and clientelism. Those wanting benefits and favours from the central government either as macro-policies or micro-favours have used Matsuoka as an instrument of delivering collective or personal gains. In turn, Matsuoka has utilised these supplicants and supporters as the means of ensuring his own continuing electoral success and financial strength.

Like politicians everywhere, Matsuoka's overwhelming concern has been the realisation of his personal political ambitions, a goal that has encompassed both electoral survival and career advancement. The details of Matsuoka's political activities in this book show that Matsuoka has been prepared to do practically whatever it takes to achieve his goals. Moreover, given the borderline criminality of a number of his activities, he has been exceedingly fortunate to escape prosecution and political demise. Reading between the lines of the book also reveals evidence of Matsuoka's character and political style—which comes across as rather overbearing, self-important and even rather bullying, but also cowardly when confronted with the prospect of being caught out at underhand activities.

Matsuoka was chosen for this study because he seemed to encapsulate, albeit as an extreme case, many of the archetypal characteristics of LDP politicians, which have been so well documented by other scholars. He is not just any politician. He is a notorious example of a particular 'genre' of politician, obsessed with money, politics, pork barrelling and the unabashed protection of vested interests. Matsuoka deserves 'thick' or 'rich' description because he is so patently illustrative of a certain political type. Matsuoka exemplifies what one might call the 'traditional paradigm' of LDP politician, which further vindicates the approach of the book as a study of Japanese politics through a focus on an individual political actor.

The question that is implicitly raised in this book is whether such a political type can survive in the brave new world of Japanese politics shaped by Prime Minister Koizumi and his successor Abe Shinzô. The projects that have resulted from Matsuoka's kind of mediation have consumed the budgets of national and local governments and wasted tax money. This 'style and structure of politics is old and is increasingly not approved of any longer'.[2] It is 'a way of politics that is now considered "old-fashioned" and is being outlawed, and seen as unpopular with voters'.[3] A shrinking pork barrel is curbing the abilities of Matsuoka and his ilk to manufacture electoral coalitions that are independent of the party by handing out economic bribes as incentives to voters. At the same time, a new policymaking process is gradually being sculpted where *zoku* politicians are being bypassed in favour of a more top-down structure where the prime minister and his enlarged executive are crafting policy initiatives and forcing them on the party and the bureaucracy.

The old LDP, of which Matsuoka is a prime example, is giving way to a new LDP, in which individual backbenchers have to yield to a more centrally directed and cohesive party policy program. The program aims to win voters' hearts and minds, not through appeals to special interests but to various policy causes that will deliver broadly based outcomes affecting all Japanese people and a more equitable distribution of scarcer public resources. Because the future contours of the LDP and its public policy philosophy remain unclear, however, relics of the old LDP such as Matsuoka may survive for a time, even in a new guise as ministers. In order to maintain his political standing and policy influence, Matsuoka has had to reinvent himself in an environment that is increasingly hostile to the old ways.

NOTES

1   Visit: http://www.nouminren.ne.jp/dat/200208/2002081202.htm
2   Hôsei University Professor Igarashi, quoted in http://www.nouminren.ne.jp/dat/200208/2002081202.htm
3   *ibid.*

BIBLIOGRAPHY

The Asahi Shimbun Company, 2005. '2005 Sôsenkyo' ['2005 General Election'], asahi.com. Available from http://www2.asahi.com/senkyo2005/kaihyo/A43003.html [Accessed: 12 September 2005]
Ajia Jinkô Kaikatsu Kaigi ni Sanka (Bangkok)' ['Participating in the Asia Population Development Conference (Bangkok)'], in *Katsudô Hôkoku [Activity Report]*, Matsuoka Toshikatsu Official Site. Available from http://matsuokatoshikatsu.org/site002//public/003.html. [Accessed: 29 April 2005]
'Anzen, Anshin de Sugureta Nihon no Nôsanbutsu o Sekai ni' ' ['Safe, Secure and Prominent Japanese Farm Products to the World'], in *Katsudô Hôkoku [Activity Report]*, Matsuoka Toshikatsu Official Site. Available from http://www.matsuokatoshikatsu.org/site002//public/051.html [Accessed: 28 March 2005]
Arai Shunzô, 1982. *Bunjin Saishô Ôhira Masayoshi [The Cultured Prime Minister Ôhira Masayoshi]*, Shunjûsha, Tokyo.
Araya Hirotake, '"Kawarimi" Matsuoka Toshikatsu: Higan no Nyûkaku wa Naru Ka?' ['Matsuoka Toshikatsu "Adapts Quickly": Will His Earnest Wish to Enter the Cabinet Eventuate?'], *Mainichi Ekonomisuto*, 18 October 2005, pp.76–7.
Asahi Shinbunsha Senkyo Honbu, 1997., *Asahi Senkyo Taikan: Dai 41-kai Shûgiin Sôsenkyo (Heisei 8-nen 10-gatsu), Dai 17-kai Sangiin Tsûjô Senkyo (Heisei 7-nen 7-gatsu) [Asahi General Survey of Election: The 41st House of Representatives General Election (October 1996), The 17th House of Councillors Regular Election (July 1995)]*, Asahi Shimbunsha, Tokyo.
——, 1993. *Asahi Senkyo Taikan: Dai 40-kai Shûgiin Sôsenkyo (Heisei 5-nen 7-gatsu), Dai 16-kai Sangiin Tsûjô Senkyo (Heisei 4-nen 7-gatsu) [Asahi General Survey of Election: The 40th House of Representatives General Election (July 1993), The 16th House of Councillors Regular Election (July 1992)]*, Asahi Shinbunsha, Tokyo.
——, 1990. *Asahi Senkyo Taikan: Dai 39-kai Shûgiin Sôsenkyo (Heisei 2-nen 2-gatsu), Dai 15-kai Sangiin Tsûjô Senkyo (Heisei Gannen 7-gatsu) [Asahi General Survey of Election: The 39th House of Representatives General Election (February 1990), The 15th House of Councillors Regular Election (July 1989)]*, Asahi Shinbunsha, Tokyo.
'Aso Shinrin Kumiai Sôdaikai' ['A Representatives' Meeting of the Aso Forestry Association'], in *Katsudô Hôkoku [Activity Report]*, 2003. Matsuoka Toshikatsu Official Site. Available from http://matsuokatoshikatsu.org/index1.html [Accessed 2 November 2005]
Ayukawa Saiji, 2002. 'Jimintô de mo Shinkô suru "Matsuoka Hazushi"'['"Removal of Matsuoka" is Even in Progress in the LDP'], *Fôsaito*, May, 2002, p. 20.
'Bêkâ Bei Chûnichi Taishi to Kaidan' ['Conversation with the U.S. Ambassador to Japan, Ambassador Baker'], in *Katsudô Hôkoku [Activity Report]*, Matsuoka Toshikatsu Official Site. Available from http://matsuokatoshikatsu.org/site002//public/054.html. [Accessed: 29 April 2005]
'Chihô Rinkatsu Giren Sôkai' ['A General Meeting of the Regional Forestry Activization Assembly Members' League'], in *Katsudô Hôkoku [Activity Report]*, Matsuoka Toshikatsu Official Site. Available from http://matsuokatoshikatsu.org/index1.html [Accessed: 2 November 2005]

'Chiikikan Kakusa o Umu Zaisei Kaikaku de yoi ka?' ['Are Fiscal Reforms that Produce Differences between Regions Good?'], in *Katsudô Hôkoku* [*Activity Report*], Matsuoka Toshikatsu Official Site. Available from http://matsuokatoshikatsu.org/index1.html [Accessed: 2 November 2005]

'Chiiki no "Sakaya San" Sonzoku no Tame ni' ['For the Existence of Regional "Sake Shops"'], in *Katsudô Hôkoku* [*Activity Report*], Matsuoka Toshikatsu Official Site. Available from http://matsuokatoshikatsu.org/site002//public/053.html [Accessed: 28 March 2005]

'Chiiki no Shakai Shihon Seibi wa Jûmin no Negai' ['The Provision of Regional Social Capital is the Hope of Local Residents'], in *Katsudô Hôkoku* [*Activity Report*], Matsuoka Toshikatsu Official Site. Available from http://matsuokatoshikatsu.org/site002//public/053.html [Accessed: 28 March 2005]

'Chûnichi Chûgoku Taishi to no Iken Kôkan' ['Exchanging Opinions with the Chinese Ambassador'], in *Katsudô Hôkoku* [*Activity Report*], Matsuoka Toshikatsu Official Site. Available from http://www.matsuokatoshikatsu.org/sit002//public/059.html [Accessed: 28 March 2005]

'Chûnichi Ôsutoraria Taishi Raisho' ['The Australian Ambassador to Japan Comes to the Office'], in *Katsudô Hôkoku* [*Activity Report*], Matsuoka Toshikatsu Official Site. Available from http://matsuokatoshikatsu.org/index1.html [Accessed 2 November 2005]

'Chûô Shôchô Saihen Sutâto', ['Reorganization of the Central Ministries and Agencies Starts'], *Nôsei Undô Jyânaru*, No.35, February 2001, pp. 12–15.

Curtis, G L., 1971. *Election Campaigning, Japanese Style*, Columbia University Press, New York.

——, 1999. *The Logic of Japanese Politics: Leaders, Institutions, and the Limits of Change*, Columbia University Press, New York.

'Dai 136-kai Kokkai Nôrinsuisan, Yosan Iinkai Shitsumon Ôtô' ['Responses to The 136th Diet Agriculture, Forestry and Fisheries and Budget Committees Responses to Question'], *Nôsei Undô Jyânaru*, No.7, May 1996, pp. 28–9.

'Dai 41-kai Shûgiin Giin Sôsenkyo: Fuken Nôsei Undô Soshiki Suishin, Shiji Tôsen Giin Ichiran' ['The 41st House of Representatives General Election: A Summary of Elected Diet Members Recommended and Supported by Prefectural Agricultural Policy Campaign Organizations'], *Nôsei Undô Jyânaru*, No.10, November 1996, pp. 19–22.

'Dai 42-kai Shûgiin Sôsenkyo ni mukete' ['With a View to the 42nd House of Representatives Election'], *Nôsei Undô Jyânaru*, No.30, April 2000, pp.2-4.

Deirii Jimin [*Daily LDP*]. Available from http://www.jimin.jp/jimin/daily/04_04/21/160421b.shtml

Editorial Group, Ushiroda Ryôe, 2002. 'Okinawa de Mitsuketa Umami' ['The Allure Found in Okinawa'], *Aera*, 11 March, pp.16–19.

'Eikoku Môri Kankyô, Shokuryô, Nôson Chiikishô Kakugai Daijin no Shôsei o Uke Hôei' ['A Visit to the United Kingdom at the Invitation of Minister Morley, Minister of State for the Department of the Environment, Food and Rural Affairs'], in *Katsudô Hôkoku* [*Activity Report*], Matsuoka Toshikatsu Official Site. Available from http://matsuokatoshikatsu.org/sit002//public/059.html. [Accessed: 28 March 2005]

'Forest Governance and Trade – Japan, the UK and EU Initiatives', Chatham House, London. Available from http://www.illegal-logging.info/events/Japan_meeting_notes.doc. [Accessed: 2 November 2005]

'"FTA Kôshô no Kantei Dokusô" o Kensei' ['Restrain the "Unilateral Action of the Prime Minister's Official Residence on FTA Negotiations"'], in *Katsudô Hôkoku* [*Activity Report*], Matsuoka Toshikatsu Official Site. Available from http://matsuokatoshikatsu.org/site002//public/048.html [Accessed: 28 March 2005]

Geertz, C., 1973. *The Interpretation of Cultures*, Basic Books, New York.

'Genchi Rupo—Kumamoto ken' ['On the Spot Report—Kumamoto Prefecture'], *Nôsei Undô Jyânaru*, No.24, March 1999, pp.26–9.

'Giin Renmei Katsudô' ['Diet League Activities'], Matsuoka Toshikatsu Official Site. Available from http://www.matsuokatoshikatsu.org/wite002//public/033.html [Accessed: 6 April 2005]

Gleneagles Plan of Action, Climate Change, Clean Energy and Sustainable Development, 2005. Available from http://www.fco.gov.uk/Files/kfile/PostG8_Gleneagles_CCChangePlanofAction.pdf [Accessed: 11 November 2005]

'GLOBE Japan (Chikyû Kankyô Kokusai Giin Renmei) Sôkai Kaisai' ['General Meeting of GLOBE (Global Legislators Organization for a Balanced Environment') Held]', in *Katsudô Hôkoku* [*Activity Report*], Matsuoka Toshikatsu Official Site. Available from http://matsuokatoshikatsu.org/sit002//public/054.html. [Accessed: 3 March 2005]

'Gomeifuku o O'inorimasu' ['I Pray for his Happiness in the Next World'], in *Katsudô Hôkoku* [*Activity Report*], Matsuoka Toshikatsu Official Site. Available from http://matsuokatoshikatsu.org/index1.html. [Accessed: 2 November 2005]

'Gurîn Kônyûhô de mo Ihô Bassai Taisaku o' ['Illegal Logging Countermeasures Even Under the Green Purchasing Law'], in *Katsudô Hôkoku* [*Activity Report*], Matsuoka Toshikatsu Official Site. Available from http://matsuokatoshikatsu.org/index1.html. [Accessed: 28 March 2005]

'Gurôsâ WTO Nôgyô Iinkai Gichô to Kaidan' ['A Talk with Groser WTO Committee on Agriculture Chairman'], in *Katsudô Hôkoku* [*Activity Report*], Matsuoka Toshikatsu Official Site. Available from http://www.matsuokatoshikatsu.org/site002//public/059.html. [Accessed: 28 March 2005]

'Gyûdon Fukatsu no Hi wa Itsu?' ['When Will the Day Come When Beef Bowl Comes Back?'], in *Katsudô Hôkoku* [*Activity Report*], Matsuoka Toshikatsu Official Site. Available from http://matsuokatoshikatsu.org/sit002//public/051.html. [Accessed: 28 March 2005]

'Hakunetsu shita Yûsei Mineika Hôan Shingi' ['Deliberations on the Postal Privatization Bills Have Reached a Climax'], in *Katsudô Hôkoku* [*Activity Report*], Matsuoka Toshikatsu Official Site. Available from http://matsuokatoshikatsu.org/index1.html. [Accessed: 2 November 2005]

'Han Koizumi Giin no "Yoru no Kao' ['The "Night Face" of a Diet Member Opposed to Koizumi'], *Shûkan Shinchô*, 13 December 2001, p.161.

Hasegawa Hiroshi, 2002a. 'Kanjûdanomi no Hazama de Shundô' ['Wriggling Through the Gaps of Bureaucratic Demands and Requests'], *Aera*, 18 February, pp.23–5.

——, 2002b. 'Nôsuishô o Haishi seyo' ['Abolish the Ministry of Agriculture, Forestry, and Fisheries'], *Aera*, 1 April, pp.35–8.

——, 2002c. 'Kokusan Gyûniku Kaiage no Nazo' ['The Mystery of the Buy-up of Domestic Beef'], *Aera*, 8 July, pp.20–2.

——, 2003. 'Jimin "Gajô" no Chikaku Hendô' ['A Tectonic Shift in an LDP "Stronghold"'], *Aera*, 24 November, pp.25–7.

Hashimoto Naoyuki, 2005. 'Letter from Yochomachi', 11 September 2005. Available from http://homepage.mac.com/naoyuki_hashimoto/iblog/C47j8131471.... [Accessed: 12 September 2005]

'Hini Kaku "Matsuoka Toshikatsu Daigishi" no Patoron no "Yappari"' ['"The Expected" from the Dignity-Lacking Patrons of "Matsuoka Toshikatsu Diet Member"'], *Shûkan Shinchô*, 13 December 2001, pp.58–9.

'Hirasawa Katsuei Vs Matsuoka Toshikatsu: "Makiko-Muneo" no Dairi Senso"' ['Hirasawa Katsuei Versus Matsuoka Toshikatsu: "The Makiko-Muneo" Proxy War'], *Shûkan Shinchô*, 21 February 2002, pp.44–5.

'Hisaichi no Genchi Chôsa ni Hairimasu' ['Participating in On-The-Spot Investigation of a Disaster Area'], in *Katsudô Hôkoku* [*Activity Report*], Matsuoka Toshikatsu Official Site. Available from http://matsuokatoshikatsu.org/index1.html. [Accessed: 2 November 2005]

Hôsaka Masayasu, 1993. *Yoshida Shigeru to iu Gyakusetsu* [*The Paradox of Yoshida Shigeru*], Chûô Kôron Shinsha, Tokyo.

Hôsei University Professor Igarashi, quoted in http://www.nouminren.ne.jp/dat/200208/2002081202.htm [Accessed: 10 May 2005]

'Inasaku Keiei Antei, Kome Seisaku no Kakuritsu e' ['Towards the Establishment of a Rice Policy and Stabilization of Rice Crop management'], *Nôsei Undô Jyânaru*, No.16, November 1997, p.20.

'IPU (Rekkoku Gikai Dômei) de Shûsan Daihyô toshite Supîchi' ['Speech as the Representative of the House of Representatives and House of Councillors at the IPU (Inter-Parliamentary Union)'], in *Katsudô Hôkoku* [*Activity Report*], Matsuoka Toshikatsu Official Site. Available from http://matsuokatoshikatsu.org/site002//public/041.html. [Accessed: 29 April 2005]

Ishii Kôki, 2000. 'Nôsuishô Osen: Amakudari Konsarutanto ga Genkyô da' ['MAFF Contamination: The Amakudari Consultants Are the Ringleaders'], *Bungei Shunjû*, May 2000, pp.192-99.

Itô Hirohide, 2000. 'Heisei Jiken Fuairu: Nôrin Jigyô Hojokin o Dokusen Suru Matsuoka Toshikatsu Shûin Nôsuiiinchô no Eikyôroku' ['Heisei Scandal File: The Influence of House of Representatives Agricultural, Forestry and Fisheries Committee Chairman Matsuoka Toshikatsu Who Monopolizes Agricultural and Forestry Works Subsidies'], *Seikai*, Vol.22, No.6, June 2000, pp.64-7.

Itô Hirotoshi, 2003. 'Matsuoka Toshikatsu Daigishi no "Maboroshi no Hon" to Nôsuishô Baiomasu Jigyô to no Fushigi na Kankei' ['Matsuoka Toshikatsu Diet Member's "Phantom Book" and Its Strange Connection to the MAFF's Biomass Business'], *Zaikai Tenbô*, January, 2003, pp.52-4.

——, 2004. '"Muneo no Meiyû" no Arata na Taidô: Matsuoka Toshikatsu ga Shikakeru Kokka Purojekuto o Oe' ['The New Movements of "Muneo's Sworn Friend": Follow the State Project that Matsuoka Toshikatsu is Going to Do'], *Gendai*, June 2004, pp.286-93.

——, 2002. 'Shinbun ga Zettai ni Hojinai "Gyuniku Giso" no Anbu' ['The Black Spots of the "Beef Camouflage" that the Newspapers Absolutely Don't Report'], *Gendai*, October 2002, pp.76-85.

Itô Terî and Editorial Department, 2002. 'O'warai Nôrinsuisanshô' ['The Comical MAFF'], *Shûkan Daiyamondo*, 20 April 2002, pp.71-5.

'Jimintô Chikusanbutsu Kakakutô Shôiinchô, Matsuoka Toshikatsu, "Chikusan Nôka no Iyoku o so ga nai Kakaku Kettei ni Zenryoku"' ['LDP Livestock Commodity Prices Etc. Subcommittee Chairman, Matsuoka Toshikatsu, "Full Power for a Price Decision that Does Not Weaken the Motivation of Livestock Farmers"'], *Nôsei Undô Jyânaru*, No.7, May 1996, p.11.

'Jimintô ni Girigiri made Yôsei, Tôhonbu ni Kesshû shi Sôkekki Shûkai' ['Taking Requests to the LDP, A General Uprising Meeting Concentrating on the Party Heaquarters'], *Nôsei Undô Jyânaru*, No.16, November 1997, p.24.

'Jimintô no "Bukai" tte??' ['What are the LDP "Divisions??"'], in *Katsudô Hôkoku* [*Activity Report*], Matsuoka Toshikatsu Official Site. Available from http://matsuokatoshikatsu.org/sit002//public/053.html. [Accessed: 28 March 2005]

'Jimintô to iu Kettei Hôhô' ['The LDP Decision Method'], in *Director's Eye*. Available from http://www.tv-asahi.co.jp/scoop/update/director/20011201_010.html.

'"Jimintô Tori Infuruenza Taisaku Honbu" Tachiageru' ['"LDP Headquarters for Avian Influenza" Established'], in *Katsudô Hôkoku* [*Activity Report*], Matsuoka Toshikatsu Official Site. Available from http://matsuokatoshikatsu.org/sit002//public/051.html. [Accessed: 28 March 2005]

Jiyû Minshutô Seimu Chôsakai (ed.), 1992. *Jiyû Minshutô Seimu Chôsakai Meibo, Heisei 4-nen, 2-gatsu, 3-nichi Genzai* [*Liberal Democratic Party Policy Affairs Research Council Membership List 3 February 1992 to the Present*], February.

Jiyû Minshutô Sômukai [*The LDP Executive Council*]. Available from http:ja.wikipedia.org/ [Accessed: 31 January 2006]

'Jizokuteki na Rakunô Shien o' ['Continuing Support for Dairy Farming'], in *Katsudô Hôkoku* [*Activity Report*], Matsuoka Toshikatsu Official Site. Available from http://matsuokatoshikatsu.org/index1.html. [Accessed: 2 November 2005]

'Jyunêbu de WTO Giin Gaikô no Hôkoku' ['Report of WTO Diet Member's Diplomacy in Geneva'], in *Katsudô Hôkoku* [*Activity Report*], Matsuoka Toshikatsu Official Site. Available from http://matsuokatoshikatsu.org/index1.html.[Accessed: 2 November 2005]

'Jyunêbu Hômon no Seika o Hôkoku' ['Reporting the Results of the Geneva Visit'], in *Katsudô Hôkoku* [*Activity Report*], Matsuoka Toshikatsu Official Site. Available from http://matsuokatoshikatsu.org/site002//public/041.html.[Accessed: 29 April 2005]

'Jyunêbu kara Kaette Kimashita' ['I Returned from Geneva'], in *Katsudô Hôkoku* [*Activity Report*], Matsuoka Toshikatsu Official Site. Available from http://matsuokatoshikatsu.org/index1.html. [Accessed: 2 November 2005]
'Kaibunsho ga Tobikau Inshitsusa Nôsui "Jinji Kôsô" no Uchimaku' ['The Insidiousness of Mysterious Documents Flying About: Inside Information on MAFF "Human Resource Battles"'], *Shûkan Daiyamondo*, 20 April 2002, pp.55–7.
Kan Naoto, 1998. *Daijin* [*Ministers*], Iwanami Shôten, Tokyo.
'Kantei Shudô no FTA Suishin ni Hanpatsu shi Kessoku mo, Kyûshinryoku Kadai' ['The Prime Minister's Official Residence Leadership's FTA Promotion, Centripetal Force Topic'], *Nôsei Undô Jyânaru*, No.53, February 2004, p.1.
'Karate 4-dan (Shinseitô) ga 2-dan (Jimintô) o Haritao shita Yoru' ['The Night that the Karate 4th Level from the Renewal Party Pushed Over the Karate 2nd Level from the LDP'], *Shûkan Asahi*, 9 December 1994, pp.32–5.
'Katô Kôichi yo Semete Muneo yori Hayaku Yamenasai' ['Katô Kôichi, You Need At Least to Resign Earlier Than Muneo'], *Shûkan Bunshun*, 21 March 2002, p.166.
'"Keiei Seisaku Taikô" o Matomaru – "Kôzô Kaikaku" no Gutaika Sutâto' ['"Management Policy Outline" Decided – The Start of Concrete Measures for "Structural Reform"'], *Nôsei Undô Jyânaru*, No. 39, October 2001, pp.20–1.
'Kinkyû Nyûin shita "Suzuki Muneo" no Funkei no Tomo: Matsuoka Toshikatsu Daigishi no Taiho Jôhô' ['"Suzuki Muneo's" Eternal Friend is Admitted to Hospital in an Emergency" A Report of Matsuoka's Arrest'], *Shûkan Shinchô*, 4 July 2002, pp.26–8.
Kitamatsu M., et al., 2002. 'Matsuoka Toshikatsu Daigishi Tettei Bunseki: Sono Ôsei naru Shûkin Nôryoku no Kiseki' ['An Exhaustive Analysis of Matsuoka Toshikatsu Diet Member: The Tracks of a Vigorous Money-Collecting Ability'], *Zaikai Tenbô*, December 2002, pp.46–9.
Kokita Kiyohito, 2002. 'Suzuki Muneo no Tsukurareta' ['How Suzuki Muneo was Made'], *Aera*, 18 February 2002, pp.19–22.
Kokkai Benran, Nihon Seikei Shinbunsha, Tokyo, various issues.
'Kokkai Kengaku' ['Diet Study Tour'], in *Katsudô Hôkoku* [*Activity Report*], Matsuoka Toshikatsu Official Site. Available from http://matsuokatoshikatsu.org/index1.html. [Accessed: 2 November 2005]
'Kôkû Gyôkai no Anteika Shien o Yôsei' ['Requesting Support for the Stabilization of the Aviation Industry'], in *Katsudô Hôkoku* [*Activity Report*], Matsuoka Toshikatsu Official Site. Available from http://matsuokatoshikatsu.org/site002//public/041.html. [Accessed: 29 April 2005]
'Kome Seisaku Kakuritsu e Zaigen Kakuho Motomeru 1300 nin Kesshû shi Zenkoku Daihyôsha Shûkai' ['A National Gathering of Representatives Uniting 1300 People Demand the Securing of a Source of Revenue for the Establishment of a Rice Policy'], *Nôsei Undô Jyânaru*, No.16, November 1997, pp.21–2.
'Kome Seisan Chôsei Menseki Sueoki o Kettei' ['The Decision to Leave the Area for Rice Production Adjustment As It Is'], in *Katsudô Hôkoku* [*Activity Report*], Matsuoka Toshikatsu Official Site. Available from http://matsuokatoshikatsu.org/site002//public/048.html. [Accessed: 29 April 2005]
Konno Kumiaki (ed), 2004. 'Results of the Regional Workshop on Strengthening the Asia Forest Partnership', *MAFF Update*, No.563, 27 October 2004. Available from http://www.maff.go.jp/mud/563.html. [Accessed: 3 November 2005]
Kôno Takeshi, and Iwasaki Masahiro, 2004. *Rieki Yûdô Seiji—Kokusai Hikaku to Mekanizumu* [*Politics That Benefit Local Interests—Mechanism and International Comparison*], Ashi Shobô, Tokyo.
Krauss, E.S., and Pekkanen, R., 2004. 'Explaining Party Adaptation to Electoral Reform: The Discreet Charm of the LDP?', *Journal of Japanese Studies*, Vol.30, No.1, Winter 2004, pp.1–34.
'Kumamoto Kenka no Zen Chôson to no Iken Kôkankai' ['Meeting to Exchange Opinions with All Towns and Villages in Kumamoto Prefecture'], in *Katsudô Hôkoku* [*Activity Report*], Matsuoka Toshikatsu Official Site. Available from http://matsuokatoshikatsu.org/site002//public/053.html. [Accessed: 28 March 2005]

Kumamoto-ken Hômu Pêji/Senkyo Kanri Iinkai, (*Dai 43-kai*) *Shûgiingiin Sôsenkyo (Shôsenkyoku) Kaihyô Kekka: Heisei 15-nen 11-gatsu 9-ka* [*(The 43rd) House of Representatives General Election (Single-Member Districts), The Results of the Vote Count: 9 November 2003*].

Kumamoto-ken Hômu Pêji/Senkyo Kanri Iinkai, (*Dai 42-kai*) *Shûgiingiin Sôsenkyo (Shôsenkyoku) Kaihyô Kekka: Heisei 12-nen 6-gatsu 25-nichi* [(*The 42nd*) *House of Representatives General Election (Single-Member Districts), The Results of the Vote Count: 25 June 2000*].

'Kumamoto-ken kara no Seisaku Teian' ['A Policy Proposal from Kumamoto Prefecture'], in *Katsudô Hôkoku* [*Activity Report*], Matsuoka Toshikatsu Official Site. Available from http://matsuokatoshikatsu.org/index1.html. [Accessed: 2 November 2005]

'Kuni no Moto, Chisan Chisui' ['The Foundation of the Country, Food Control and Riparian Works'], in *Katsudô Hôkoku* [*Activity Report*], Matsuoka Toshikatsu Official Site. Available from http://matsuokatoshikatsu.org/site002//public/041.html. [Accessed: 2 November 2005]

'Kyûshû no Shûchû Gôu Higai Taisaku on Kinkyû Giron ['Urgent Deliberation on Localized Torrential Downpour Damage Countermeasures in Kyushu'], in *Katsudô Hôkoku* [*Activity Report*], Matsuoka Toshikatsu Official Site. Available from http://matsuokatoshikatsu.org/site002//public/041.html [Accessed: 29 April 2005]

'LDP Claims Survival of Japanese at Stake in WTO Farm Talks', *Kyodo News*. Available from http://www.japantoday.com/e/?content=news&cat=1&id=251497. [Accessed: 6 April 2005]

Le Blanc, R., 2005. 'The Small Director System'. Available from http://ssj.iss.u-tokyo.ac.jp/archives/2005/10/ssj_3934_re_the.html

Machidori Satoshi, 2005. 'The 1990s Reforms Have Transformed Japanese Politics', *Japan Echo*, June 2005, pp. 38–43.

'Matsuoka Toshikatsu, Jimintô Nôgyô Kihon Seisaku Iinkai Iinchô ni Kiku' ['Listening to Matsuoka Toshikatsu, Chairman of the Agricultural Basic Policy Subcommittee'], *Nôsei Undô Jyânaru*, No.23, February 1999, pp.12–13.

'Matsuoka Toshikatsu Daigishi Hannan nado Giwaku Sanseki' ['Suspicions Accumulate About Diet Member Matsuoka Toshikatsu Such As Hannan'], *Kokumin Shinbun*, July 2002. Available from http://www5f.biglobe.ne.jp/~kokumin-shinbun/H14/1407/140769matsuoka.html. [Accessed: 10 October 2005]

'Matsuoka Toshikatsu Daigishi ni Hisho no Taishokukin & Kyûyo Pinhane Giwaku' ['Suspicion that Matsuoka Toshikatsu is Raking Off His Secretary's Retirement Money & Allowances'], *Flash*, 5 February 2002, pp.14–15.

'Matsuoka Toshikatsu Daigishi kara Minasama e' ['To Everyone from Matsuoka Toshikatsu Diet Member'], Matsuoka Toshikatsu Official Site. Available from http://www.matsuokatoshikatsu.org/site003//public/077.html. [Accessed: 21 September 2005]

'Matsuoka Toshikatsu, Nishi no Muneo no Gyôten Sukyandaru: Mitsukoshi "Nisenmanen Bîruken" Fumitaoshi Kosaku' ['Matsuoka Toshikatsu, Muneo of the West's Astonishing Scandal: The Plot Not to Pay for the Mitsukoshi "¥20 Million Beer Coupons"'], *Shûkan Bunshun*, 5 September 2002:1, pp.68–71.

'Matsuoka Toshikatsu no Rirekisho' ['Curriculum Vitae of Matsuoka Toshikatsu'], Matsuoka Toshikatsu Official Site. Available from http://www.matsuokatoshikatsu.org/site002//public/008.html. [Accessed: 6 April 2005]

'Matsuoka Toshikatsu: Purofuiru' ['Profile'], in Seisaku Jihôsha, *Seikan Yôran* [*A Handbook of Politicians and Bureaucrats*], August 1998, Latter Half Year Edition. Seisaku Jihôsha, Tokyo.

'Matsuoka Toshikatsu: Purofuiru' ['Profile'], in Seisaku Jihôsha, *Seikan Yôran* [*A Handbook of Politicians and Bureaucrats*], First Half Year Edition, March 1995, Seisaku Jihôsha, Tokyo.

'Matsuoka Toshikatsu: Purofuiru' ['Profile'], in Seisaku Jihôsha, *Seikan Yôran* [*A Handbook of Politicians and Bureaucrats*], First Half Year Edition, March 1990, Seisaku Jihôsha, Tokyo.

'Matsuoka Toshikatsu Shi ni Kiku (Aso Gun Ishi Renmei Shiryoo no Peiji)' ['Ask Mr Toshikatsu Matsuoka (The Data Page of Aso County Doctors Federation)']. Available from http://www.geocities.jp/e_osan/ishirenmei_aso03_T_Matsuoka.html [Accessed: 13 May 2005]

Matsuoka Toshikatsu, 'Shinsan - Kôtoku Chiiki no Shinkihon Keikaku no Gaiyô' ['Outline of the New Industry-Industrial Special Regions New Basic Plan'], *Rinya Jihô*, April 1977:34–7.

——, 2005. 'Kinkyô Hôkoku' ['A Report on Recent Doings'], Matsuoka Toshikatsu Official Site. Available from http://matsuokatoshikatsu.org/index1.html. [Accessed: 20 June 2005]
'Midori no Enerugî Kakumei ga Zenshin' ['The Green Energy Revolution Advances'], in *Katsudô Hôkoku* [*Activity Report*], Matsuoka Toshikatsu Official Site. Available from http://matsuokatoshikatsu.org/index1.html. [Accessed: 2 November 2005]
'Midori no Enerugî Kakumei Suishin Giin Renmei Hossoku!!' ['Launching the Diet Members' League to Promote the Green Energy Revolution!!'], in *Katsudô Hôkoku* [*Activity Report*], Matsuoka Toshikatsu Official Site. Available from http://www.matsuokatoshikatsu.org/sit002//public/003.html. [Accessed: 29 April 2005]
'Midori no Enerugî Kaikaku Suishin Giren Dai-2 Kai Benkyôkai' ['The Second Study Group of the Diet Members' League for Promoting the Green Energy Revolution'], in *Katsudô Hôkoku* [*Activity Report*], Matsuoka Toshikatsu Official Site. Available from http://matsuokatoshikatsu.org/site002//public/041.html. [Accessed: 29 April 2005]
'"Mizu to Shinrin no Taisetsusa" Sekai e Uttaeru' ['An Appeal on the "Importance of Water and Forests" to the World'], in *Katsudô Hôkoku* [*Activity Report*], Matsuoka Toshikatsu Official Site. Available from http://matsuokatoshikatsu.org/sit002//public/041.html. [Accessed: 29 April 2005]
'Mokuzai ni mo Torêsabiritei o' ['Traceability Even for Timber'], in *Katsudô Hôkoku* [*Activity Report*], Matsuoka Toshikatsu Official Site. Available from http://matsuokatoshikatsu.org/index1.html. [Accessed: 2 November 2005]
'"Muneo no Bôrei"'ni Maketa Meiyû "Matsuoka Toshikatsu"' ['The Sworn Friend "Matsuoka Toshikatsu" Who Lost to "Muneo's Ghost"'], *Shûkan Shinchô*, 20 November 2003:28–9.
'Muneo no Meiyû "Matsuoka Toshikatsu" Jr. no Igai na Kekkon Aite' ['The Unexpected Marriage Partner of Muneo's Sworn Friend "Matsuoka Toshikatsu" Jr.'], *Shûkan Shinchô*, 28 November 2002, p.32.
'Muneo "Waido": "Usotsuki Toshu Otoko" to Kakarete Honshi o Utaeta "Suzuki Muneo" no Sente Shisaku' ['Muneo "Wide Show": Having Written That He Was an "Habitual Liar", The Failed First Move of "Suzuki Muneo" in Suing Our Magazine'], *Shûkan Shinchô*, 14 March 2002, pp.47–9.
'Naka Kyûshû Ôdan Dôrô' ['For the Early Realization of the Central Kyushu Crossing Road'], in *Katsudô Hôkoku* [*Activity Report*], Matsuoka Toshikatsu Official Site. Available from http://matsuokatoshikatsu.org/site002//public/048.html. [Accessed: 6 April 2005]
Nakanishi Akihiko and Journal Reporter Group, 'Matsuoka Toshikatsu to Iu Giwaku Nin' ['The Suspicious Person Called Matsuoka Toshikatsu'], *Bungei Shunjû*, 1 September 2002:178–87.
Nakanishi Akihiko and Special Reporting Group, 'Suzuki Muneo, Matsuoka Toshikatsu: Riken no Kyôbô' ['Suzuki Muneo and Matsuoka Toshikatsu: Conspiracy for Concessions'], *Bungei Shunjû*, May 2000, pp.94–105.
Nakanishi Tomiki, 2002. 'Matsuoka Toshikatsu (Jimin Daigishi): Igai na Sugao to Shûkin Ryoku' ['Matsuoka Toshikatsu (LDP Diet Member): An Exceptional "Warts and All" and Money Collecting Power'], *Shûkan Asahi*, February 2002, pp.28–9.
Nakano Minoru, 1992. *Gendai Nihon no Seisaku Katei* [*Policy-Making Process in Contemporary Japan*], Tôkyô Daigaku Shuppankai, Tokyo.
'Ni Chaneru Kako Rogu' ['Channel 2, Previous Entries/Log'], Giin Section. Available from http://piza.2ch.net/giin/kako/987/987905181.html. [Accessed: 13 May 2005]
'Nihongata Chokusetsu Shiharai o dô subeki ka' ['How Should Japan-Style Direct Payments Be?'], in *Katsudô Hôkoku* [*Activity Report*], Matsuoka Toshikatsu Official Site. Available from http://matsuokatoshikatsu.org/index1.html. [Accessed: 2 November 2005]
'Nihonhan Chokusetsu Shiharai no Saishûan Happyô', ['Announcement of the Final Draft of the Japanese Edition Direct Payments'], in *Katsudô Hôkoku* [*Activity Report*], Matsuoka Toshikatsu Official Site. Available from http://matsuokatoshikatsu.org/index1.html. [Accessed: 2 November 2005]
'Nihon, Mekishiko FTA (Jijû Bôeki Kyôtei) Gôi' ['The Japan-Mexico FTA (Free Trade Agreement) Reached'], in *Katsudô Hôkoku* [*Activity Report*], Matsuoka Toshikatsu Official Site. Available from http://matsuokatoshikatsu.org/sit002//public/053.html. [Accessed: 28 March 2005]

'Nihon no Kiki o Sukui Shin no Kaikaku o Jitsugen shi Akarui Mirai o Sôzô suru Giin Renmei (Mirai Sôzô Giren) Sankasha Risuto' ['Participants List of Diet Members' League to Create a Bright Future by Saving Japan and Realizing True Reform (Future Creating Giin League)']. Available from http://www.bea.hi-ho.ne.jp/naito38/mirailist.htm [Accessed: 2 November 2005].

Embassy of Japan in China, 'Nihon to Chûkoku no Kankei: Saikin no Ugoki—Matsuoka Nôrinsuisan Fukudaijin ga Hôchû, Yunyû Yasai ya Ryokka Kyôryoku ni Tsuite Kaidan' ['Japan-China Relations: Recent Activities—Deputy Agriculture, Forestry and Fisheries Minister Matsuoka on a Visit to China, Talks concerning Vegetable Imports and Cooperation in Tree-Planting']. Available from http://www.cn.emb-japan.go.jp/bilateral_j/j-c010320-2j [Accessed: 29 April 2005]

'Nikkanchû Kokusai Nôgyô Kaigi o Kaisai' ['Holding a Japan-Korea-China International Agricultural Conference'], in *Katsudô Hôkoku* [*Activity Report*], Matsuoka Toshikatsu Official Site. Available from http://matsuokatoshikatsu.org/site002//public/003.html. [Accessed: 29 April 2005]

Nishikawa Shinichi, 'Tako Tsubo ni Tojikomotte Ôkoku o Gyûjiru Nôgyô Doboku' ['The Agriculture Civil Engineering Technical Bureaucrats Stuck in a Foxhole and Controlling Their Kingdom'], *Shûkan Daiyamondo*, 20 April 2002, p.48.

'"Nishi no Muneo" Matsuoka Toshikatsu no Sokkin Hisho mo Yukue o Kuramashita' ['"Muneo of the West" Matsuoka Toshikatsu's Close Associate and Secretary Also Disappears'], *Shûkan Bunshun*, 4 July 2002, p.38–9.

'Nishi no Muneo: Matsuoka Toshikatsu o Torimaku amari ni Kuroi Jinmyaku' ['Muneo of the West: The Extremely Evil Personal Connections that Surround Matsuoka Toshikatsu'], *Shûkan Bunshun*, 21 March 2002, p.167.

'Nishi no "Muneo" Matsuoka Toshikatsu wa Kisha "I" o "Gokiburi ika" to Kimetsuketa' ['"Muneo" of the West Matsuoka Toshikatsu Asserts Journalist "I" is "Lower than a Cockroach"], *Shûkan Bunshun*, 27 March 2003, p.29.

'Nitchû Ryokuka Suishin Giin Renmei, Yakuinkai nite Konnendo no Shien Katsudô' ['The Diet Members' League to Promote Japan-China Tree-Planting Checks This Fiscal Year's Support Activities at an Executive Committee'], in *Katsudô Hôkoku* [*Activity Report*], Matsuoka Toshikatsu Official Site. Available from http://www.matsuokatoshikatsu.org/sit002//public/003.html. [Accessed: 29 April 2005]

'Nôgyô Kankei Seisaku Kettei no Ashidori' ['The Steps of Agriculture-Related Policy Decisions'], *Nôsei Undô Jyânaru*, No.35, February 2001, p.29.

'Nôgyô Kankei Seisaku Kettei no Ashidori', ['The Steps of Agriculture-Related Policy Decisions'], *Nôsei Undô Jyânaru*, No.37, June 2001, p.30.

'Nôgyô Kankei Seisaku Kettei no Ashidori' ['The Steps of Agriculture-Related Policy Decisions'], *Nôsei Undô Jyânaru*, No.42, April 2002, p.29.

'Nôgyô Kankei Seisaku Kettei no Ashidori', ['The Steps of Agriculture-Related Policy Decisions'], *Nôsei Undô Jyânaru*, No.46, December 2002, p.29.

'Nôgyô Kankei Seisaku Kettei no Ashidori', ['The Steps of Agriculture-Related Policy Decisions'], *Nôsei Undô Jyânaru*, No.51, October 2003, p.29.

'Nôgyô Kankei Seisaku Kettei no Ashidori' ['The Steps of Agriculture-Related Policy Decisions'], *Nôsei Undô Jyânaru*, No. 52, December 2003, p.31.

'Nôgyô Kankei Seisaku Kettei no Ashidori' ['The Steps of Agriculture-Related Policy Decisions'], *Nôsei Undô Jyânaru*, No.53, February 2004, p.23.

'Nôgyô Kankei Seisaku Kettei no Ashidori' ['The Steps of Agriculture-Related Policy Decisions'], *Nôsei Undô Jyânaru*, No. 54, April 2004, p.29.

'Nôgyô Kankei Seisaku Kettei no Ashidori' ['The Steps of Agriculture-Related Policy Decisions'], *Nôsei Undô Jyânaru*, No.55, June 2004, p.29.

'Nôgyô Kankei Seisaku Kettei no Ashidori' ['The Steps of Agriculture-Related Policy Decisions'], *Nôsei Undô Jyânaru*, No.57, October 2004, p.31.

'Nôgyô Kankei Seisaku Kettei no Ashidori' ['The Steps of Agriculture-Related Policy Decisions'], *Nôsei Undô Jyânaru*, No.63, October 2005, p.31.
'"Nôrin Giin" mo Kôkeisha Fusoku?' ['"Agriculture and Forestry Diet Members" Also Lack Successors?'], *Nôsei Undô Jyânaru*, No.30, April 2000, p.1.
Nôrinsuisanshô, Tôkei Jôhôbu, 1992. *Dai-66 Nôrinsuisanshô Tôkeihyô*, [*The 66th Statistical Yearbook of Ministry of Agriculture, Forestry and Fisheries*], 1989-90, Nôrin Tôkei Kyôkai, Tokyo.
——, 2003. *Dai-77 Nôrinsuisanshô Tôkeihyô* [*The 77th Yearbook of Ministry of Agriculture, Forestry and Fisheries, Japan*], 2000-2001], Nôrin Tôkei Kyôkai, Tokyo.
'Nôsuishô "Chikusan Riken": Inamikitta "Niku" to "Uma" Gyôsei' ['The Ministry of Agriculture and Fisheries' "Livestock Concessions"" The Totally Denied "Meat" and "Horse" Administration'], *Sentaku*, May 2002, pp.126-29.
'Oishikute Anshin na Nippon no Nôrinsuisanbutsu o Kaigai e!' ['Delicious and Quality Assured Japanese Agriculture, Forestry and Fisheries Products for the Overseas Market!'], in *Katsudô Hôkoku* [*Activity Report*], Matsuoka Toshikatsu Official Site. Available from http:// www.matsuokatoshikatsu.org/site002//public/053.html. [Accessed: 28 March 2005]
'Ondanka Bôshi no Kagi o Nigiru "Shinrin Seibi" no Kakujû o' ['Expansion of "Forestry Maintenance" Holds the Key to Preventing Global Warming'], in *Katsudô Hôkoku* [*Activity Report*], Matsouka Toshikatsu Office. Available from http://matsuokatoshikatsu.org/sit002// public/041.html [Accessed: 29 April 2005]
'Oishikute Anshin na Nippon no Nôrinsuisanbutsu o Kaigai e!' ['Delicious and Quality Assured Japanese Agriculture, Forestry and Fisheries Products for the Overseas Market!'], in *Katsudô Hôkoku* [*Activity Report*], Matsuoka Toshikatsu Official Site. Available from http:// www.matsuokatoshikatsu.org/site002//public/053.html. [Accessed: 29 April 2005]
'Oguni Machi de Genchi Chôsa' ['On-the-Spot Investigation' in Oguni Town'], in *Katsudô Hôkoku* [*Activity Report*], Matsuoka Toshikatsu Official Site. Available from http:// matsuokatoshikatsu.org/site002//public/060.html. [Accessed: 24 November 2005]
'Oguni-machi Gikai' ['Oguni Town Assembly'], in *Katsudô Hôkoku* [*Activity Report*], Matsuoka Toshikatsu Official Site. Available from http://matsuokatoshikatsu.org/index1.html. [Accessed: 2 November 2005]
Okano Sadahiko, 2005. 'Progress Toward a New Policy-Making Process', *Japan Echo*, Vol.10, October 2005, pp.40-4.
'Onsen de Kokusai Kyôsôryoku Kyôka Nôgyô Yosan Muda Tsukai no Kôzu' ['Internationalization Strengthened Through Hot Springs – The Composition of Wasteful Spending in the Agricultural Budget'], *Shûkan Daiyamondo*, 20 April 2002, pp.52–3.
'Ôzume no WTO Kôshô e Shûgiin yori Daihyôdan Haken' ['Dispatch of the Delegation from the House of Representatives for the Final Phase of the WTO Negotiations'], in *Katsudô Hôkoku* [*Activity Report*], Matsuoka Toshikatsu Official Site. Available from http:// matsuokatoshikatsu.org/site002//public/041.html. [Accessed: 29 April 2005]
Pempel, T.J., 1998. *Regime Shift: Comparative Dynamics of the Japanese Political Economy*, Cornell University Press, Ithaca.
'Posuto "Muneo Sôsa" no Shôten e: Tokusôbu ga Kanshin o Motsu Matsuoka Toshikatsu no "Kôdô"' ['The Focus of the Post "Muneo Investigation": The "Action" of Matsuoka Toshikatsu in Which the Special Investigation Department has an Interest'], *Themis*, July 2002, pp.34–5.
Reed, S.R., 2002a. 'More Muneo', 20 March 2002. Available from http://ssj.iss.u-tokyo.ac.jp/ archives/2002/03/ssj_2654_re_mor.html [Accessed: 6 October 2005]
——, 2002b. 'Revelations About Suzuki Muneo', 13 March 2002. Available from Available from http:/ /ssj.iss.u-tokyo.ac.jp/archives/2002/03/ssj_2644_re_rev.html [Accessed: 6 October 2005]
Rensai Kikaku [*Serial Project*], 2002. Available from http://www.nca.or.jp/shinbun/20040213/ nouiin040213_2_rensai.html. [Accessed: 14 November 2005]
Rhodes R.A.W. and Weller P., 2001. *The Changing World of Top Officials: Mandarins or Valets?*, Open University Press, Buckingham and Philadelphia.
'Rieki Yûdô' ['Guiding Benefits'], *Wikipedia*, http://ja.wikipedia.org/wiki/ %E5%88%A9%E7%9B%8A%E8%AA%98%E5%B0%8E. [Accessed: 13 May 2005]

'Rinya Gyôsei ga Baramaki Tsuzukeru Dôtô Kensetsu Hojokin to Sugi Kafun' ['Forestry Administration Continues to Scatter Road Construction Subsidies and Cedar Pollen'], *Shûkan Daiyamondo*, 20 April 2002, pp.66–9.
Rondan, Kisha Kurabu [*Discussion, Press Club*]. Available from http://www.rondan.co.jp/html/kisha/0206/020625-3.html [Accessed: 13 May 2005]
Sakamoto Tetsushi Official Site, 'Sakamoto Tetsushi o Sasaerukai Nyûkai no Goannai' ['Information for Becoming a Member of Sakamoto Tetsushi's Support Association']. Available from http://www.tetusi.com/sasaeru/index.html [Accessed: 18 May 2005]
Satô Masaru, 2005. *Kokka no Wana: Gaimushô no Rasupuchi to Yobarete* [*National Trap: The So-Called Rasputin of the Foreign Ministry*], Shinchôsha, Tokyo.
Scheiner E., 2006. *Democracy Without Competition in Japan: Opposition Failure in a One-Party Dominant State*, Cambridge University Press, New York.
Schlesinger J.M., 1999. *Shadow Shoguns: The Rise and Fall of Japan's Postwar Political Machine*, Stanford University Press, Stanford.
'Seidoteki na Shikumi wa Dekita. Nôgyô Genba no Jikkô ni Kitai Shitai' ['A Systematic Framework has Emerged. I Expect that This Will be Executed at the Agricultural Grass Roots'], *Nôsei Undô Jyânaru*, No.16, November 1997, p.18.
'Seijika o Kattei ni Kenkyû suru' ['Researching a Politician Off My Own Bat'], http://www.geocities.co.jp/WattStreet-Stock/4518/matsuoka_t.htm [Accessed: 13 May 2005]
'Seijika o Katte ni Kenkyû Suru: Matsuoka Toshikatsu' ['Researching Politicians Arbitrarily: Toshikatsu Matsuoka'], *Yûkan Fuji News*, 9 August 2000. Available from http://ww.fujinews.com/today/2000-08/200000809/0809-08.htm [Accessed: 13 May 2005]
'Seiji Shikin Pâtî ga Dai Seikyô' ['Great Success of Political Funds Parties']. Available from http://www.kenkin.com/etcetra/sikinparty.html [Accessed: 2 November 2005]
'Seiji Shikin Zenkoku Chôsa Kekka: 96-nen Sôsenkyô, Shôsenkyoku Kanren Tôsensha 384 Ninbun o Kôkai' ['Results of the National Investigation of Political Funding': Disclosure of 384 Successfully Elected Persons in the 1996 General Election']. Available from http://www.asahi.com/paper/special/shikin/ [Accessed: 13 May 2005]
Self, P., 1993. *Government by the Market? The Politics of Public Choice*, Macmillan, Houndmills.
'Sessoku na Shichôson Gappei no Saikô o' ['Reconsidering the Hasty Municipality Mergers'], in *Katsudô Hôkoku* [*Activity Report*], Matsuoka Toshikatsu Official Site. Available from http://matsuokatoshikatsu.org/site002//public/003.html. [Accessed: 29 April 2005]
Shigeki Nishihira, 1995. 'Shosenkyoku Bunrui Kijun no Teian' ['Proposals for a Classification Standard for the Single-Member Electorate System'], *Chûô Chôsahô*, No.449, March 1995, pp.1–5.
'Shinbun Kiji ni Miru Matsuoka Toshikatsu Daigishi no Seiji Katsudô' ['Matsuoka Toshikatsu Diet Members' Political Activities Seen in a Newspaper Article']. Available from http://www.matsuokatoshikatsu.org/site002//public/069.html [Accessed: 1 September 2005]
'Shinshokuryôhô Ketchaku' ['The New Food Law Launched'], *Nôsei Undô Jyânaru*, No.5, January 1996, pp.12–14.
Shiota Ushio, 1996. *Kishi Nobusuke*, Kôdansha, Tokyo. 'UR Taisaku 6 Chô 100 Oku En tô Nôgyô Kankei Yosan Kakuho e Jimintô Sôgô Nôsei Chôsakai, Nôrin Bukai Zenryoku' ['The LDP's Comprehensive Agricultural Policy Investigation Committee and Agriculture and Forestry Division Put All Power Into Securing the UR Countermeasures ¥6.01 Trillion Agriculture-Related Budget'], *Nôsei Undô Jyânaru*, No.13, May 1997, pp.4–15.
'"Shôni Kyûkyû Denwa Sôdan Jigyô" Jitsugen e!!'['Realising "The Paediatric Emergency Telephone Consultation Project"!!'], in *Katsudô Hôkoku* [*Activity Report*], Matsuoka Toshikatsu Official Site. Available from http://matsuokatoshikatsu.org/sit002//public/053.html. [Accessed: 28 March 2005]
'Shûchû Gôu no Higaichi o Chôsa Shimashita' ['I Investigated Areas Damaged by Concentrated Heavy Rain'], in *Katsudô Hôkoku* [*Activity Report*], Matsuoka Toshikatsu Official Site. Available from http://matsuokatoshikatsu.org/index1.html. [Accessed: 2 November 2005]

'Shûgiin Yosan Iinkai "Riji" ni Shûnin' ['Taking up the Post of "Director" of the Budget Committee in the House of Representatives'], in *Katsudô Hôkoku* [*Activity Report*], Matsuoka Toshikatsu Official Site. Available from http://matsuokatoshikatsu.org/site002//public/048.html. [Accessed: 29 September 2005]

Sômuchô, Tôkei Kyoku, 1991. *Heisei 2-nen Kokusei Chôsa Saishû Hôkoku Dai 3-kan Dai 2-ji Kihon Shûkei Kekka Sono 2 Todôfuken, Shichôson-hen 43 Kumamoto-ken* [*1990 National Census Report, Vol. 2, Second Basic Statistical Results 2 Prefectures and Municipalities, Edition 43 Kumamoto Prefecture*], Sômuchô, Tôkei Kyoku, Statistics Bureau, Tokyo.

——, Statistics Bureau, 1991. *Heisei 2-nen Kokusei Chôsa Saishû Hôkokusho Nihon no Jinkô (Shiryô Hen)* [*1990 National Census Closing Report Japanese Population (Data Edition)*], Sômuchô, Tôkei Kyoku, Tokyo.

——, 2001. *Heisei 12-nen Kokusei Chôsa Hôkoku Dai 2-kan Dai I-ji Kihon Shûkei Kekka Sono 2 Todôfuken, Shichôson Hen-43 Kumamoto-ken* [*Year 2000 National Census Report Vol. 2 Primary Basic Statistical Results 2 Prefectures and Municipalities Edition 43 Kumamoto Prefecture*], Sômuchô, Tôkei Kyoku, Tokyo.

'Sôsa Shinsa Repôto: Suzuki Muneo, Matsuoka Toshikatsu — Kesareta Shokuniku Rûto' ['Report from the Depths of the Criminal Investigation: Suzuki Muneo, Matsuoka Toshikatsu — The Meat Route that Has Been Wiped Out'], *Shûkan Bunshun*, 19 September 2002:150–53.

'Sôsaisen, Sôsenkyo, Tsugi no Rîdâ...' ['Election for the President, General Election, The Next Leaders...'], *Nôsei Undô Jyânaru*, No.50, August 2003, p.1.

'Sukûpu! Mitsui Bussan Kanbu Shain o Taiho: Chiken Tokusôbu ga Jimintô Daigishi Futari o Chôshu' ['Scoop! The Arrest of Mitsui Co. Executives: The Special Investigation Department of the Public Prosecutor's Office Listens to Two LDP Diet Members'], *Shûkan Gendai*, 20 July, 2002, pp.55–6.

'Sutoppu the Kankyô Hakai, Kokusan Zai o Tsukaeba Mondai Kaiketsu' ['Stop the Destruction of the Environment, Solve the Problem by Using Domestic Timber'], in *Katsudô Hôkoku* [*Activity Report*], Matsuoka Toshikatsu Official Site. Available from http://matsuokatoshikatsu.org/sit002//public/054.html. [Accessed: 28 March 2005]

'"Stop the Illegal Logging" Summary', International Symposium on Countermeasures for Illegal Logging, <http://www.zenmoku.jp/sinrin/symp2003_rep_e.html>.

'Suishin Giin to Meiyû no Paipu o Futoku' ['Fattening the Staunch Friend Pipe with Recommended Diet Members'], *Nôsei Undô Jyânaru*, No.15, September 1996, pp.15–23.

'Supachai WTO Jimukyokuchô to Mendan' ['Talking Personally with WTO Director-General Supachai'], in *Katsudô Hôkoku* [*Activity Report*], Matsuoka Toshikatsu Official Site. Available from http://matsuokatoshikatsu.org/site002//public/003.html. [Accessed: 29 April 2005]

'Suzuki Muneo Giin no Ritô to Sono Yoha' ['Suzuki Muneo Diet Member's Split From the Party and its Aftermath'], *Nôsei Undô Jyânaru*, No.42, April 2002, p.1.

'Suzuki Muneo ni Dôkatsu sareta Jisatsu shita Nôsuishô Kyaria Kanryô' ['The MAFF Career Bureaucrat Threatened by Suzuki Muneo Who Committed Suicide'], *Shûkan Bunshun*, 21 February 2002, pp.26–9.

Tachibana Takashi, 1976. *Tanaka Kakuei Kenkyû: Zenkiroku* [*Tanaka Kakuei Research: A Total Record*], Kôdansha, Tokyo.

——,1980. *Nôkyô: Kyodai na Chôsen* [*Nokyo: The Enormous Challenge*], Asahi Shinbunsha, Tokyo

'Taifû to Jishin de Nôgyô ni Daihigai' ['Great Damage to Agriculture in the Typhoons and Earthquake'], *Nôsei Undô Jyânaru*, No.58, December 2004, p.20.

Tamagawa Tôru, 'Matsuoka Toshikatsu Giin no "Senryaku"' ['Matsuoka Toshikatsu Diet Member's "Strategy"']. Available from http://www.tv-asahi.co.jp/scoop/update/director/20011201_010.html

Tatebayashi Masuhiko, 2004. *Giin Kôdô no Seiji Keizaigaku: Jimintô Shihai no Seido Bunseki* [*The Political Economy of Diet Members' Activities: An Analysis of the System of LDP Rule*], Yuhikaku, Tokyo.

'Tokubetsu Kokkai ga Owarimashita' ['The Special Diet Ended'], in *Katsudô Hôkoku* [*Activity Report*], Matsuoka Toshikatsu Official Site. Available from http://matsuokatoshikatsu.org/index1.html. [Accessed: 2 November 2005]

'Tokubetsu Kokkai Kaikai kara Jyunêbu e' ['From the Opening Session of a Special Diet to Geneva'], in *Katsudô Hôkoku* [*Activity Report*], Matsuoka Toshikatsu Official Site. Available from http://matsuokatoshikatsu.org/index1.html. [Accessed: 2 November 2005]
'Tokubetsu Kokkai no Kaikaishiki' ['Opening Ceremony of the Special Diet'], in *Katsudô Hôkoku* [*Activity Report*], Matsuoka Toshikatsu Official Site. Available from http://matsuokatoshikatsu.org/index1.html. [Accessed: 2 November 2005]
'Tori Infuruenza Yosan Iinkai de Shitsumon' ['Interpellating on Avian Influenza in the Budget Committee'], in *Katsudô Hôkoku* [*Activity Report*], Matsuoka Toshikatsu Official Site. Available from http://matsuokatoshikatsu.org/sit002//public/051.html. [Accessed: 28 Marchr 2005]
'Tô no Nôrin Kankei Kaigi ga Meijirooshi' ['The Party's Agriculture and Forestry-Related Council Stands Close Side by Side'], in *Katsudô Hôkoku* [*Activity Report*], Matsuoka Toshikatsu Official Site. Available from http://matsuokatoshikatsu.org/index1.html. [Accessed: 2 November 2005]
'Tô Nôrinsuisanbutsu Bôeki Chôsakai de WTO Kôshô no Jôkyô Kaiseki' ['Situational Analysis of WTO Negotiations at the Party's Agriculture, Forestry and Fishery Products Trade Investigation Committee'], in *Katsudô Hôkoku* [*Activity Report*], Matsuoka Toshikatsu Official Site. Available from http://matsuokatoshikatsu.org/index1.html. [Accessed: 2 November 2005]
'Tori Infuruenza wa Banzen no Taisaku de' ['All Possible Countermeasures Against Bird Flu'], in *Katsudô Hôkoku* [*Activity Report*], Matsuoka Toshikatsu Official Site. Available from http://matsuokatoshikatsu.org/index1.html. [Accessed: 2 November 2005]
'"Tô Tori Infuruenza Taisaku Honbu" Kachiku Densenbyôo Yôbôhô Kaiseian o Ryôshô' ['"The LDP Avian Influenza Countermeasures Headquarters" Approves the Revised Bill for the Livestock Infectious Diseases Prevention Law'], in *Katsudô Hôkoku* [*Activity Report*], Matsuoka Toshikatsu Official Site. Available from http://matsuokatoshikatsu.org/sit002//public/053.html. [Accessed: 28 March 2005]
Wada Yoshitaka, 2002. 'Kenshô: Sêfugâdo wa Naze Hatsudô sareta ka?' ['Investigation: Why Were the Safeguards Invoked?'], *Ekonomisuto*, 23 April 2002:88–93.
'WTO Hi Nôsanhin Shijô no tame Saido Hôô' ['Revisiting Europe for WTO Non-Farm Products Market Negotiations'], in *Katsudô Hôkoku* [*Activity Report*], Matsuoka Toshikatsu Official Site. Available from http://matsuokatoshikatsu.org/site002//public/041.html. [Accessed: 29 April 2005]
'WTO Jimukyokuchô ra to Kaidan' ['Talks with the WTO Director-General and Others'], in *Katsudô Hôkoku* [*Activity Report*], Matsuoka Toshikatsu Official Site. Available from http://matsuokatoshikatsu.org/site002//public/048.html. [Accessed: 29 April 2005]
'WTO Kôshô e muke Nanbei Shokoku e Giin Gaikô Tenkai' ['Development of Diet Members' Diplomacy to the Countries of South America for the WTO Negotiations'], in *Katsudô Hôkoku* [*Activity Report*], Matsuoka Toshikatsu Official Site. Available from http://matsuokatoshikatsu.org/site002//public/054.html. [Accessed: 28 March 2005]
'WTO Kôshô, Nanbei Shokoku Giin Gaikô Gaiyô' ['Summary of Diet Members' Diplomacy to the Countries of South America for the WTO Negotiations'], in *Katsudô Hôkoku* [*Activity Report*], Matsuoka Toshikatsu Official Site. Available from http://matsuokatoshikatsu.org/site002//public/056.html. [Accessed: 6 June 2005]
'WTO Kôshô Nikkan Daihyô Giin Jyunêbu e' ['Japan and South Korea Assembly Delegation Group for WTO Negotiations Go to Geneva'] in *Katsudô Hôkoku* [*Activity Report*], Matsuoka Toshikatsu Official Site. Available from http://matsuokatoshikatsu.org/site002//public/003.html. [Accessed: 29 April 2005]
'WTO Kôshô ni tsuite Chûnichi Ôsutoraria Taishi to Iken Kôkan' ['Exchanging Opinions with the Australian Ambassador to Japan About the WTO Negotiations'], in *Katsudô Hôkoku* [*Activity Report*], Matsuoka Toshikatsu Official Site. Available from http://matsuokatoshikatsu.org/site002//public/003.html. [Accessed: 29 April 2005]
'WTO mo Chokusetsu Shiharai mo Ôzume' ['The Final Wrap Up of Both the WTO and the Direct Payments'], in *Katsudô Hôkoku* [*Activity Report*], Matsuoka Toshikatsu Official Site. Available from http://matsuokatoshikatsu.org/index1.html [Accessed: 2 November 2005]

'"WTO ni kansuru Giin Kaigi" Unei Iinkai e Nihon kara Shûgiin Daihyôdan o Haken' ['The Dispatch of the House of Representatives' Delegation from Japan to the Steering Committee of the "Parliamentary Conference on the WTO"'], in *Katsudô Hôkoku* [*Activity Report*], Matsuoka Toshikatsu Official Site. Available from http://matsuokatoshikatsu.org/site002//public/053.html [Accessed: 28 March 2005]

'WTO Nôgyô Kôshô ni mukete Tô Hakken ni yoru Giin Gaikô o Tenkai' ['Developing Diet Members' Diplomacy Through Dispatch by the Party for the WTO Agriculture Negotiations'], in *Katsudô Hôkoku* [*Activity Report*], Matsuoka Toshikatsu Official Site. Available from http://matsuokatoshikatsu.org/site002//public/054.html [Accessed: 28 March 2005]

'Yatsu Yoshio Shûgiin Giin, Matsuoka Toshikatsu Shûgiin Giin to Jirâru Gichô to no Kaidan no Kekka Gaiyô' ['A Summary of the Results of Talks amongst House of Representatives Member Yatsu Yoshio, House of Representatives Member Matsuoka Toshikatsu and Chair Girard'], 13 May 2003. Available from http://www.rinya.maff.go.jp/kouhousitu/wto/files/0305ym.htm [Accessed: 27 March 2005]

'Yûsei Mineikahô Yoyatôan no Shingi' ['Deliberations of Government and Opposition Party Drafts of the Postal Privatization Laws'], in *Katsudô Hôkoku* [*Activity Report*], Matsuoka Toshikatsu Official Site. Available from http://matsuokatoshikatsu.org/index1.html [Accessed: 2 November 2005]

'Yushutsu Sokushin de Juyô Kakudai' ['Expanding Demand by Promoting Exports'], in *Katsudô Hôkoku* [*Activity Report*], Matsuoka Toshikatsu Official Site. Available from http://matsuokatoshikatsu.org/index1.html [Accessed: 2 November 2005]

'Zaisei Kôzô Kaikaku ni kakawaru Nôgyô Kankei Yosan Kakuho Taisaku no Torikumi ni tsuite' ['About Grappling with Countermeasures to Secure the Agriculture-Related Budget Endangered by Fiscal Structural Reform'], *Nôsei Undô Jyânaru*, No.14, August 1997, p.5.

'Za Sankuchuari: Jimintô "Nôrin Zoku"' ['The Sanctuary: LDP "Agriculture and Forestry Tribe"'], *Sentaku*, Vol.30, No.2, February 2004, pp.58–61

INDEX

Abe Shintarô, 154
Abe Shinzô, 226, 243, 253
Agriculture and Livestock Industries Corporation (ALIC), 150–1
Aoki Mikio, 90, 206
Arai Hiroyuki, 223
Arai Satoshi, 11, 141
Araki Katsutoshi, 192–3, 234
Arima Harumi, 168
Asada Mitsuru, 177–80, 189
Aso City 239, 242
Aso County, 6, 12, 14, 19, 23, 36, 51, 57, 60, 98, 105, 139, 142, 226
Asô Tarô, 103
Aso Town, 6, 12, 51, 53–4, 64, 69, 81, 139, 227, 234

Baker, Howard, 132
BSE scandal, 174–81, 190, 216

Cabinet Office, 103
clientelistic interests, 41–3, 69–70, 123
Comprehensive Agricultural Policy Investigation Committee (CAPIC) (Sôgô Nôsei Chôsakai) *see under* Liberal Democratic Party, Matsuoka Toshikatsu
Conservative Party, 225
dairy farmers' political league (*rakunô seiji renmei*), 27

Democratic Party of Japan (DPJ), 11, 64, 65, 122, 140, 162, 166, 229, 238, 240
Diet, 1, 2, 19, 70, 126
  standing committees (Kokkai *iinkai*), 38, 64, 76–7, 78, 102–7, 121
  members' leagues (*giin renmei*), 38, 107–11, 184, 205–6, 207, 208, 221–2, 223–4
  see also electoral system, elections, Liberal Democratic Party (LDP), *nôrin giin, nôrin zoku*
electoral system (Lower House) 13, 16, 50–1, 51–6, 65, 68–9, 141, 227
elections
  Lower House election (1986), 12
  Lower House election (1990), 16–25, 178
  Lower House election (1993), 26, 30–2
  Lower House election (1996), 57–61, 84
  Lower House election (2000), 61–3, 169
  Lower House election (2003), 140, 204, 225–35
  Lower House election (2005), 193, 238–42
  Upper House election (1989), 13, 16

Upper House election (2001), 89, 207
Etô Takami, 56, 123–4, 154, 169, 175, 176, 205–6, 223, 224

factions
  see under Liberal Democratic Party, Etô Takami etc
Free Trade Agreements (FTAs), 95–6, 133, 210–12, 243
Fuji Bank scandal, 173–4
Fujimura Yoshiharu, 179
Fujinami Takao, 189
Fujita Hoshimitsu, 28
Fukuda Takeo, 154
Fukuda Yasuo, 211
Furuya Keiyû, 124
Furuya Keiji, 124
Futada Kôji, 133, 157

giin gaikô (Diet members' diplomacy)
  see under Matsuoka Toshikatsu
gikan (specialist)
  see under Ministry of Agriculture, Forestry and Fisheries (MAFF)
Gotô Keiki, 140
Groser, Tim, 132, 133, 134

Hamaguchi Kazuhisa, 64
Hanada Toshikatsu, 173
Harbinson, Stuart, 128
Hashimoto Ryûtarô, 82, 166, 169, 170, 237
Hata Tsutomu, 154
Hiranuma Takeshi, 166
Hirasawa Katsuei, 155, 156
Hirota Daisuke, 29–30
Horinuchi Hisao, 94–5
Horiuchi Mitsuo, 166
Hosokawa Morihiro, 28, 30–1, 52, 57, 61, 229

Ibuki Bunmei, 242
Ichihara Norita, 142
Ichinomiya Town, 142–3
Ikeda Kazutaka, 213, 216
Inose Naoki, 206
Ishiba Shigeru, 154
Ishihara Mamoru, 124
Ishii Kôki, 162
Iwanaga Mineichi, 237
Izeri Seigo, 55
Izumi Hideki, 182–4

Japan Communist Party (JCP), 13, 17, 23, 81
Japan New Party, 28, 30
Japan Socialist Party (JSP), 13, 16, 17, 18, 28, 31, 52, 77, 207
jiban (local support base), 12–4, 20, 26, 35, 52–5, 57, 60, 64, 66, 226, 233
jimukan (generalist)
  see under Ministry of Agriculture, Forestry and Fisheries (MAFF)

kaban (political funds), 9–11, 15–6, 29–30, 55–6, 60, 92, 158–68, 170–4, 178, 184–91, 193
Kamei Shizuka, 56, 90, 123–4, 140, 153, 155, 166, 169, 170, 180, 205–6, 237, 242
Kamei Yoshiyuki, 131
Kamoto County, 57, 58, 60
kanban (name recognition), 14–5, 26, 60, 153
Kaneda Hideyuki, 90
Kaneko Yasushi, 156
Kanemaru Shin, 180
Kantei (Prime Minister's Official Residence), 90, 186, 211, 225
Katô Kôichi, 154, 166, 172
Kawamura Takeo, 242
Kawasaki Atsuo, 53

Kikuchi City, 58, 64, 67, 68
Kikuchi County, 20, 53, 55, 57, 58, 60, 67, 81, 193
Kikuyo Town, 53, 55, 58, 67, 81, 82, 226
Kitaguchi Hiroshi II 17, 19, 20, 23, 26
Kitamura Naoto, 187
Kobayashi Yoshio, 175
kôenkai (personal support group), 12, 29, 30, 37, 42–3, 53–6, 142, 159, 185, 192, 232, 233
Koga Makoto, 225
Koizumi Junichirô, 5, 89, 90, 93, 94, 127, 136, 163, 169, 191, 204–12, 214, 220, 225, 226, 229, 233–8, 239, 241–2, 253
Komachi Kyôgi, 182–3
Kômeitô, 13, 19, 29, 31, 52, 226, 229, 239, 241
Kôno Yôhei, 156
Kumamoto City, 12, 14, 17, 19, 24, 30, 31, 36, 82, 192
Kumamoto Prefecture, 6, 9, 10, 12, 14, 27, 39, 40, 50, 51, 67, 92, 122, 136, 140, 153, 156, 164, 177, 212, 214, 226
Kunii Masayuki, 90
Kurata Eiki, 29

Liberal Democratic Party (LDP), 1, 12, 220, 228, 229, 232, 233–4, 252
 Agricultural Basic Policy Subcommittee (Nôgyô Kihon Seisaku Shôiinkai), 79–84, 94–5, 101
 Comprehensive Agricultural Policy Investigation Committee (CAPIC) (Sôgô Nôsei Chôsakai), 73–5, 79–90, 94, 99–102, 121, 176, 211, 217

 Executive Council 18, 83, 107, 156, 204–5
 factions, 25, 56, 83, 123–4, 160, 166
 see also Etô Takami etc
 and Ministry of Agriculture, Forestry and Fisheries (MAFF) 14, 17, 127
 Policy Affairs Research Council (PARC) (Seimu Chôsakai) 4, 18, 38, 73, 74–5, 83, 107, 133, 143, 169, 170, 210, 217
 policymaking, 3–4, 5, 28, 83–5, 127, 210, 235, 242
 split, 26, 28, 31, 154
 see also elections, Matsuoka Toshikatsu (and Liberal Democratic Party)
Livestock Industry Promotion Corporation (LIPC), 180

Matsuno Raizô. 18, 20, 23–4
Matsuoka Toshikatsu
 and Agricultural Basic Policy Subcommittee (Nôgyô Kihon Seisaku Shôiinkai), 79–84, 94–5, 101, 125, 135, 152, 154, 188
 and avian flu, 96–7
 and biomass, 212–6, 223
 and BSE scandal, 174–81, 190, 216
 clientelistic interests, 41–3, 69–70, 123, 139, 171–3, 182–4, 209, 252
 and Comprehensive Agricultural Policy Investigation Committee (CAPIC) (Sôgô Nôsei Chôsakai), 73–5, 79–90, 94, 99–102, 217–8
 and construction companies, 10, 36, 42, 70, 82, 156, 157–8, 163–7, 222, 226, 228, 234

and dairy farmers, 27, 85, 108–9
Deputy Minister (MAFF), 90–4
and Diet committees, 64, 76–7, 78, 102–7, 121
and Diet members' leagues, 107–11, 184, 205–6, 207, 208, 221–2, 223–4
education, 6–7, 9, 14, 24
early career, 6–16
effects of electoral reform (1994), 51–6, 65, 68–9, 141
and Executive Council, 204–5
and farm interests, 4, 14–5, 27–8, 37–40, 52, 67–9, 74, 89, 99–102, 104, 107–11, 126, 135–7, 222, 242
and Forestry Agency, 7–10, 12, 27, 123, 151, 160–2, 173–4, 184–7, 191, 217, 221, 222
and forestry interests, 4, 27–8, 37–40, 52, 67–9, 74, 99, 104, 109, 110–11, 123, 126, 160–3, 166, 184–91, 217–23
and Free Trade Agreements (FTAs), 95–6, 133, 210–12, 242–3
and Fuji Bank scandal, 173–4
and *giin gaikô* (Diet members' diplomacy), 88, 127–35
jiban (local support base), 12–4, 20, 26, 35, 52–5, 57, 60, 64, 66, 226, 233
kaban (political funds), 9–11, 15–6, 29–30, 55–6, 60, 92, 158–68, 170–4, 178, 181, 184–91, 193
kanban (name recognition), 14–5, 26, 60, 153
kôenkai (personal support group), 12, 29, 30, 37, 42–3, 53–6, 142, 159, 192, 232, 233
and Land Agency, 8, 9, 11

and Liberal Democratic Party (LDP), 11, 13–4, 15, 19–20, 25–6, 28, 38, 55, 56, 64, 64, 66, 107, 135, 165, 181, 232–3, 237, 242
see also Comprehensive Agricultural Policy Investigation Committee (CAPIC), Policy Affairs Research Council (PARC)
and local interests, 35–7, 42, 53, 65–7, 69, 206, 209, 226, 252
and Lower House election (1990), 16–25, 178
and Lower House election (1993), 26–35, 229
and Lower House election (1996), 57–61
and Lower House election (2000), 61–3
and Lower House election (2003), 140, 204, 225–35
and Lower House election (2005), 238–42
and media, 150–1, 156, 168
and Ministry of Agriculture, Forestry and Fisheries (MAFF) (Ministry of Agriculture and Forestry (MAF)), 7–9, 11, 15, 17, 20–3, 37, 75, 77–9, 87–8, 90–3, 100, 103, 122, 123–6, 138–43, 151, 154, 155, 157–8, 159, 163, 175–81, 187, 237–8, 242–3, 252
and Ministry of Construction, 37
and Ministry of Foreign Affairs (MoFA), 182–4, 219
and Ministry of Internal Affairs and Communications (MIAC), 97, 103
and Ministry of Land, Transport and Infrastructure (MLIT), 98, 103, 139, 163

and Ministry of Transport 37
and National Land Agency, 8, 106
and Nokyo, 17, 19, 27, 28, 39, 60–1, 65, 68, 80, 84–7, 92, 95, 101, 122, 240
nôrin giin (agriculture and forestry Diet member), 38, 52, 53, 64, 68, 73–111, 154, 215
nôrin zoku (agriculture and forestry 'tribe'), 64, 75, 121–43, 152, 153–8, 159, 204, 209, 225, 235
parliamentary vice-minister, 77–9
and Policy Affairs Research Council (PARC) (Seimu Chôsakai) and committees, 73–5, 76, 77, 79–111, 121, 143, 217–8, 251
and postal privatisation, 236–7, 238–40
and public works, 4, 9, 26, 35–7, 42, 66, 70, 83, 88, 98–9, 102–4, 111, 121–2, 138–43, 153, 158, 163, 207, 208–9, 218, 234
and rice price, 86–7, 89, 95, 100, 155
and sectional interests, 37–40, 138–9, 209, 252
see also farm interests etc
and Suzuki Muneo, 11, 15–6, 25, 56, 81, 105, 151–8, 166, 168–71, 175–81, 183, 184, 185–91, 215–6, 228, 237
and Syrian Embassy affair, 181–4
and *teikô seiryoku* (resistance forces), 191, 204–12, 225–6
and tobacco farmers, 27, 89–90, 96, 155
and Uruguay Round (UR) Agriculture Countermeasures Expenditure (*UR Nôgyô Taisakuhi*), 37, 80–88, 121, 124, 125, 140, 159
and Uruguay Round negotiations, 107–8
and World Trade Organization (WTO), 87–8, 91–4, 95–6, 127–35, 243
and Yamarin scandal, 184–91, 192, 216, 217, 210, 228, 237–8
Matsushita Tadahirô, 186–7, 219
McCarthy, John, 127
McLean, Murray, 133–4
methodology, 2–5, 251–2
Ministry of Agriculture and Forestry (MAF)
see Ministry of Agriculture, Forestry and Fisheries (MAFF)
Ministry of Agriculture, Forestry and Fisheries (MAFF), 137, 154, 157, 174–5, 191, 211, 214, 223–4, 225
BSE scandal, 174–81
Fisheries Agency, 155
Food Control system, 78–9
Forestry Agency, 7–10, 12, 99, 123, 151, 160–2, 173–4, 184–7, 191, 217, 219, 221, 222
gikan (specialist), 7, 8, 17, 23, 125, 225
jimukan (generalist), 7, 17, 23, 124, 225
and Liberal Democratic Party (LDP), 14, 17, 127
Minister's Secretariat, 8
National Land Agency, 8, 9
and Nokyo, 17, 19, 27
Old Boys (OBs), 17, 20–3, 25, 84, 123, 125, 141, 154, 161–2, 174, 180
see also Matsuoka Toshikatsu (and Ministry of Agriculture, Forestry and Fisheries)

Ministry of Construction, 37
Ministry of Economy, Trade and Industry (METI), 92, 130, 137, 238
Ministry of Education Science and Technology, 175
Ministry of Finance (MOF), 175, 225
Ministry of Foreign Affairs (MoFA), 92, 126, 137, 155, 156, 157, 182–4, 191, 194, 219
Ministry of Internal Affairs and Communications (MIAC), 98, 103, 170
Ministry of Health, Labour and Welfare (Ministry of Health and Welfare), 27, 175
Ministry of Home Affairs, 15
Ministry of Internal Affairs and Communications, 190
Ministry of Justice, 186
Ministry of Land, Transport and Infrastructure (MLIT), 98, 103, 139, 163
Ministry of Transport 37
Mitsuzuka Hiroshi, 25, 154
Mitsuzuka Kôkichi, 213–4
Miyazawa Kiichi, 108
'money politics', 4, 190–1
see also *kaban* (political funds)
Mori Yoshirô, 56, 90, 159, 166, 169, 213
Murata Kazuyoshi, 238
Murayama Tomiichi, 77

Nagamura Takemi, 124
Nakagawa Hidenao, 166
Nakagawa Ichirô, 11, 12, 15, 154, 177, 187, 189
Nakagawa Shôichi, 94, 151, 189, 238, 242
Nakagawa Yoshio, 211

Nakamura Yasushi, 162
Nakasone Yasuhirô, 56
National Land Agency, 8, 106
New Frontier Party, 53, 57, 61, 187
New Party Harbinger (Shintô Sakigake), 159
New Party Mother Earth (Shintô Daichi), 193
Noda Masaharu, 29
Noda Takeshi, 17, 19, 28, 29, 31, 52, 225
Nokyo, 17, 19, 27, 28, 39, 60–1, 65, 68, 80, 84–7, 92, 95, 101, 122, 179, 193, 240
see also Zenchû
Nonaka Hiromu, 154, 155, 156, 170, 180, 210
Norota Hôsei, 133, 181, 211
nôrin giin (agriculture and forestry Diet member), 38, 52, 53, 64, 67, 73–111, 154, 157, 191, 228
nôrin zoku (agriculture and forestry 'tribe'), 11, 64, 75, 121–43, 152, 153–8, 159, 174, 175, 204, 210–12, 225, 235, 242
Nukaga Fukushirô, 210

Obuchi Keizô, 210, 221
Ôgata Yasuo, 161
Ogawa Yasuo, 124
Oguni Town, 19, 20, 59, 67, 98, 103
Old Boys (OBs) see *under* Ministry of Agriculture, Forestry and Fisheries (MAFF)
Omori Town, 67
Ôshima Tadamori, 211
Ozaki Mitsurô, 29
Ozawa Ichirô, 36
Ozu Town, 20, 58

People's New Party (Kokumin Shintô), 242
Policy Affairs Research Council (Seimu Chôsakai) *see under* Liberal Democratic Party, Matsuoka Toshikatsu
postal privatisation, 236–7, 238–40
public works, 4, 26, 35–7, 42, 66, 69, 83, 88, 98–9, 102–4, 111, 121–2, 138–43, 153, 158, 163, 207, 208–9, 218, 234

Recruit scandal 16
Renewal Party, 26

Sakamoto Tetsushi, 229, 232–3, 238–41
Sakurai Shin, 128–34, 211
Sasaki Kichinosuke, 10, 172
Satô Masato, 124–5
Satô Saburô, 172
Seirankai (Young Storm Society), 11
Shigeie Toshinori, 182–3
Shimamura Yoshinobu, 237
Shisuikai 169, 242
Social Democratic Party (SDP), 229
Sonoda Hiroyuki, 173
Sonoda Toshiyuki, 240
Soyo Town, 140
Sugata Kikuhito, 150
Supachai, Panitchpakdi, 128, 129–30
Suzuki Muneo, 11, 15–6, 25, 56, 81, 105, 151–8, 159, 160, 165–6, 168–71, 175–81, 183, 184, 185–91, 193–4, 207, 216, 225, 228, 237, 238
Suzuki Zenkô, 124
Syrian Embassy affair, 181–4

Takagi Yûki, 142
Takebe Tsutomu, 174–5, 176, 178–9

Takeshita Noboru, 16, 180
Tamaki Kazuo, 11, 12, 121
Tanaka Hironao, 124
Tanaka Kakuei, 2, 153
Tanaka Makiko, 183
Tanaka Naoki, 90, 91
Tanaka Shôichi 17, 29
Tanikawa Kazuo, 182
teikô seiryoku (resistance forces), 191, 204–12
Tominaga Kiyotsugu, 82
Tottori University, 7, 9, 23, 37, 187–8

Ueki Town, 57, 58
Uemura Tsukasa, 183
Uozumi Hirohide 18, 19, 26, 27, 28, 31, 36, 52, 53, 57, 64
Uruguay Round Agriculture Countermeasures Expenditure (*UR Nôgyô Taisakuhi*), 37, 79–86, 121, 124, 125, 140, 159

Watanabe Michio, 11, 154
Watanabe Yoshinori, 180
World Trade Organization (WTO), 87–8, 91–4, 95–6, 127–35, 217, 243

Yamada Isao, 185–9
Yamada Rintarô, 188
Yamada Satoshi, 187
Yamaga City, 53, 58, 60, 64, 68
Yamaguchi Rikio, 234
Yamamoto Tôru, 150–1
Yamarin scandal, 184–91, 192, 216, 217, 219, 228, 232
Yamasaki Taku, 169, 206, 232
Yatsu Yoshio, 91–4, 123, 130, 132, 134, 177, 211, 242
Yoshii Junichi, 166, 168
Zenchû, 134

www.ingramcontent.com/pod-product-compliance
Lightning Source LLC
Chambersburg PA
CBHW050901240426
43672CB00025B/2990